DESIGN AND FABRICATION OF SELF-POWERED MICRO-HARVESTERS

DESIGN AND FABRICATION OF SELF-POWERED MICRO-HARVESTERS

ROTATING AND VIBRATING MICRO-POWER SYSTEMS

C. T. Pan

National Sun Yat-Sen University, Taiwan

Y. M. Hwang

National Sun Yat-Sen University, Taiwan

Liwei Lin

University of California, Berkeley, USA

Ying-Chung Chen

National Sun Yat-Sen University, Taiwan

IEEE

WILEY

This edition first published 2014

Registered office
John Wiley & Sons Singapore Pte. Ltd., 1 Fusionopolis Walk, #07-01 Solaris South Tower, Singapore 138628.

For details of our global editorial offices, for customer services and for information about how to apply for permission to reuse the copyright material in this book please see our website at www.wiley.com.

Library of Congress Cataloging-in-Publication Data applied for.

A catalogue record for this book is available from the British Library.

ISBN: 978-1-118-48779-2

Typeset in 11/13pt Times by Laserwords Private Limited, Chennai, India
Printed and bound in Singapore by Markono Print Media Pte Ltd

1 2014

Contents

About the Authors

Dr. C.T. Pan was born in Nauto, Taiwan, in 1969. He received master and doctoral engineering degrees in 1993 and 1998 respectively, from the Power Mechanical Engineering Department of National Tsing Hua University in Hsinchu, Taiwan. He was a researcher in the field of laser machining polymer at TU Berlin (IWF) in Germany from 1997 to 1998 and a researcher in the MEMS Division of MIRL/ITRI, Hsinchu, Taiwan from 1998 to 2003. He joined National Sun Yat-Sen University, Kaohsiung, Taiwan, as an Assistant Professor in 2003, then earned his associate professorship and full professorship in 2005 and 2008, respectively. He won the Outstanding Professor Award (2009–2013) from National Sun Yat-Sen University. From June 2009 to June 2010, he was a visiting professor at the department of ME in UC Berkeley. His current research interests focus on MEMS, nanofabrication, micro-scale energy, and LIGA process.

Dr. Y.M. Hwang was born in Chanhwa, Taiwan, Republic of China in 1958. He received his Bachelor's (1981) and Master's (1983) degrees in power mechanical engineering from National Tsing Hua University in Hsinchu, Taiwan. He earned his Doctor's degree (1990) in industrial mechanical engineering from Tokyo University in Japan. He has been a professor at the Department of Mechanical and Electro-Mechanical Engineering (MEME), National Sun Yat-Sen University (NSYSU), Kaohsiung, Taiwan, since 1996. He has served as the department chair (2002–2005) of MEME. His research interests have been in the area of metal forming, machine design and mechanics. He won the Best Paper Award (1992) and Outstanding Engineering Professor Award (2007) from the Chinese Society of Mechanical

Engineers in Taiwan. He earned the Fellow title from Japan Society for Technology of Plasticity (JSTP), Japan (2012) and Distinguished Professor of NSYSU (2012).

Dr. Liwei Lin is a Professor at the Department of Mechanical Engineering at the University of California at Berkeley, and Co-Director of the Berkeley Sensor and Actuator Center. He received his B.S. degree (1986) in Power Mechanical Engineering from the National Tsing Hua University, Taiwan, and M.S. (1991) and Ph.D. (1993) degrees from UC Berkeley in Mechanical Engineering. After graduation, Professor Lin held the position of Senior Research Scientist at BEI Electronics Inc., Associate Professor at the National Taiwan University, Taiwan and Assistant Professor at the University of Michigan, Ann Arbor, USA before joining the faculty at UC Berkeley in 1999. His research interests and activities include MEMS, NEMS, Nanotechnology, design and manufacturing of microsensors and microactuators, development of micromachining processes by silicon surface/bulk micromachining, micro molding process, and mechanical issues in MEMS such as heat transfer, solid/fluid mechanics and dynamics. Professor Lin is the co-inventor of 16 US patents in MEMS and has co-authored more than 130 journal and 200 refereed conference papers. He has supervised 29 Ph.D. and 30 M.S. students.

Dr. Ying-Chung Chen was born in Tainan, Taiwan, R.O.C., on 4 November 1956. He received the M.S. and Ph.D. degree in electrical engineering from National Cheng Kung University, Tainan, Taiwan, in 1981 and 1985 respectively. Since 1983, he has been at National Sun Yat-Sen University (NSYSU), Kaohsiung, Taiwan, where he is a professor of electrical engineering. Previously, he served as the department chair (2006–2009) of EE, the Dean of the College of Engineering (2011–2014) and won the Distinguished Professor Award and Outstanding Professor Award (2009–2013) from National Sun Yat-Sen University. His current research interests are in the areas of electronic devices, surface acoustic wave devices, thin-film technology, and electronic ceramics. He is a member of the Chinese Society for Materials Science and a registered electrical engineer in Taiwan.

Preface

Energy harvesting is known as power harvesting or energy scavenging to store and capture ambient energy which is natural, self-regenerating, or renewable. For example, ambient energy may include wind, solar, hydro, geothermal, and tide. Energy harvesting takes advantage of these sources to provide energy that is renewable and eco-friendly as compared with energy derived from fossil fuels. As an enabling technology from ambient vibrational energy sources, energy harvesting could find potential applications in low-power devices such as sensors, actuators, and electronics with ecological advantages in reducing chemical wastes from batteries and is maintenance-free.

This book covers recent advances in energy harvesting using different transduction mechanisms, including mechanics, semiconductor process, and electrical circuitry. The dissemination of this technology is important for the industry but there are only a limited number of introductory books or handbooks in the global community. Compared to other semiconductor disciplines, such as MEMS and LIGA, the gap in advanced knowledge of energy harvesting has yet to be filled. This book partially fills this gap by documenting the latest and most frequently cited research results of a few key energy harvesting processes. Scientists and researchers from various disciplines have contributed heavily to the related energy harvesting literature. Our hope with the current book is to provide reliable and practical techniques for analytical models of piezoelectric and electromagnetic energy harvesters and their relevant phenomena. This book presents a state-of-the-art understanding of diverse aspects of energy harvesting with a focus on broadband energy conversion, as well as new concepts in designs and fabrication processes.

The book is arranged in five chapters to describe the research and development of energy harvesters. Chapter 1 introduces the background of energy harvesting, as well as the development of electromagnetic and piezoelectric energy harvesters. Chapter 2 describes the development of the single ZnO energy harvesters, broad bandwidth vibrational energy harvesting systems, and double-sided piezoelectric energy harvesters. The design and fabrication of vibration-induced electromagnetic generators is presented in Chapter 3. Chapter 4 focuses on the design, fabrication, test, and application of in-plane rotary electromagnetic micro-generator. Chapter 5 describes the design and fabrication of PVDF electrospun piezo-energy harvesters with interdigital electrodes, details of the fabrication process involved in flexible piezoelectric composites and the demonstration of energy harvesters from NFES (Near-Field Electrospinning) PVDF fibers.

C. T. Pan, Ying-Chung Chen, Y. M. Hwang, and Liwei Lin
September 2013

Acknowledgments

Although four authors worked on this text, it would not have been written without support from various sources. We express our thanks to Z. H. Liu and Y. C. Candice Chen, undergraduate students who helped with edits and figures.

1

Introduction

1.1 Background

Various approaches in developing clean and green energy systems have been extensively explored in the past few years. Much research has focused on properties such as small size, light weight, high density in power, economically efficient, and environmentally friendly. Possible sources of ambient energy include light energy, wind power, kinetic energy, and thermal energy.

In many applications, researchers have been investigating the use of microactuators and microsensors in MEMS (microelectromechanical systems) where an independent power source is needed. A possible solution is to design the power supply at the same scale as actuators, sensors, and electronics. The conventional solution is to use batteries, but batteries can be undesirable for many reasons: they tend to be quite bulky, contain a finite amount of energy, have a limited life, and contain chemicals that could cause a hazard. Vibration energy harvesters and energy scavengers recover mechanical energy from their surrounding environment and convert it into usable electricity as sustainable self-sufficient power sources to drive micro-to milli-watt-scale powered instruments independently.

Mechanical kinetic energy is ubiquitous in real environments. The conversion of ambient mechanical vibration to electrical energy is considered one of the likely methods of powering a wireless sensor, without hazardous byproducts related to

Design and Fabrication of Self-Powered Micro-Harvesters: Rotating and Vibrating Micro-Power Systems, First Edition.
C. T. Pan, Y. M. Hwang, Liwei Lin and Ying-Chung Chen.

power generation. The power source does not need to be replaced and fuel does not need to be replenished from time to time like batteries.

There are two types of harvesting systems of mechanical kinetic energy under investigation in this book, as follows.

1. **Electromagnetic microgenerator:** based on Faraday's law to harvest energy.
2. **Piezoelectric energy harvester:** using piezoelectric material to convert strain energy into electricity.

Microsystems have to be self-powered, so efficient energy scavenging is crucial. The self-powered microsystems are designed to avoid the replacement of energy cells and miniature sensing devices. Vibration-based energy harvesting is a process of capturing ambient kinetic energy and converting it into usable electricity. The growing demand of cell phone devices such as miniature wireless sensor networks and the recent advent of extremely low-power controlled circuit and MEMS devices make such renewable power sources very attractive.

In addition, the energy harvesting process must be compatible with environmental vibrations such as running machines and human body movement. However, the wide range of environmental vibration frequencies means the harvester operates at low efficiency when deployed in a stochastic surrounding vibration.

1.2 Energy Harvesters

Microtransducers have been of the focus of much attention since the rapid development of MEMS technology based on micro-wireless sensors and actuators with different applications in the communication, military, and biomedical industries. Research in the field of kinetic energy harvesters using piezoelectric materials and magnetic modules that harvest environmentally random energy, such as irregular vibrations, light airflow, and human activity, have attracted considerable attention. This type of mechanical energy available in our environment has a wide spectrum of frequencies and time-dependent amplitudes, however. The design of high-sensitivity energy harvesters is therefore crucial, including electromechanical conversion of materials, efficient energy transfer of mechanical structures, and controlled circuit to obtain good performance.

In this book, electromagnetic transduction mechanisms include vibrational and rotary power generators. Microgenerators to harvest vibration-induced and rotary-induced energy such as human motion are an attractive study topic. According to Faraday's law, a permanent magnet moving relative to a coil can induce an EMF (electromotive force). Conventional microcoil fabrication processes include LIGA (Lithographie, Galvanoformung, Abformung), LIGA-like, filament winding, printed circuit board etching, and LTCC (low-temperature co-fired ceramic). LIGA and LIGA-like processes consist of expensive methods such as thick photoresistant exposure, thin film, and a high-aspect-ratio electroplating process. The latest research in the field of LTCC technology applications for the field of microgenerators are included in this book.

Much research has focused on electromechanical coupling effects of various ambient energy sources using PZT (lead zirconate titanate), ZnO (zinc oxide), PVDF (polyvinylidene fluoride), and AlN (aluminum nitride) -based piezoelectric thin films. The use of piezoelectric PZT devices has been restricted because of environmental related issues. Using ZnO and PVDF piezoelectric materials is attractive in terms of their low cost, high resistance to fatigue, and environmentally friendly applications. Significantly, the deposition process of sputtering ZnO thin films with high c-axis preferred orientation and electrospun PVDF fibers with high piezoelectric β-phase crystallization are controlled at room temperature; they therefore do not require post-annealed and electrical repoling processes to obtain an excellent piezoelectricity. The sputtering ZnO thin films and well-aligned electrospun piezoelectric PVDF fibers can be directly deposited on flexible substrates at room temperature, such as PET (polyethylene terephthalate) and PI (polyimide). Utilizing ZnO and PVDF as the main elements of piezoelectric materials has the following additional advantages:

1. they have high piezoelectric coupling coefficients,
2. they do not cause environmental pollution; and
3. the deposition process is controlled at room temperature, which is suitable for all flexible substrates.

1.2.1 Piezoelectric ZnO Energy Harvester

In the fabrication process, the piezoelectric ZnO thin film is deposited between the electrodes. ZnO thin film with excellent distribution of crystal grains with a high diffraction peak (002) was deposited using radio frequency (RF) magnetron sputtering. To fabricate the selectively deposited UV-curable resin lump structures as a proof mass, electrospinning techniques were used to construct various patterns on the back of the PET-based composite plate to control the resonant frequencies. Piezoelectric ZnO thin-film harvesters can therefore harvest energy in small working frequencies. This ZnO piezoelectric energy harvester consists of a piezoelectric laminated cantilever and a mass, and harvests vibrational energy. Through the resonant inertial oscillation of the cantilever, the power of a piezoelectric generator is generated by the inherent piezoelectric effect. The application of stress on a piezoelectric material generates a corresponding electric charge. The mechanical energy of a piezoelectric energy harvester is converted into electrical energy. A piezoelectric energy harvester must operate at the resonance frequency to harvest higher power; the friction and damping reduce its output. These obstacles are addressed in Chapter 2.

1.2.2 Vibrational Electromagnetic Generators

Microelectromagnetic generators demonstrate the advantages of power output and integrated circuit (IC) package processing, thereby saving energy and improving generator efficiency. An electromagnetic generator converts mechanical energy into electrical energy. According to Faraday's law, a permanent magnet moving relative to a coil creates an EMF. Electromagnetic transduction mechanisms include vibrational and rotational power generators. When a permanent magnet installed on a mass of a vibrational electromagnetic generator moves relative to a coil, it induces an EMF. The magnetic microgenerator includes an LTCC microinductor, a magnetic core element, a magnet module, an oscillation mass, and an attached adjustable resistor. The magnet module is composed of two NdFeB magnets of thickness 2 mm attached on the upper and lower surfaces of the LTCC microinductor. NdFeB magnets are commercially available and contribute towards cost-reduced mass production. A cylindrical hole is located at its center to accommodate the microspring oscillation. The magnetic elements also provide the supporting frame for mechanical oscillation. The power of the vibrational generator is limited by natural resonance frequencies and the operation of the vibrational source.

1.2.3 Rotary Electromagnetic Generators

The recycling of mechanical energy has been the focus of many previous studies. Rotary electromagnetic generators show potential for the development of self-sufficient energy sources. Machines using planar rotary permanent magnets have been developed for various applications due to their ability to eliminate field excitation losses. The frequencies of rotary electromagnetic generators do not have specific restrictions, making them environmentally friendly. Furthermore, as long as there is movement between the coil and the magnet, energy can be continuously created, ensuring performance in machines of all sizes. The microgenerator comprises a multilayer planar LTCC silver (Ag) or copper (Cu) microcoil and multipole hard magnets of neodymium/iron/boron (Nd/Fe/B). Finite-element simulations and analytical solutions are performed to predict the induced voltage. Various configurations of planar microcoils were investigated: sector-shaped, circular, and square microcoils. Recently, hub dynamos have become popular for powering bicycle lights. Rotary electromagnetic generators are suitable for use in bicycle dynamos. The principle of a rotary electromagnetic generator, which can be easily installed due to its rotary motion, is similar to that of a vibrational electromagnetic generator, and is not limited to use at the resonance frequency.

1.2.4 NFES Piezoelectric PVDF Energy Harvester

Other research efforts have focused on energy harvest and actuation properties of the modified NFES (near-field electrospinning) piezoelectric PVDF fibers. A direct-write electrospinning technique by means of NFES was developed to produce the controllable nanofiber deposition. Compared to the conventional electrospinning process, it shows that decreasing electrical field in continuous NFES results in smaller line-width fiber deposition. Piezoelectric fibers have the characteristics of electromechanical energy conversion and offer the advantages of nanoscale size. Piezoelectric fibers can harvest relatively low-frequency kinetic energy, such as body movement and muscle stretching. Researchers have published many studies on piezoelectric fibers, wires, and rods. A direct-write electrospinning techniques using NFES was developed to achieve controllable fiber deposition for various materials. Unlike the conventional electrospinning process, NFES only needs a small electric field to produce continuous fibers with fine diameters. The fabrication process of PVDF fibers in this study is compatible with a flexible substrate and does not involve any complex processes. PVDF is a potential piezoelectric polymer because of its high flexibility, biocompatibility, and low cost. These features make PVDF attractive for energy conversion applications involving microelectromechanical devices, electromechanical actuators, and energy harvesters.

1.3 Overview

The book is arranged in five chapters to describe the research and development of energy harvesters. Chapter 1 describes the background of energy harvesting, the development of electromagnetic generators and piezoelectric materials for flexible energy harvesters, and the use of electrospun piezoelectric nanofibers.

Chapter 2 describes the design and fabrication of flexible piezoelectric generators based on ZnO thin films. The development of a single ZnO broad-bandwidth vibrational energy harvesting system and double-sided piezoelectric energy harvester are described, including experiments, measurements, and interfacial adhesion of PET-based substrate.

In Chapter 3 the design and fabrication of vibration-induced electromagnetic microgenerators are discussed. The manufacturing technology of LTCC applied to the construction of multilayer silver microinduction coils and spiral ceramic microspring in microgeneration is also introduced. Compared to silicon-based processes, ceramic and silver are not easily broken or fractured after sintering. Microgeneration has high-current output because of the low resistance of the multilayer structures.

Chapter 4 focuses on the design, fabrication, testing, and application of an in-plane rotary electromagnetic microgenerator to obtain a high-power output. Finite-element simulations have been performed to observe electromagnetic information. The study also establishes analytical solutions for the microgenerator to predict the induced voltage for three different configurations of planar microcoils.

Chapter 5 describes the design and fabrication of a PVDF electrospun piezo energy harvester with interdigital electrode and the fabrication processes of flexible piezoelectric composites and NFES PVDF fibers. Furthermore, adding modified CNTs (carbon nanotubes) to reinforce PVDF nanofibers can enhance the crystallinity of the β phase and increase its ability to store charges. NFES can potentially be reduced to the nanometer scale and form any shape for various sensing and actuation applications.

2

Design and Fabrication of Flexible Piezoelectric Generators Based on ZnO Thin Films

2.1 Introduction

The study of energy has recently been gaining more attention. Various clean, green, and renewable energy systems have been extensively developed and explored to help address the problem of energy shortage. Alternative energy systems have been actively researched worldwide while regarding considerations such as compact size, light weight, high density in power, economy, safety, and environmental concerns. The numerous renewable sources of green energy include hydroelectric [1], wind [2], oceanic [3], solar [4–6], and vibration [7, 8]. However, environmental constraints generally limit recycling power applications. For instance, hydroelectric power plants are always built with turbines, such as the Pelton turbine or the chain turbine.

Solar power is the most frequently used energy source. However, applying solar power is restricted by day-and-night limitations [9]. The development of green energy

Design and Fabrication of Self-Powered Micro-Harvesters: Rotating and Vibrating Micro-Power Systems, First Edition.
C. T. Pan, Y. M. Hwang, Liwei Lin and Ying-Chung Chen.

should therefore address the disadvantages of solar power. In many applications, for example, solar power has been applied in microactuators and microsensors in microelectromechanical systems (MEMS) that require an independent and embedded power source without an outside connection. A possible solution is to design the power supply to the same scale as actuators, sensors, and electronics [10]. The conventional solution involves using batteries however, which can be undesirable for many reasons: they tend to be relatively bulky, contain a finite amount of energy, have a limited life, and contain hazardous chemicals [11].

In many applications, microsystems must be self-powered; efficient energy scavenging is therefore crucial. Mechanical vibration is ubiquitous in real environments. Converting ambient mechanical vibration to electrical energy is considered one of the likely methods for powering wireless sensors without hazardous byproducts related to power generation. In addition, the power source of wireless sensors does not need to be replaced and the fuel does not require periodical replenishment as for batteries [12]. Piezoelectric materials have electrical-mechanical coupling effects, and are the leading candidates for converting mechanical energy into electricity [13]. White *et al.* [14] developed an inertial piezoelectric generator that uses 7-μm-thick film of lead zirconate titanate (PZT) with a 100-μm-thick steel substrate. The device produces an output of 2.1 μW from vibrations in the environment. Roundy *et al.* [15] improved the geometry of the scavenger's piezoelectric bimorph. Using the same volume of PZT and a trapezoidal geometry can supply more than twice the energy of rectangular geometry. Fang *et al.* [16] used a generator structure of a nickel-metal composite cantilever. At a resonant frequency of approximately 608 Hz, 608 mV of AC voltage value with a 2.16 μW power level was obtained. Minazara *et al.* [17] used a mechanically excited unimorph piezoelectric membrane transducer. A power output of 0.65 mW was generated at the resonance frequency (1.71 kHz) across a 5.6 kΩ optimal resistor.

Several harvesting systems of resonant mechanical vibration energy are currently under investigation. Three considerations potentially affect transducer technology selection:

- **Electromagnetic**: a coil attached to a mass that vibrates through a magnetic field to induce voltage according to Faraday's law [18–23].
- **Piezoelectric**: using piezoelectric material to convert strain energy into electricity [9–18, 24–26].
- **Electrostatic**: inducing capacitor voltage through the movement of a mass that has its permanent charges electrically arranged [15, 27].

Dutoit *et al.* [27] compared these three generators according to energy density. Of these three mechanisms, the electrostatic transducer has the lowest energy storage density and conversion efficiency. Further, electromagnetic generators usually require a large operating space, complex instrumentation, and continual management of facilities, all of which are expensive. Many researchers have therefore examined micropiezoelectric power generators and their fabrication from mostly piezoelectric materials (Pb,Zr)TiO_3 (PZT), aluminum nitride (AlN), and zinc oxide (ZnO). Table 2.1 lists the general characteristics and properties of PZT, AlN, and ZnO films [28].

The production of PZT piezoelectric material causes considerable environmental pollution. This study adopted ZnO as a piezoelectric thin film for three main reasons: (1) it has relatively high piezoelectric coefficients compared to piezoelectric AlN [24]; (2) producing ZnO thin film does not cause substantial environmental pollution compared to piezoelectric PZT thin film; and (3) it is a non-ferroelectric material that requires no poling or post-deposition annealing. These factors make ZnO piezoelectric material useful for elucidating the application of piezoelectric energy harvester in environments.

This chapter is organized in four sections to describe the research and developmental efforts. Section 2.1 briefly introduces the background of green and renewable energy systems, the development of generator technologies, and the materials of piezoelectric thin films for piezoelectric generators.

Section 2.2 focuses on the theoretical analysis and simulations of flexible piezoelectric generators based on ZnO thin films on polyethylene terephthalate (PET) substrates. A piezoelectric cantilever plate was designed and simulated using commercial software ANSYS FEA (finite element analysis) to determine the optimal thickness of the PET substrate, internal stress distribution, operation frequency, and electric potential.

Section 2.3 describes the relationship between the model solution of piezoelectric cantilever plate equation, vibration-induced electric potential, and electric power. Section 2.4 discusses the development of high-performance piezoelectric generators using single-sided and double-sided ZnO thin films on a flexible stainless steel substrate (SUS304). Relevant fabrication processes, experiments, measurements, and piezoelectric responses are addressed.

Table 2.1 General characteristics and properties of piezoelectric materials [28]

	Mass density (kg m^{-3})	Electromechanical coupling coefficient k_t^2 (%)	Dielectric constant ε_r
PZT	1550	20.25	309.8
ZnO	5680	7.5	8.85
AlN	3270	6	9.14

2.2 Characterization and Theoretical Analysis of Flexible ZnO-Based Piezoelectric Harvesters

2.2.1 Vibration Energy Conversion Model of Film-Based Flexible Piezoelectric Energy Harvester

Figure 2.1a shows the composite cantilever plate used as a harvester with a rectifying circuit (B) and a capacitor (C) as the energy storage module. Ambient vibration amplitude on the anchor side induced vertical deflection at the front end of the cantilever plate. The piezoelectric layer on the composite structure deformed when the cantilever plate oscillated with mass. The bending results in a mechanical strain distributed along the cantilever plate, which is then converted to alternating voltage through the transverse-mode (d_{31}) piezoelectric effect (the generated strain is perpendicular to the electric field, which forms the d_{31} mode of the piezoelectric element), and overcomes the forward bias of the rectifying diodes. Subsequently, the full-wave rectified potential can quickly charge the storage module.

Figure 2.1b shows a schematic model of vibration energy conversion. The mechanical behavior of a vibrating piezoelectric harvester can be modeled using a seismic mass m bonded on a spring k corresponding to the stiffness and a damping factor b_m, which in turns corresponds to the mechanical losses of the structure. A displacement $y(t)$ is used to vibrate the rigid house, resulting in differential movement $x(t)$ between the mass and the house; the relative displacement of the mass is therefore $U_z(t)$.

This section reports the analysis of a new self-powered flexible Cu/ZnO/Al/PET-based piezoelectric energy harvester with a storage module. To verify the optimal thickness of the PET substrate evaluated by performing finite element model (FEM) analysis, an accurate analytical formula was developed. The optimal thickness of the substrate was calculated. Thus, the tensile and compressive strains do not occur

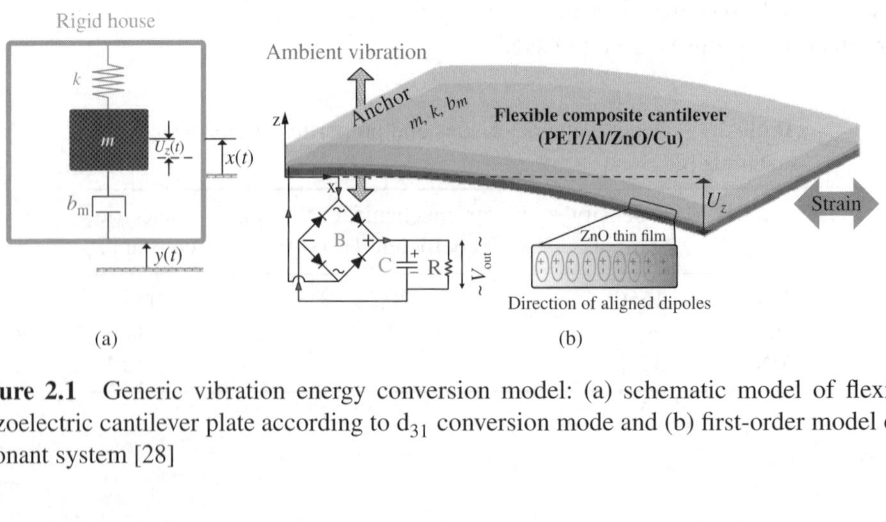

Figure 2.1 Generic vibration energy conversion model: (a) schematic model of flexible piezoelectric cantilever plate according to d_{31} conversion mode and (b) first-order model of a resonant system [28]

on the piezoelectric ZnO layer simultaneously. Static analysis, modal analysis, and harmonic response analysis were performed to determine a suitable thickness of the PET substrate, structural modal parameters, frequency response functions, and electric potential.

In the fabrication process, the piezoelectric ZnO thin film was deposited between the Al and Cu electrodes. The Al thin film used as the bottom electrode was deposited on the PET substrate by using a sputtering deposition method because of its superior adhesion to the PET substrate and lattice constant matching with ZnO. The ZnO thin film with superior compactness and more even distribution of crystal grains with a high (002) diffraction peak at $2\theta = 34.44°$ and full-width at half-maximum (FWHM) of 0.422° was deposited using RF magnetron sputtering. To fabricate the selectively deposited ultraviolet- (UV)-curable resin lump structures as a proof mass, electrospinning and stereolithography techniques were used to construct various patterns on the back of the PET-based piezoelectric composite plate, which can adjust the resonant frequencies of the composite cantilever plate to achieve larger deflection. Piezoelectric ZnO thin-film harvesters can therefore harvest energy in small work spaces. In addition, they demonstrate long-term stability, have a low operating frequency with low-level excitation and a high deflection response compared to traditional PZT-based bulk ceramic vibrational structures, and can be used at low cost in clean energy applications. The schematic layout of the self-powered storage system and the design parameters of the flexible piezoelectric composite plate are shown in Figure 2.2.

In this study, a broad bandwidth harvesting system with a wide bandwidth of 100–400 Hz was designed and fabricated. Four individual, flexible PET-based ZnO piezoelectric harvesters were assembled in parallel with a rectifying circuit

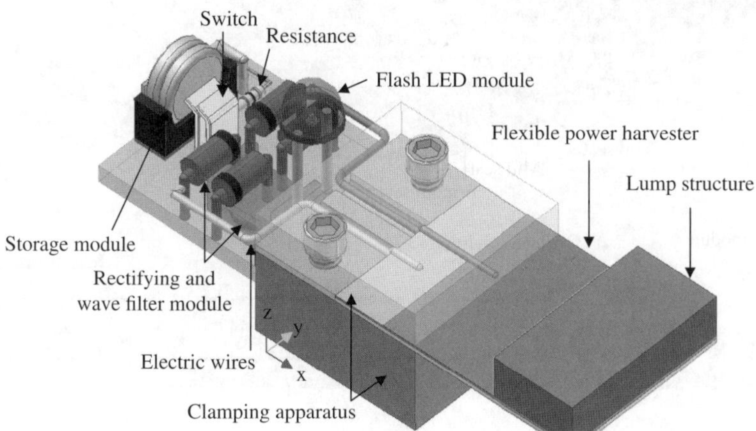

Figure 2.2 Schematic illustration of self-powered storage system and flexible composite structure [28]

and storage module. A schematic illustration of a broad bandwidth piezoelectric harvesting system with the rectifying storage module is shown in Figure 2.3. There are four rectifying circuits in parallel connection and four capacitors are assembled into a storage module [29].

Finally, the effect of vibration on the mechanical and electrical properties of multilayer flexible electronics composites such as ITO (indium tin oxide)/PET, Al/PET, ZnO/ITO/PET, and ZnO/Al/PET was examined to analyze the coating adhesion, mechanical strength, and conductivity. Nanoscratching tests (Nanoindenter XP system) were conducted to examine the adhesion of multilayer interfaces. Field-emission scanning electron microscopy (FE-SEM) and optical microscopy (OM) were employed to observe the surface morphology following the test. In addition, scratching experiments using Berkovich and conical probes on multilayered structures were compared and investigated.

2.2.2 *Piezoelectricity and Polarity Test of Piezoelectric ZnO Thin Film*

Figure 2.4 shows the response of induced electric potential when a periodic force is applied to the piezoelectric cantilever structure (Cu/ZnO/Al/PET). The bending

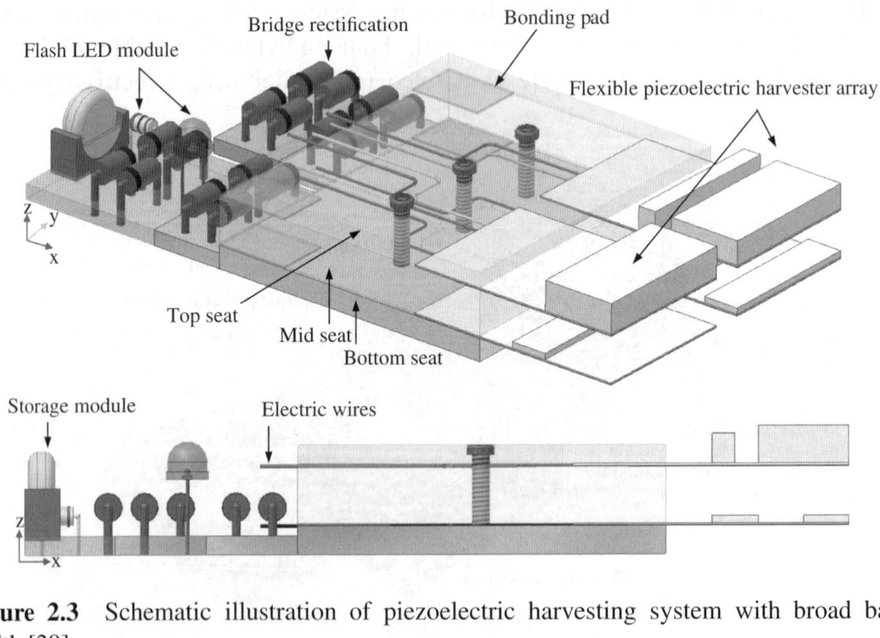

Figure 2.3 Schematic illustration of piezoelectric harvesting system with broad bandwidth [29]

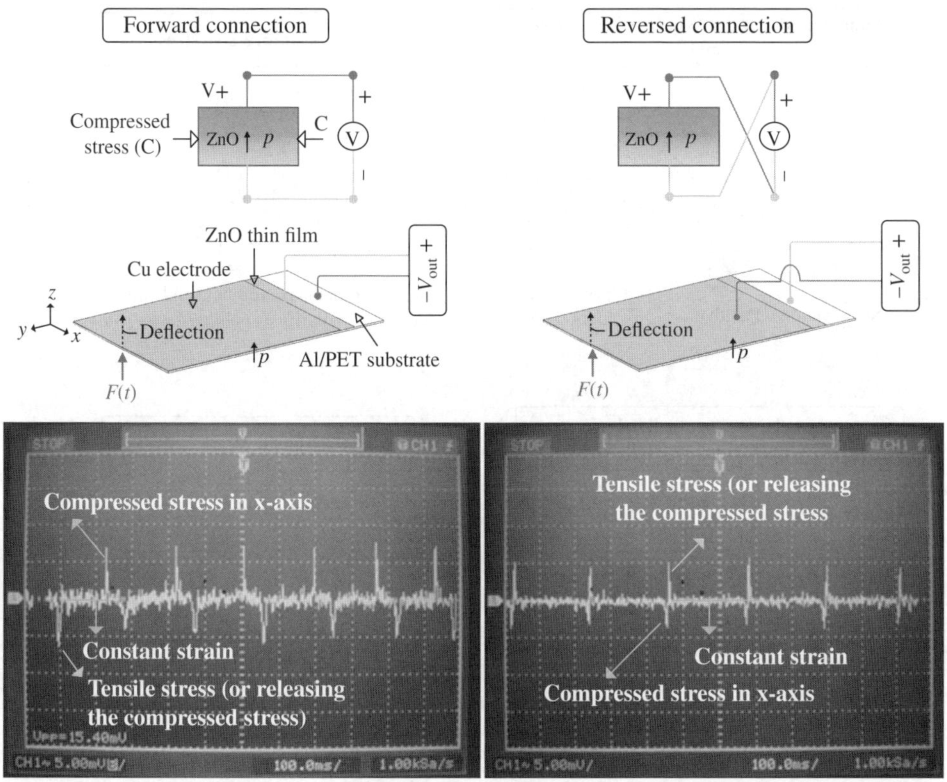

Figure 2.4 Polarity test of piezoelectric ZnO thin film can be observed by piezoelectricity responses: (a) forward-connected to the measurement system and (b) reverse-connected to the measurement system [28]

results in a mechanical strain distributed along the ZnO thin film, which is then converted to alternating voltage through the transverse-mode (d_{31}) piezoelectric effect.

The center of the cations (Zn^{2+}) and the center of the anions (O^{2-}) relatively displaces in the Wurtzite-structured ZnO crystal by applying stress and result in an electric dipole moment (Figure 2.5a). A constructive summation of the dipole moments created by all units in the crystal creates a piezoelectric field (E_z) due to the polarization of atoms, resulting in a macroscopic potential drop between the two sides along the crystal z-axis. This is the piezoelectric potential, or piezopotential, which can serve as the driving force for the flow of electrons in the external load once it is subjected to mechanical deformation, the fundamental working principle of the ZnO film-based energy harvester. Figure 2.5b shows a simplified model of the piezoelectricity of ZnO film. In the absence of external stress, the ionic charge distribution is symmetric and the net electric dipole moment is zero.

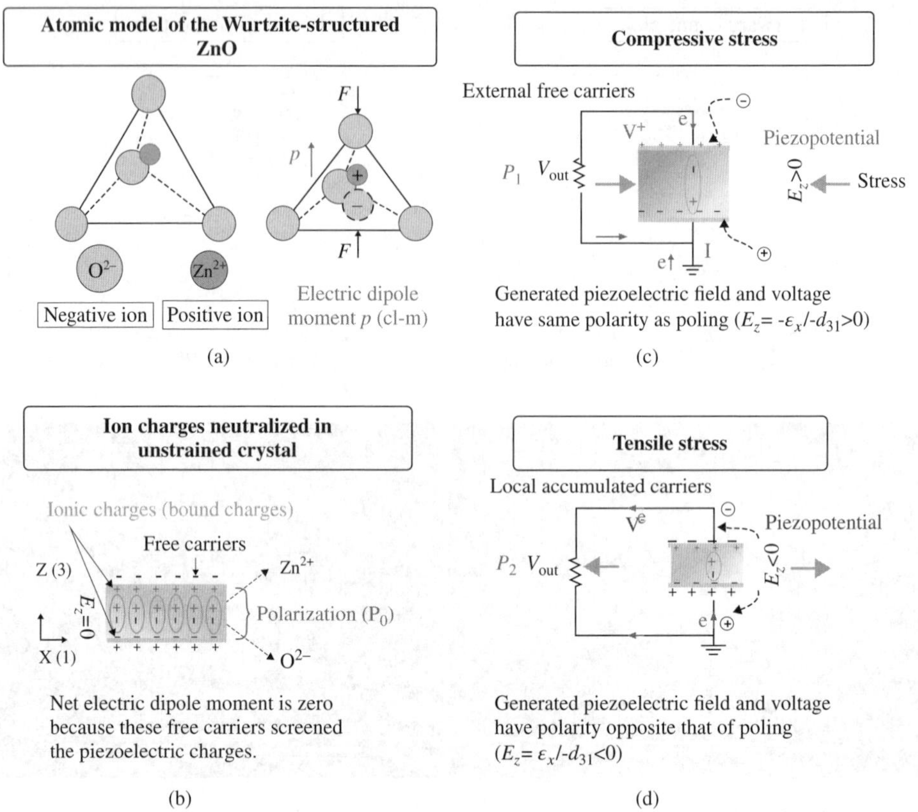

Figure 2.5 Mechanical bending of the substrate creates tensile strain and a corresponding piezoelectric potential in ZnO thin film: (a) atomic model of the wurtzite-structured ZnO; (b) the net electric dipole moment is zero in unstrained ZnO thin film; (c) a compressive stress is applied to the ZnO thin film in x-axis, leading to a drop in the piezoelectric potential along z-axis, and forcing electrons to flow along an external circuit to charge the film; and (d) when the stress is released, electrons flow back in the opposite direction [28]

When compressive stress is applied to the crystal on the x-axis (Figure 2.5c), a piezoelectric field ($E_z > 0$) is created in the ZnO film because of the induced ionic charges (or polarization P_1) in the crystal, and induces a higher piezopotential V^+ on the $+c$-axis side. These piezoelectric charges, which are non-mobile, remain in the ZnO thin film for an extended period without being depleted by the free carriers as long as the stress is preserved [30]. Because the charges associated with the ions are rigid and affixed to the atoms, they cannot move freely. The electrons can be driven forward by the positive piezopotential through the external circuit (known as the piezoelectric-potential-driven flow of external electrons), forming the positive potential signal as shown in the red line area of Figure 2.4a. In this study, this process has to be faster than the charge-accumulation process because the external free carriers are also driven to balance this piezopotential. Under a constant strain, the piezopotential gradually vanished because of a balance between external free carriers and this potential (known as the screening effect of the charge carriers). The piezopotential in piezoelectric materials is therefore a function of the changing stress. As the tensile stress is applied (the compressed stress on the x-axis is released) (Figure 2.5d), the piezopotential V^- reverses with a higher potential at the c-axis side. The local accumulated carriers at both ends of the ZnO film quickly flow to the top electrode, creating a circular flow of the electrons in the external circuit and producing a negative signal caused by the reverse-flowing carriers shown in the green line area. This process is considerably faster than the charge-accumulation process. Furthermore, when the harvester is connected in reverse polarity, all of the response signals exhibit reversed output (Figure 2.4b). It is proposed that satisfying the polarity and criteria is a necessary and sufficient condition for confirming that the output is generated by the flexible ZnO energy harvester.

2.2.3 *Optimal Thickness of PET Substrate*

The neutral axis (NA) position in a composite plate can influence stress distribution [30]. The optimal thickness of a PET substrate is calculated using the geometric parameters on the y–z-plane, as shown in Figure 2.6. The neutral axis position is situated between the PET layer and the Cu foil layer. The desirable thickness of the PET substrate $\bar{t}_{\mathrm{PET}}$ is defined

$$\bar{t}_{\mathrm{PET}} \geq t_{\mathrm{Cu}}\sqrt{\eta_2} \pm 0.5 t_{\mathrm{ZnO}} \tag{2.1}$$

where t_i is the thickness of each layer ($i = 1, 2, 3$), η_i is the ratio of the elastic constant and $\eta_1 = E_{\mathrm{ZnO}}/E_{\mathrm{PET}}$ and $\eta_2 = E_{\mathrm{Cu}}/E_{\mathrm{PET}}$, where E_{PET}, E_{ZnO}, and E_{Cu} are the elastic constant of the PET substrate, the ZnO piezoelectric material, and the Cu foil.

In the composite plate, the tensile and compressive strain should not occur on the piezoelectric ZnO layer as the plate experiences a time-varying vibration. By contrast, the one-way mechanical strain is sufficient to enhance the electric charge output, which is generated within the top and bottom electrodes of the ZnO layer.

2.2.4 *Model Solution of Cantilever Plate Equation*

Figure 2.6 depicts schematic diagrams of the composite cantilever plate with geometric variables. The deflection at each position x of the composite plate is defined by Equation (2.2).

$$U_z(x) = \frac{M(l-x)l^3}{3E_{\mathrm{PET}}Ix} \tag{2.2}$$

where $U_z(x)$ is derived from an equivalent moment $M(l-x) = F_z x$, which is caused by an externally applied force F_z. The maximum vertical deflection at the free end of

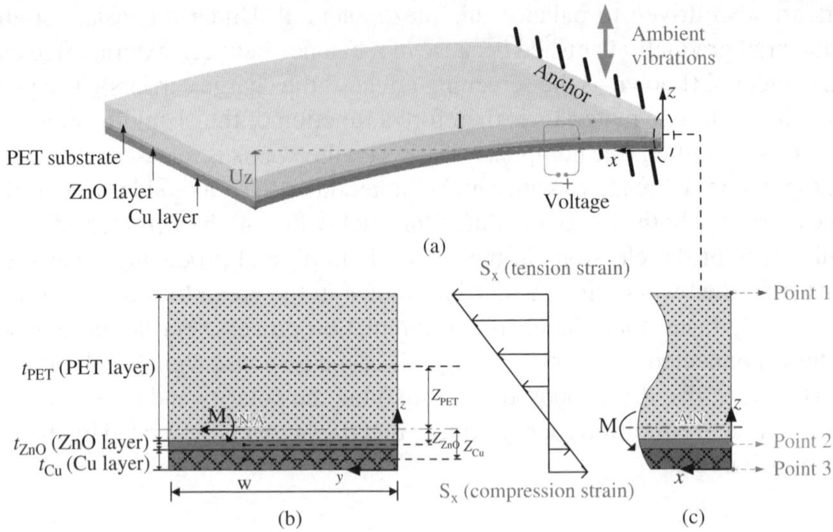

Figure 2.6 Schematic layout of the composite cantilever plate: (a) the deflection (U_z) in the free end of the cantilever, (b) the geometry of ZY plane, and (c) the position of neutral axis and the strain distribution [28]

the cantilever plate is given by Equation (2.3):

$$U_z(l) = \frac{M(0)l^2}{3E_{\mathrm{PET}}I} \tag{2.3}$$

The effective moment of the inertia of the composite plate is developed using Equation 2.4:

$$I = \frac{wt_{\mathrm{PET}}{}^3}{12} + wt_{\mathrm{PET}}z_{\mathrm{PET}}{}^2 + \frac{\eta_1 wt_{\mathrm{ZnO}}{}^3}{12} + \eta_1 wt_{\mathrm{ZnO}}z_{\mathrm{ZnO}}{}^2 + \frac{\eta_2 wt_{\mathrm{Cu}}{}^3}{12} + \eta_2 wt_{\mathrm{Cu}}z_{\mathrm{Cu}}{}^2 \tag{2.4}$$

where z_i is the distance between the center of the geometry and the neutral axis of the three layers (z_{PET}, z_{ZnO}, and z_{Cu}), and w is the width of the plate.

The large deflection theory of thin plate can be employed when the deflection is small relative to the width and length of the plate [31]. The internal strain on the x-axis for a point within the plate can be expressed [32]:

$$S_x = -z\frac{\partial^2 U_z(x)}{\partial x^2} \tag{2.5}$$

The axial strain on the ZnO layer is defined by Equation (2.6).

$$S_{x,\mathrm{ZnO}} = -z_{\mathrm{ZnO}}\frac{\partial^2 U_z(x)}{\partial x^2} \tag{2.6}$$

The distance z is from the neutral axis to each plane within the thickness of the composite plate. The internal stress distribution is related to the ratio of the elastic constant, and the stress in the plate is defined by Equation (2.7) [33]:

$$T_x = -\frac{E}{1-\upsilon} z \frac{\partial^2 U_z(x)}{\partial x^2} \tag{2.7}$$

where υ is Poisson's ratio.

The axial stress on the ZnO layer must be modified according to the ratio of the elastic constant η_1:

$$T_{x,\mathrm{ZnO}} = -\eta_1 \frac{E_{\mathrm{PET}}}{1-\upsilon_{\mathrm{PET}}} z_{\mathrm{ZnO}} \frac{\partial^2 U_z(x)}{\partial x^2} \tag{2.8}$$

The neutral axis $\overline{Z}$ of the composite plate is calculated via:

$$Z = \frac{(wt_{\mathrm{PET}})\left(\frac{t_{\mathrm{PET}}}{2} + t_{\mathrm{ZnO}} + t_{\mathrm{Cu}}\right) + (\eta_1 w t_{\mathrm{ZnO}})\left(\frac{t_{\mathrm{ZnO}}}{2} + t_{\mathrm{Cu}}\right) + (\eta_2 w t_{\mathrm{Cu}})\left(\frac{t_{\mathrm{Cu}}}{2}\right)}{(wt_{\mathrm{PET}}) + (\eta_1 w t_{\mathrm{ZnO}}) + (\eta_2 w t_{\mathrm{Cu}})} \tag{2.9}$$

The maximum axial stress and strain distributed on the ZnO layer are defined by Equations (2.10) and (2.11), respectively:

$$T_{x,\mathrm{ZnO,max}} = \frac{M(0) z_{\mathrm{ZnO}}}{I} = \frac{3\eta_1 E_{\mathrm{PET}} z_{\mathrm{ZnO}} U_z(l)}{l^2} \tag{2.10}$$

$$S_{x,\mathrm{ZnO,max}} = \frac{3(1-\upsilon_{\mathrm{PET}}) z_{\mathrm{ZnO}} U_z(l)}{l^2} \tag{2.11}$$

Improving the axial stress on the ZnO layer is achieved by analysis of these parameters. To obtain the electrical circuit equation with mechanical coupling, the following piezoelectric constitutive relationship should be considered:

$$D_z = d_{31} T_x + \varepsilon_{33}^T E_z \tag{2.12}$$

where D_z is the electric displacement in the direction of the z-axis, d_{31} is the piezoelectric strain coefficient that couples the electric field in the z-axis with the stress T_x, ε_{33}^T is the permittivity at a constant stress, and E_z is the external electric field in the ZnO in terms of the electric potential across the ZnO layer (i.e. $E_z = v/t_{\mathrm{ZnO}}$ where v is electric potential). If Equation (2.12) is rearranged to express the maximum axial stress T_x of the ZnO layer, then the external electric field is assumed to be zero and the electric displacement of the ZnO layer can be expressed:

$$D_z = d_{31} \frac{3\eta_1 E_{\mathrm{PET}} z_{\mathrm{ZnO}} U_z(l)}{l^2} \tag{2.13}$$

The total strain-induced charge Q on the ZnO layer is defined:

$$Q = \int D_z dA = \int_0^l D_z w \, dx = \frac{3\eta_1 E_{\mathrm{PET}} z_{\mathrm{ZnO}} U_z(l) w d_{31}}{l} \tag{2.14}$$

The maximum induced electric potential between the top and bottom surfaces of the ZnO layer is defined:

$$V_{\max} = \frac{Q}{C_{\mathrm{ZnO}}} = \frac{3\eta_1 E_{\mathrm{PET}} z_{\mathrm{ZnO}} U_z(l) t_{\mathrm{ZnO}} d_{31}}{l^2 \varepsilon_{33}} \tag{2.15}$$

where C_{ZnO} and ε_{33} represent the capacitance and the permittivity of the ZnO, respectively.

Detwiler *et al.* [34] demonstrated a cantilevered piezoelectric composite plate with a continuous electrode layer, in which the induced electric potential is an average value. The electric potential is dependent on the integrated strain rates over the electrode area, which is known as the integrated effect of the piezoelectric sensor [35]. The designed electrodes of the Cu foil and Al film are continuous electrodes on both sides of the piezoelectric ZnO layer. The electric potential is an average value. Equation (2.15) must therefore be corrected by using the half-maximum axial strain on the ZnO layer. The integrated effect is also applied in establishing the boundary condition of the piezoelectric cantilevered ZnO layer in the simulation process.

$$V' = \frac{3\eta_1 E_{\mathrm{PET}} z_{\mathrm{ZnO}} U_z(l) t_{\mathrm{ZnO}} d_{31}}{2 l^2 \varepsilon_{33}} \tag{2.16}$$

The axial stress is applied repeatedly with a resonant frequency of ω_n (Hz), and the maximum electric power of the ZnO layer is obtained without any resistive load [36]:

$$P = V' Q \omega = \frac{9\eta_1^2 E_{\mathrm{PET}}^2 z_{\mathrm{ZnO}}^2 U_z^2(l) t_{\mathrm{ZnO}} w d_{31}^2 \omega_n}{2 l^3 \varepsilon_{33}} \tag{2.17}$$

2.2.5 *Vibration-Induced Electric Potential and Electric Power*

In this study, a sinusoidal signal $y(t) = u_0 \cos(\omega\, t)$ excitation was applied persistently at the anchor of the cantilever. The deflection $U_z(x,t)$ was assumed, and the Euler–Bernoulli beam model was adopted for simulation. The continuous system dynamic behavior can be modeled:

$$\frac{\partial^2}{\partial x^2}\left[EI \frac{\partial^2 U_z(x,t)}{\partial x^2}\right] + m \frac{\partial^2 U_z(x,t)}{\partial t^2} + b_{\mathrm{m}} \frac{\partial U_z(x,t)}{\partial t} = F_z(x,t) = -m u_0 \omega^2 \cos(\omega t) \tag{2.18}$$

where EI is the flexural stiffness of the plate, $U_z(x,t)$ denotes the vertical deflection of the mass point, m is the mass per unit length of the plate, b_{m} is the damping per unit length, $F_z(x,t)$ is the force per unit length, and u_0 is the amplitude of shaker vibration.

The maximum vertical deflection in the free end of the cantilever, when the excited frequency is equal to the resonant frequency (i.e. $\omega = \omega_n$), is defined [37]:

$$U_z(l,\omega_n) = \frac{A_n(l) m u_0}{\xi} \frac{[\cos(k_n l)\sinh(k_n l) - \cosh(k_n l)\sin(k_n l)]}{\sin(k_n l) + \sinh(k_n l)} \tag{2.19}$$

where ξ is the damping ratio of the cantilevered structure, defined

$$\xi = b_{\mathrm{m}}/c_{\mathrm{c}} = b_{\mathrm{m}}/2\sqrt{km} \tag{2.20}$$

where c_{c} is the critical damping coefficient, k is the equivalent spring constant ($k = 3E_{\mathrm{PET}}I/l^3$), $m = lw(\rho_{\mathrm{PET}}t_{\mathrm{PET}} + \rho_{\mathrm{ZnO}}t_{\mathrm{ZnO}} + \rho_{\mathrm{Cu}}t_{\mathrm{Cu}})$ denotes the mass of the composite plate [38], and A_n is an integration constant.

By combining Equations (2.16), (2.17), and (2.19), the maximum electric potential ($\overline{V}$) and maximum electric power ($\overline{P}$) in the resonant state are defined:

$$V = \frac{3\eta_1 E_{\mathrm{PET}} z_{\mathrm{ZnO}} t_{\mathrm{ZnO}} d_{31} A_n(l) m u_0}{2l^2 \varepsilon_{33} \xi} \frac{[\cos(k_n l)\sinh(k_n l) - \cosh(k_n l)\sin(k_n l)]}{\sin(k_n l) + \sinh(k_n l)} \tag{2.21}$$

$$P = \frac{9\eta_1^2 E_{\mathrm{PET}}^2 z_{\mathrm{ZnO}}^2 t_{\mathrm{ZnO}} w d_{31}^2 \omega A_n^2(l) m^2 u_0^2}{2l^3 \varepsilon_{33} \xi^2} \left\{ \frac{[\cos(k_n l)\sinh(k_n l) - \cosh(k_n l)\sin(k_n l)]}{\sin(k_n l) + \sinh(k_n l)} \right\}^2 \tag{2.22}$$

The analytical model shows that the harvested vibration-induced electric potential and electric power are proportional to the ratio of the elastic constant η_1, elastic constant for the PET substrate E_{PET}, the distance z_{ZnO} from the geometric center of the ZnO layer to the neutral axis, the thickness of ZnO thin film t_{ZnO}, the transverse piezoelectric strain coefficient d_{31}, the mass density of the composite plate m, and the amplitude of shaker vibration u_0. However, the harvested vibration-induced electric potential and electric power are inversely proportional to the length of the cantilevered plate l, the dielectric coefficient of ZnO thin film ε_{33}, and the damping ratio ξ.

2.2.6 *Static Analysis to Calculate the Optimal Thickness of the PET Substrate*

To calculate the optimal thickness of the PET substrate, the FEA package ANSYS is used to perform static analysis, the results of which are compared to those obtained using Equation (2.1). The composite cantilever plate is then loaded with a constant deflection ($U_z = -2$ mm) as shown in Figure 2.6a and the optimal thickness of the PET substrate is corrected gradually. In the simulation process, a square composite structure with an effective area of 0.01×0.01 m^2 is constructed using different thicknesses of the PET substrate. The variation of the substrate thickness and the electric potential of the ZnO layer are shown in Figure 2.7. The optimal thickness of the PET substrate cannot be around 212–213 μm. Figure 2.7 shows that the resonant frequency is a function of the thickness of the PET substrate according to model analysis, as the resonant frequency depends on the substrate stiffness. Three points in the anchor of the cantilever plate are selected to analyze the internal stress and strain on the x-axis, as shown in Figure 2.6b. The thickness of the PET substrate is gradually accumulated with the internal strain on the x-axis, as shown in Figure 2.8. Point 1 indicates a long

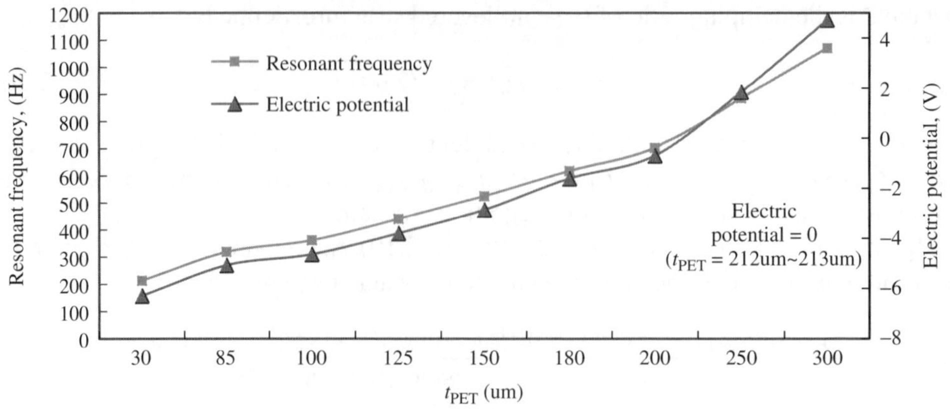

Figure 2.7 Variation of the accumulated thickness of PET substrate with the electric potential of ZnO layer and resonant frequency [29]

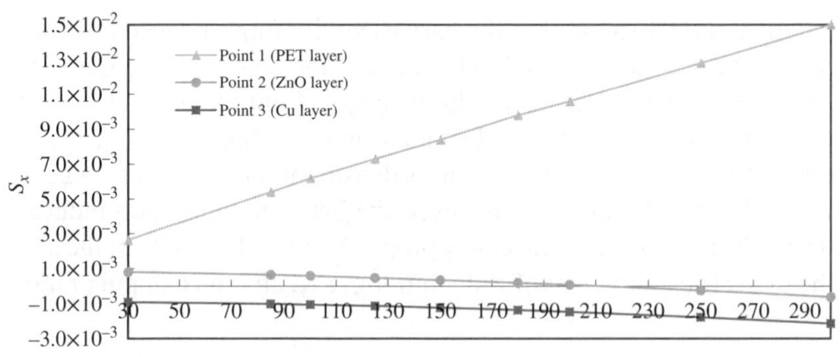

Figure 2.8 Variation of the accumulated thickness of PET substrate with the internal strain in the x-axis [29]

distance from the neutral axis to the top of the PET substrate, compared to the other two points. Therefore, the top of the PET layer receives a maximum tensile strain. The different thicknesses of the PET substrate with the internal stress in the x-axis are shown in Figure 2.9. There is a noticeable trend in the x-axis stress distribution of the three layers. The internal stress of the ZnO layer and Cu foil layer are enhanced because of the ratio of the elastic constants η_1 and η_2. The internal stress distribution of the ZnO layer and Cu foil layer is therefore obvious compared to that of the PET layer.

The simulation result shows that the neutral axis is within $\overline{Z} = 10–20\,\mu m$ when the thickness of the PET substrate is 30 μm. The neutral axis is within $\overline{Z}$= 32–57 μm when the thickness of the PET substrate is 300 μm. When the thickness of the PET substrate

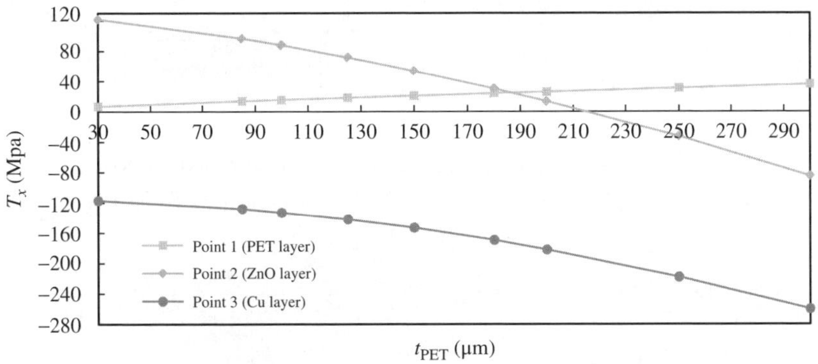

Figure 2.9 Variation of the accumulated thickness of PET substrate with the internal stress in the x-axis [29]

is smaller than 212.5 μm, the tensile strain and stress are within the piezoelectric ZnO layer. The compressive strain and stress are within the piezoelectric ZnO layer when the thickness is greater than 212.5 μm; consequently, a positive electric potential is induced. If we select a material that is of a higher elastic constant than the PET, the elastic constant ratio η_1 of the substrate would be lower. Thus, the x-axis stress and the induced electric potential on the ZnO layer would be reduced, as seen from Equations (2.10) and (2.15). In addition, a thicker and stiffer substrate creates a lower vertical tip deflection and a lower electric potential. A thin substrate with stiffness that is too weak to counteract the input vibrational energy increases the mechanical damping ratio, as described by Equation (2.20). The optimal thickness of the PET substrate is therefore 100 μm.

2.2.7 Model Analysis and Harmonic Analysis

A square structure with an effective area of $0.01 \times 0.01\ m^2$ was developed during the simulation process. Modal analysis and harmonic response analysis were performed to obtain the structural modal parameters, frequency response functions, and the induced voltage from the ZnO layer.

The layout of the flexible piezoelectric harvester is shown in Figure 2.10, depicting the schematic piezoelectric composite plate and the designed dimensions. Table 2.2 lists the design parameters of these five harvesters, the lengths of the lump structure W_1 and W_2, the gaps L_1 and L_2 between W_1 and W_2, and the thickness of the lump structures p.

Three-dimensional (3D) FEMs of the flexible piezoelectric cantilever plate were constructed using a Solid5 3D-coupled field element. The FEMs are shown in Figure 2.11. The physical properties of the materials and the piezoelectric matrix of

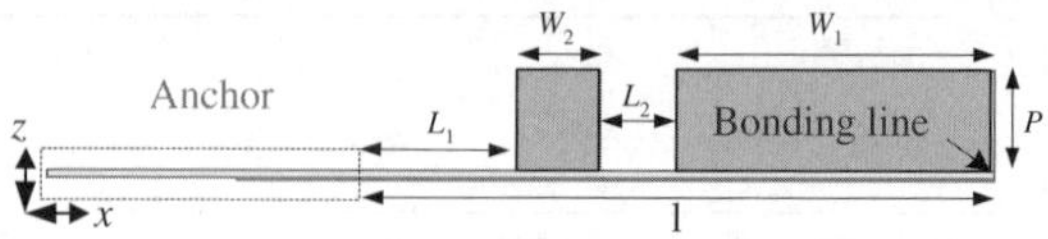

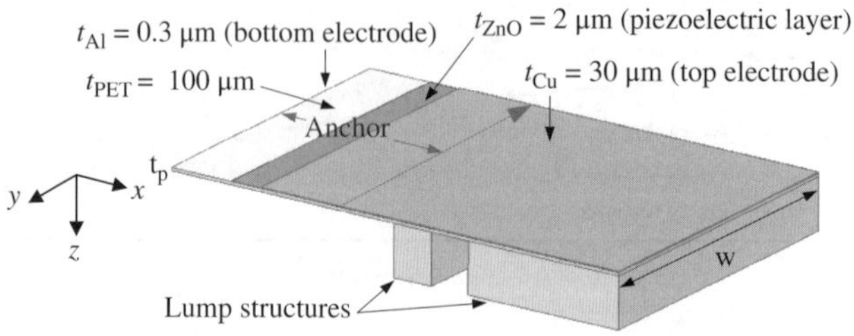

Figure 2.10 Schematic layout of flexible piezoelectric harvester and designed parameters [29]

Table 2.2 Properties of five harvesters [28]

Case No.	Length of lump structure W_1 (m)	Length of lump structure W_2 (m)	Gap between W_1 and W_2 (m)	Thickness of lump structure p (μm)
1	0.005	0	$L_1 = 0.005$ $L_2 = 0$	2000
2	0.005	0.0025	$L_1 = 0.00125$ $L_2 = 0.00125$	1600
3	0.0025	0.0025	$L_1 = 0.0025$ $L_2 = 0.0025$	400
4	0	0.0025	$L_1 = 0.005$ $L_2 = 0.0025$	400
5 (no lump)	–	–	–	–

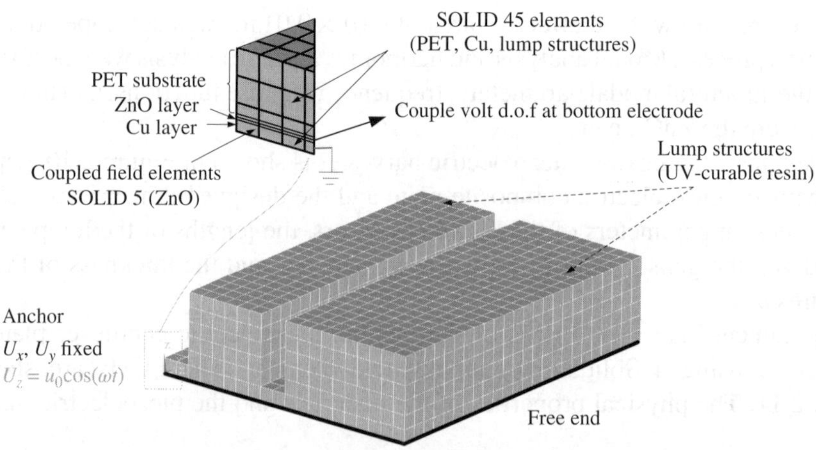

Figure 2.11 Plot of the finite element mesh model of Case 2

Table 2.3 Physical properties of the four materials [28]

Material properties	PET substrate	ZnO	Cu electrode	Exposed UV-curable resin
Young's modulus E ($N\,m^{-2}$)	2.4×10^{9}	$[c]_{6\times6}$ (see Table 2.4)	1.19×10^{11}	2.4×10^{9}
Density ρ ($kg\,m^{-3}$)	1400	5665	8900	1037.78
Poisson ratio υ	0.36	–	0.326	0.34

ZnO thin film are listed in Tables 2.3 and 2.4. Modal analysis is performed to obtain the structural modal parameters. Furthermore, for the harmonic response analysis process, $y(t) = u_0 \cos(\omega\, t)$ sinusoidal signals were applied on the anchor side of the cantilever plate. The amplitude of the vibration was set at a fixed displacement.

2.2.8 *Results of Model Analysis and Harmonic Analysis*

The result shows that the first mode of Case 2 (Table 2.2) is the uniform bending mode. Therefore, the maximum electric potential of Case 2 can also be obtained in the first mode. The first two resonant frequency mode shapes of Case 2 are shown in Figure 2.12; the first resonant frequency $\omega_n(1)$ was 225.173 Hz. In the second mode however (resonant frequency at 1593 Hz), an inflection point occurred in the middle of the composite cantilever plate which eliminated the induced voltage on the ZnO layer because of the charge cancellation. The vibration amplitude and displacement response when the cantilever plate operated at resonance frequency are shown in Figure 2.13. The induced maximum upward and downward deflection was observed when the vibration amplitude shifted to origin (Figure 2.13).

Figure 2.14 shows the resonant frequency mode shapes of five cases, which could be obtained at 151.432, 225.173, 315.605, 401.145, and 356.744 Hz. Table 2.5 lists the resonant frequencies, tip deflection, damping ratio, and electric potential of the five cases. The results show that the damping ratio of Cases 1 and 2 is smaller than that of the other three cases. The vertical deflection and maximum electric potential have the same trend. The stiffness of the five cases is the same because they were all designed using the same composite substrate (l, w, t_{PET}, t_{ZnO}, and t_{Cu} are constant). The selectively deposited UV-curable resin lump structures used as an additional mass were added to the cantilever plate to change the resonant frequency, reduce the damping ratio, and achieve a large tip deflection of a single harvester, as expressed in Equations (2.19) and (2.20). Figure 2.15a shows that Case 1 has the maximum deflection of 2123.3 μm; Case 2 has 1456.2 μm, Case 3 has 769.4 μm, Case 4 has 454.4 μm, and Case 5 has 404.4 μm. Figure 2.15b shows that Case 1 has the maximum electric potential of 3.998 V; Case 2 has 2.521 V, Case 3 has 1.47 V, Case 4 has 0.858 V, and Case 5 has 0.697 V (Table 2.2).

Table 2.4 Material properties matrix of piezoelectric ZnO thin film [28]

ZnO stiffness matrix $[c]$ $(\mathrm{N\,m^{-2}})$

$$[c] = \begin{bmatrix} 20.97 & 12.11 & 10.51 & 0 & 0 & 0 \\ 12.11 & 20.97 & 10.51 & 0 & 0 & 0 \\ 10.51 & 10.51 & 21.09 & 0 & 0 & 0 \\ 0 & 0 & 0 & 4.43 & 0 & 0 \\ 0 & 0 & 0 & 0 & 4.24 & 0 \\ 0 & 0 & 0 & 0 & 0 & 4.24 \end{bmatrix}_{6\times6} \times 10^{10}$$

ZnO compliance matrix $[s]$ $(\mathrm{m^2\,N^{-1}})$

$$[s] = \begin{bmatrix} 7.85525 & -3.43143 & -2.20457 & 0 & 0 & 0 \\ -3.43143 & 7.85525 & -2.20457 & 0 & 0 & 0 \\ -2.20457 & -2.20457 & 6.93884 & 0 & 0 & 0 \\ 0 & 0 & 0 & 22.5734 & 0 & 0 \\ 0 & 0 & 0 & 0 & 23.5849 & 0 \\ 0 & 0 & 0 & 0 & 0 & 23.5849 \end{bmatrix}_{6\times6} \times 10^{-12}$$

ZnO piezoelectric stress matrix $[e]$ $(\mathrm{C\,m^{-2}})$

$$[e] = \begin{bmatrix} 0 & 0 & -0.570878 \\ 0 & 0 & -0.570878 \\ 0 & 0 & 0.428446 \\ 0 & 0 & 0 \\ 0 & -0.480816 & 0 \\ -0.480816 & 0 & 0 \end{bmatrix}_{6\times3}$$

ZnO piezoelectric strain matrix $[d]$ $(\mathrm{C\,N^{-1}})$

$$[d] = \begin{bmatrix} 0 & 0 & -3.47 \\ 0 & 0 & -3.47 \\ 0 & 0 & 5.49 \\ 0 & 0 & 0 \\ 0 & -11.3 & 0 \\ -11.3 & 0 & 0 \end{bmatrix}_{6\times3} \times 10^{-12}$$

ZnO dielectric matrix $[\varepsilon_r]$ under the constant strain

$$[\varepsilon_r] = \begin{bmatrix} 7.57 & 0 & 0 \\ 0 & 7.57 & 0 \\ 0 & 0 & 8.31 \end{bmatrix}_{3\times3}$$

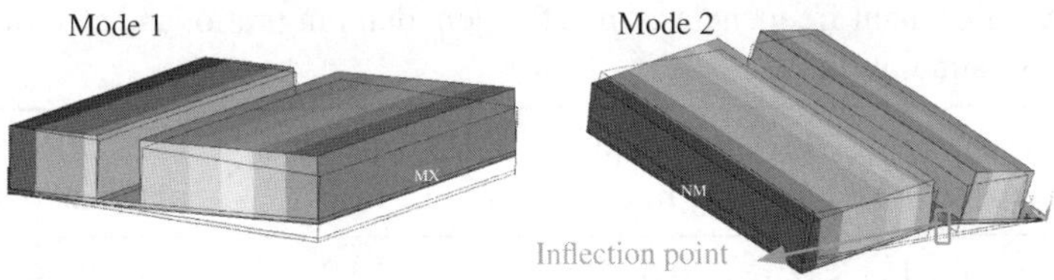

Figure 2.12 (a,b) The first and second resonance mode shapes of Case 2

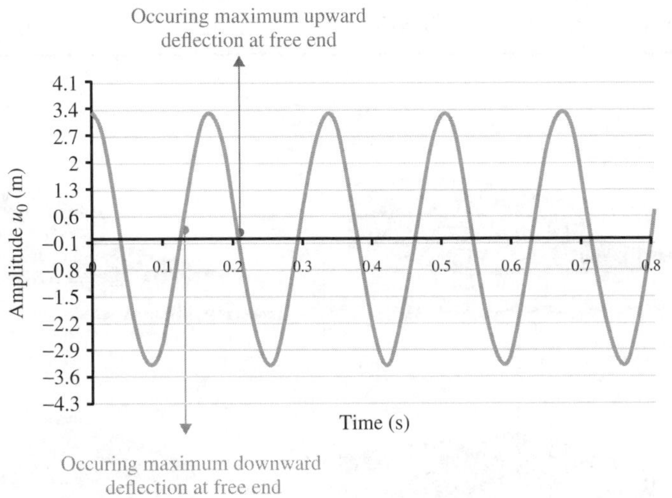

Figure 2.13 The harmonic response analysis of Case 2: vibration process (excited frequency of 225.173 Hz and a fixed sinusoidal vibration amplitude of 33 μm at the anchor side)

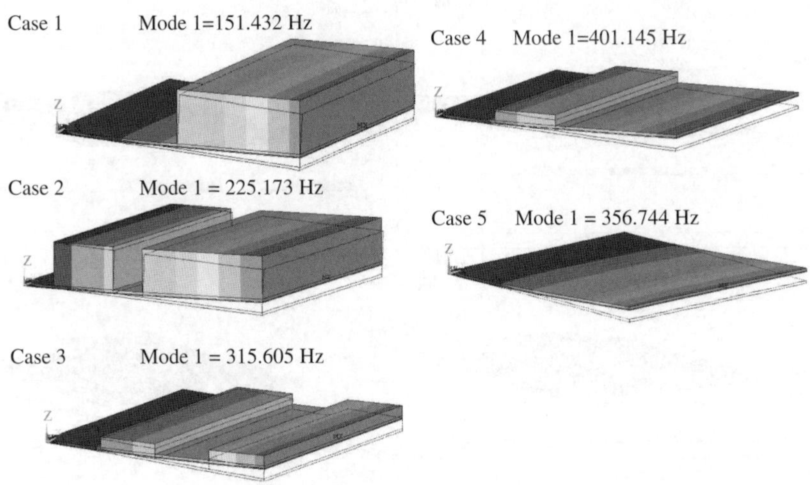

Figure 2.14 (a–e) The simulated results of the first resonance mode of five cases [29]

Table 2.5 Resonant frequencies, tip deflection, damping ratio, and electric potential for simulation [28]

Case No.	$\omega_{n(1)}$ (Hz)	Damping ratio ξ	Tip deflection U_z (μm)	Electric potential (OCV) (V)
1	151.432	0.01043	2123.3	3.990
2	225.173	0.01634	1456.2	2.523
3	315.605	0.03198	769.4	1.473
4	401.145	0.06128	454.4	0.853
5 (no lump)	356.744	0.0653	404.4	0.698

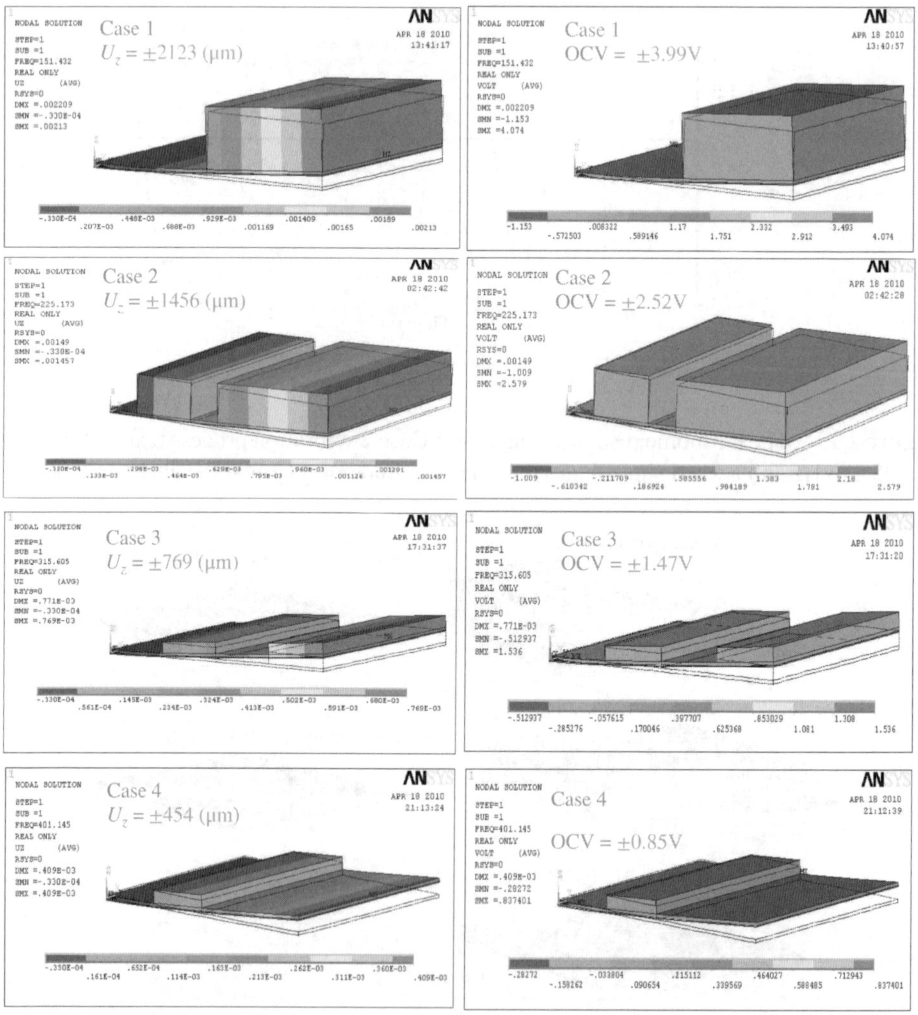

Figure 2.15 The simulated results of four cases: (a) plots of the deflection and (b) plots of the electric potential [28]

2.3 The Fabrication of Flexible Piezoelectric ZnO Harvesters on PET Substrates

2.3.1 Bonding Process to Fabricate UV-Curable Resin Lump Structures on PET Substrates

The harvester fabrication procedure using the bonding process is described in the following.

2.3.1.1 Fabrication of a PMMA Template

A micro computer numerical control (CNC) machining center is used to machine a polymethyl methacrylate (PMMA) billet into desired shapes. When the template is completed (Figures 2.16 and 2.17a), the template is sprayed with release agents.

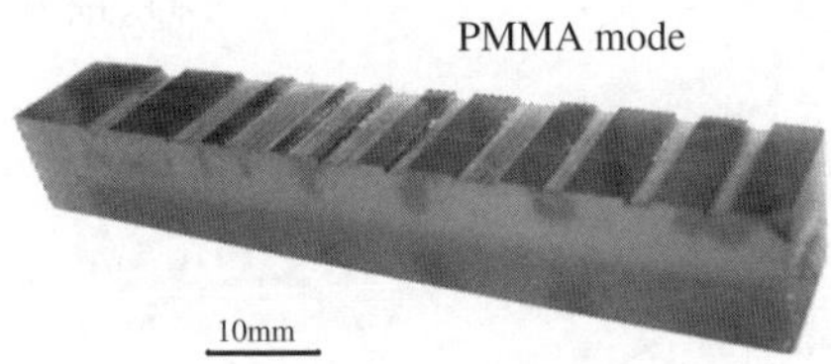

Figure 2.16 PMMA template machined by end mill

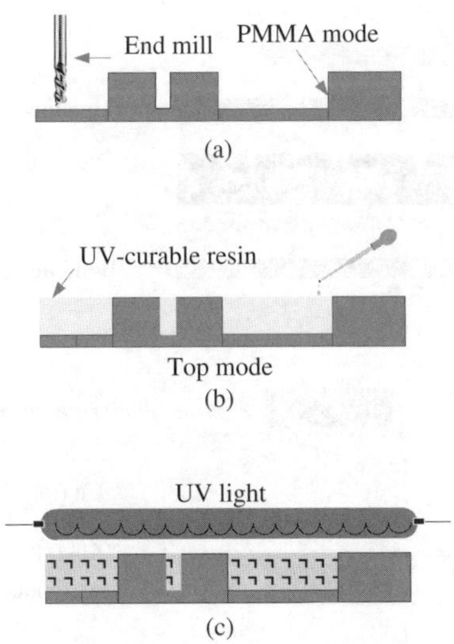

Figure 2.17 Illustration of fabrication process of top mode (a) the template is machined, (b) the UV curable resin is cast in the template, and (c) UV light exposes the resin

The UV-curable resin is injected uniformly into the cavities (Figure 2.17b). Exposure to UV light (the exposure dosage is from 350 to 450 mJ cm^{-2}) solidifies the resin (Figure 2.17c).

2.3.1.2 Definition of Bonding Pad

The PET flexible substrate is coated with a thick photoresistant SU-8 and the bonding pad region for subsequent bonding is defined, followed by an exposure process and a developmental process. It is then injected with thin UV-curable resin to complete the bonding pad for the bonding process (Figure 2.18a–c).

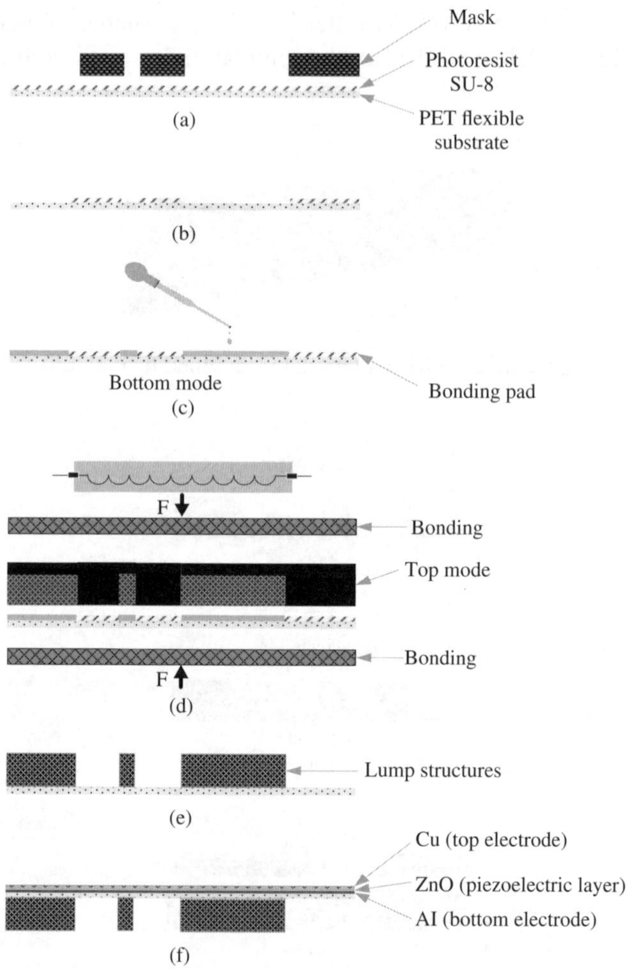

Figure 2.18 Illustration of fabrication process of lump structures on flexible multilayer plate (a) exposure process, (b) patterning on PET flexible substrate, (c) bonding process, (d) UV curing, (e) lump structures making, and (f) Cu/ZnO/Al layer with lump structures

2.3.1.3 Bonding Process

The bonding process is used to transfer the exposed UV-curable resin to the selected area on the PET flexible substrate. First, the top mode is flipped with the exposed UV-curable resin. Both the top and bottom modes are aligned and pressed. The UV light then causes the lump structures to transfer to the PET subtract accurately, and the exposure dosage is from 350 to 450 mJ cm^{-2}. Secondly, the dilute acetone is used to remove the residual photoresistant SU-8 with an ultrasonic cleaner (Branson Model 1510). Thirdly, the PET substrate with lump structures is easily separated from the template when the template is sprayed with some release agents. Finally, the PET substrate with lump structures is flipped, and the Al, ZnO, and Cu layers are continuously fabricated on the PET substrate surface (Figure 2.18d–f).

2.3.2 Near-Field Electro-Spinning with Stereolithography Technique to Directly Write 3D UV-Curable Resin Patterns on PET Substrates

To fabricate the selectively deposited UV-curable resin lump structures as a proof mass, electrospinning [30] and stereolithography [31] techniques were used to directly draw patterns on the back of the PET-based piezoelectric composite substrate, as shown in Figure 2.19a. This lithography-free method can improve the imprinting process, and can be applied to transfer UV-curable resin configurations onto a PET substrate. Electrospinning equipment in tandem with a UV light source (Hamamatsu co. LCS LightningCure^{TM}), high-voltage power supply (HP 6515A), X-Y stage (Tanlian electro optics co. TL-ST9090-50), and infusion pump (KDS 101) were assembled, as shown in Figure 2.19b. The UV-curable resin was selectively spun on the PET flexible substrate by using a high electric field (5×10^6–1.6×10^7 V m^{-1}) and controllable infusion pump (1 mL min^{-1}). Subsequently, the UV light exposed the patterns to immediately solidify the resin (exposure dosage at 400 mJ cm^{-2}), based on layer-by-layer processing.

2.3.3 Sputtering of Al and ITO Conductive Thin Films on PET Substrates

The conductive Al and ITO films were deposited directly onto the PET substrate by using RF magnetron sputtering at room temperature (Al/PET and ITO/PET) and subsequent ZnO deposition (ZnO/Al/PET and ZnO/ITO/PET). The fabrication procedures include PET substrate cleaning followed by sputtering Al and ITO as conductive substrates.

The PET substrates were cleaned ultrasonically in toluene, isopropyl alcohol, and methanol for 10 minutes each. The pre-deposited PET substrate surfaces were then cleaned using plasma in a vacuum.

After cleaning the PET substrates, an Al thin film (200 nm) was selected as the conductive electrode by using sputtering deposition; its low sheet resistance and matching

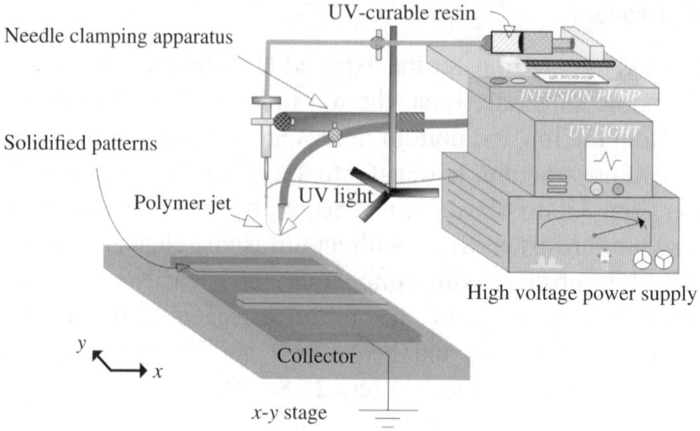

(a)

(b)

Figure 2.19 Fabrication processes of electrospinning with stereolithography technique based on UV-curable resin to fabricate 3D patterns on PET substrate: (a) schematic diagram of equipment and (b) experimental apparatuses

lattice constants with ZnO thin film provide favorable adhesion. The Al film sheet resistance was below 9.7 Ω/sq (ohms per square). The process parameters are shown in Table 2.6. Conversely, ITO film (200 nm) was also deposited onto the PET substrate as the conductive electrode. ITO films were deposited using RF reactive magnetron sputtering [32]. The sheet resistance of the ITO films was below 117.5 Ω/sq. Table 2.7 lists the process parameters.

Table 2.6 Al sputtering deposition parameters [28]

Target	Substrate temperature	Ar flow rate (sccm)	Sputtering pressure (mTorr)	DC power (W)
Al (99.9%)	RT	7.6	3	300

Table 2.7 ITO sputtering deposition parameters [28]

Alloy target	Substrate temperature (°C)	Ar flow rate (sccm)	Sputtering pressure (mTorr)	RF power (W)
In_2O_3:SnO_3 (90:10) (99.9%)	30	6.5	15	30

Table 2.8 Sputtering parameters of ZnO thin film [28]

Target	Substrate temperature	$O_2/Ar+O_2$ (%)	Sputtering pressure (mTorr)	RF power (W)
Zn (99.995%)	RT	60	12	75

2.3.4 Deposition of Piezoelectric ZnO Thin Films by Using RF Magnetron Sputtering

An RF reactive magnetron sputtering method was employed to deposit ZnO piezoelectric thin films. A metallic zinc disk with a 5 cm (2 inches) diameter and a thickness of 0.6 cm (0.25 inch) with 99.995% purity was used as the target. After the base pressure was evacuated below 6×10^{-6} Torr, O_2 and Ar gases were introduced into the chamber as a gas atmosphere during sputtering. The shutter was opened to start sputtering for 1 hour after 30 minutes of pre-sputtering. Table 2.8 lists the optimal sputtering parameters of work pressure, RF power, and O_2 concentration. The ZnO thin film of 2 μm was then deposited on the Al/PET and ITO/PET at room temperature.

For energy harvester applications, the physical properties of piezoelectric ZnO thin films should theoretically have precise surface structures and high *c*-axis orientations. The flexible multilayer structures (Al/PET, ITO/PET, ZnO/ITO/PET, and ZnO/Al/PET) are shown in Figure 2.20a. The surface morphologies of the deposited thin films were investigated. The SEM images of films are shown in Figure 2.20b–e. The crystallite size of Al films deposited on the PET substrate (Figure 2.20b) is uniform and clear compared with the ITO film on the PET substrate (Figure 2.20c).

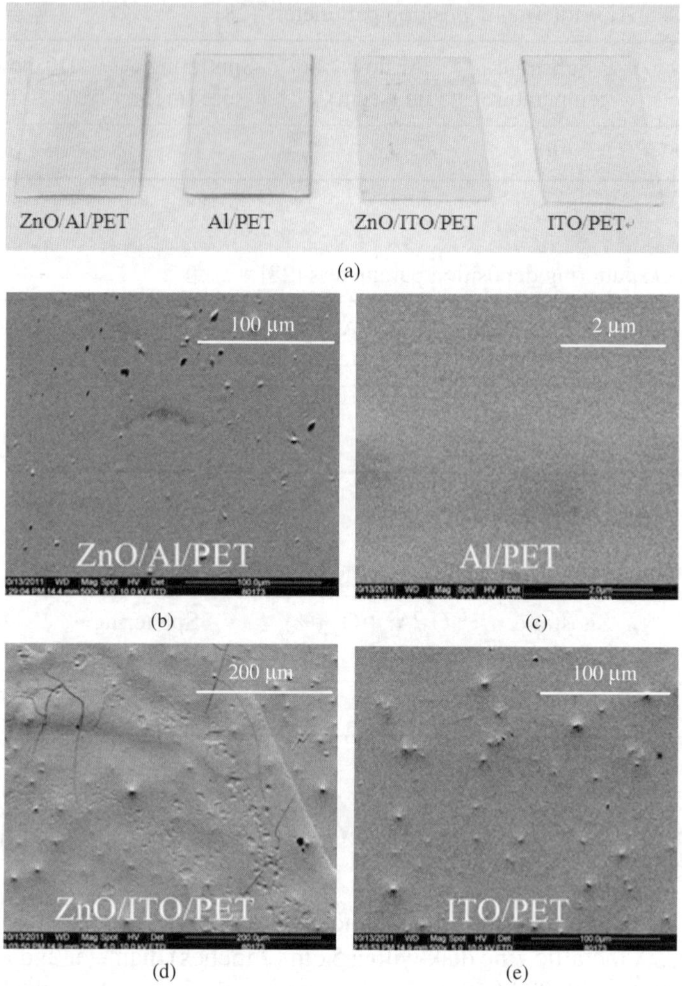

Figure 2.20 Flexible multilayer plate prototypes and images of the coating layer surfaces deposited on PET substrate: (a) four prototypes, (b) ZnO/Al/PET, (c) Al/PET, (d) ZnO/ITO/PET, and (e) ITO/PET

Furthermore, when the ZnO thin film was deposited on the Al/PET conductive substrate (Figure 2.20d), the crystallite structure was composed of many circle-like grains [33, 34] as shown in Figure 2.21a. The uniform grain size distribution was homogeneous with a smooth surface compared with the ZnO thin film deposited on the ITO/PET conductive substrate because the ITO film surface is relatively rough (Figure 2.20e). The microstructure and thickness of various films were evaluated, and the cross-sectional SEM images of ZnO/Al/PET composite structures are shown in Figure 2.21b.

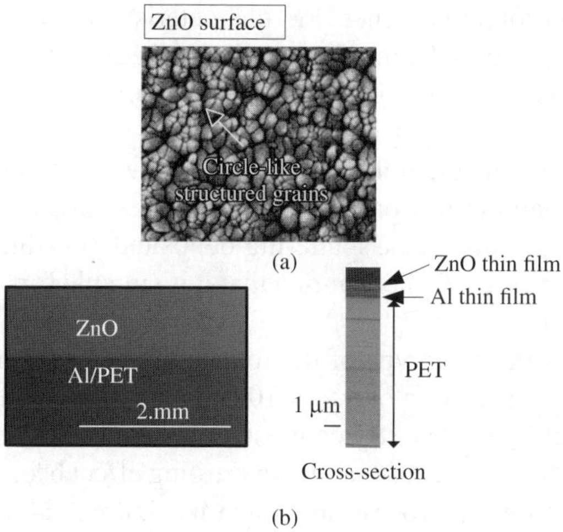

Figure 2.21 (a) SEM observations of the surface morphology of ZnO thin film on Al/PET substrate and (b) the cross-sectional SEM image of ZnO/Al/PET composite structures

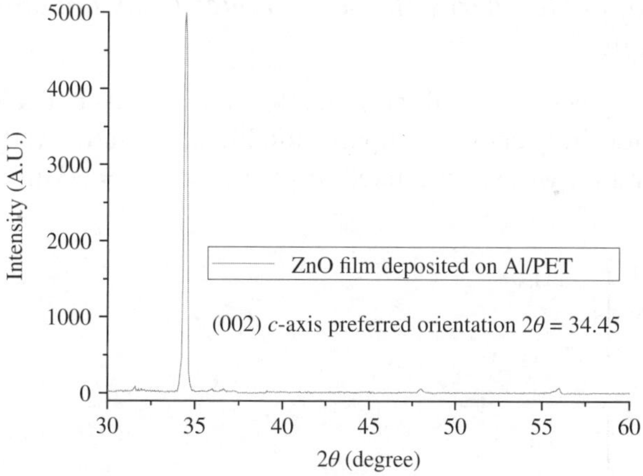

Figure 2.22 XRD patterns of ZnO thin film deposited on Al/PET (ZnO thin film was developed at room temperature with RF power of 75 W, a mixed gas $O_2/Ar + O_2$ of 60%, sputtering pressures of 15 m Torr, and base pressure of 6×10^{-6} Torr) [29]

The preferred orientation and crystal properties of the ZnO thin films were evaluated using XRD (X-ray diffraction) (a Shimadzu XRD-6000 with CuKα radiation). Figure 2.22 shows a high (002) diffraction peak at $2\theta = 34.45°$ of the ZnO thin film. The precise circle-like grains with such orientation of ZnO thin film exhibit excellent piezoelectricity.

The effects of RF magnetron sputtering parameters on the luminescent characteristics of ZnO thin films were also investigated. The optimal sputtering parameters for the most favorable photoluminescence (PL) characteristics were determined. In the PL spectrum, only UV light of 379 nm wavelength (UV region) appeared, and no visible light could be detected. The photon energy of the characteristic UV light is 3.27 eV, which corresponds to the band gap of ZnO as shown in Figure 2.23. The luminescence mechanisms of UV and visible light emissions of ZnO thin films were considered in the case of band-to-band transition and existing defects involved in ZnO. From the results of XRD and PL analyses the sputtering-deposited ZnO thin film exhibits excellent piezoelectric properties, demonstrating that it is suitable for use as a piezoelectric material for generators.

Figure 2.24a shows the prototype of the four flexible energy harvesters. It reveals that a commercial PET substrate of $t_{PET} = 100\,\mu m$ can reduce the warping effect and obtain a uniform cantilever plate. By contrast, a thin PET substrate of approximately $t_{PET} = 30–80\,\mu m$ in thickness causes a severe warping effect because of thermal deformation during the sputtering process, as shown in Figure 2.24b. Selecting a suitable thickness of the PET substrate provides more robustness in controlling the warping deformation.

2.3.5 *Testing a Single Energy Harvester under Resonant and Non-Resonant Conditions*

The harvester was placed on a vibration platform of a shaker (B&K BK4800) at an adjustable vibration frequency. To supply suitable acceleration to the harvester, the amplitude of vibration was set at a fixed displacement ($u_0 = 33\,\mu m$). The shaker can

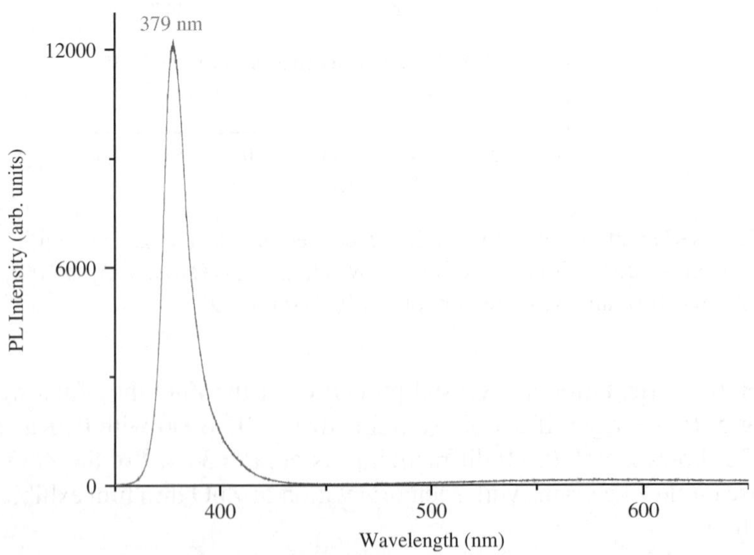

Figure 2.23 The PL spectra of ZnO thin film with optimal sputtering parameters

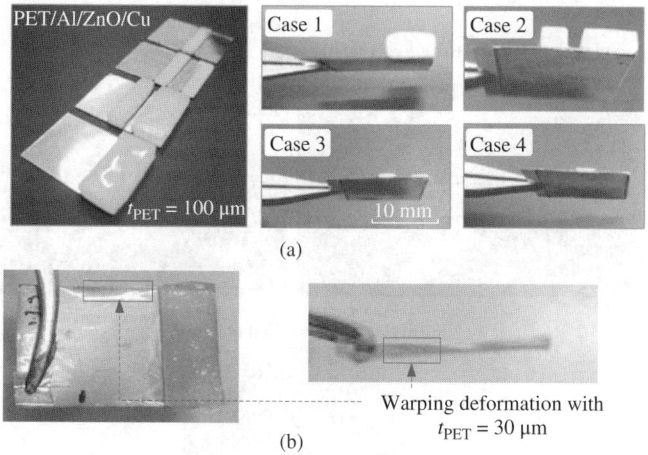

Figure 2.24 Prototype of energy harvesters: (a) the four energy harvesters with PET thickness of 100 μm and (b) a severely warping cantilever plate with $t_p = 30$ μm [29]

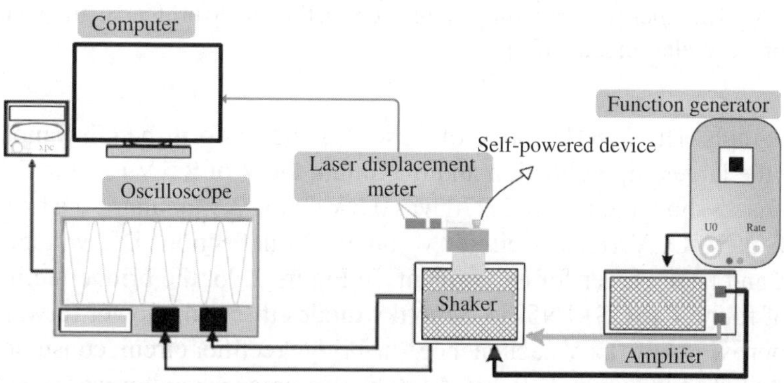

Figure 2.25 The measurement system for the self-powered flash signal LED module

be controlled by the function generator. The induced voltage signal from the harvester was measured using an oscilloscope (Agilent 54622D). A laser displacement meter (Keyence LK-G5001V) can be used to measure the vertical deflection at the free end of the cantilever plate. The measurement system is schematically shown in Figure 2.25. Figure 2.26 shows the energy-harvesting system, which consists of the rectifying bridge and a storage electrolytic capacitor.

Case 2 (Table 2.2) was directly excited at a fixed amplitude of the vibration source and scanned using the frequency 200–250 Hz. Case 5 was scanned from 340 to 370 Hz. The cantilever tip deflection was measured using a laser displacement meter. The vibration amplitude at the anchor side was measured as shown in Figure 2.27a. The resonant frequency and OCV (output circuit voltage) signal were recorded using

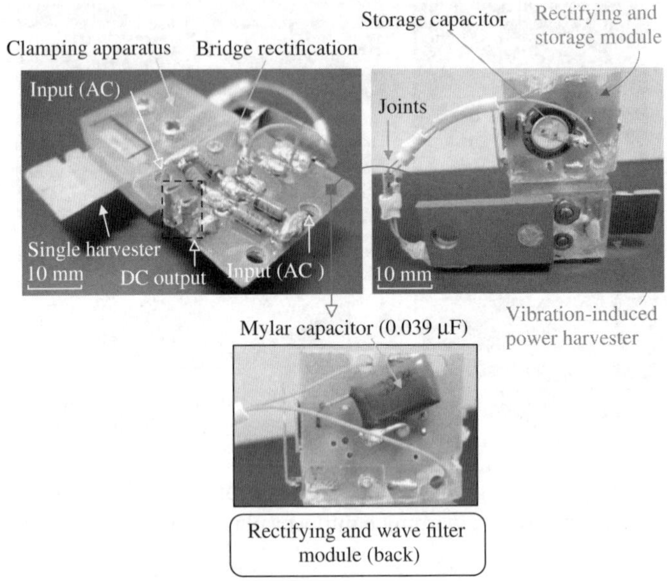

Figure 2.26 The energy harvesting system consisting of a bridge rectifier and storage capacitor or wave filter module [29]

an oscillograph. The 218 Hz mode of Case 2 is the uniform bending mode, which exhibited the largest tip deflection and maximum OCV of 2.5 V (Figure 2.27b). The OCV of Case 5 shown in Figure 2.27d was 0.6 V. Table 2.9 tabulates the experimental results of OCV, CCV (closed circuit voltage), tip deflection, DC voltage across a capacitor, and peak power for experiment. In Figure 2.26, the bridge rectifier circuit consists of four ST (SGS) 1N5711 Schottky diodes that had a lower forward voltage drop of approximately 0.2 V each; hence, the bridge rectifier circuit consumes an electric potential of approximately 0.4 V. A 0.039 μF storage capacitor was selected as the wave filter because it charges faster than storage electrolytic capacitors do. As shown in Figure 2.27c, the DC voltage of Case 2 was 2 V across a Mylar capacitor after rectification, whereas the DC voltage was 0.14 V for Case 5, as shown in Figure 2.27e.

To measure the power output, the load resistance was connected in parallel to convert the mechanical vibration to the AC load voltage. Subsequently, Case 2 was connected to an external resistance load between the two electrodes. The CCV was measured at the first mode resonance as a function of load resistance. According to the current–voltage (*I*–*V*) test, the internal resistance of various ZnO thicknesses was within 2–5 MΩ [39]. The external load must be adjusted to match the output voltage; the matching of the external load and the optimal power density was achieved by fabricating ZnO-based harvesters with a ZnO thin film thickness of 2 μm. The measurement results show that the load voltage increased with the load resistance; Case 2 reached 2.16 V, and Case 5 reached 0.24 V (Figure 2.28). In addition, Figure 2.29 shows the maximum power output of 0.556 μW from input vibrations of 61.8 m s^{-2} at 218 Hz for

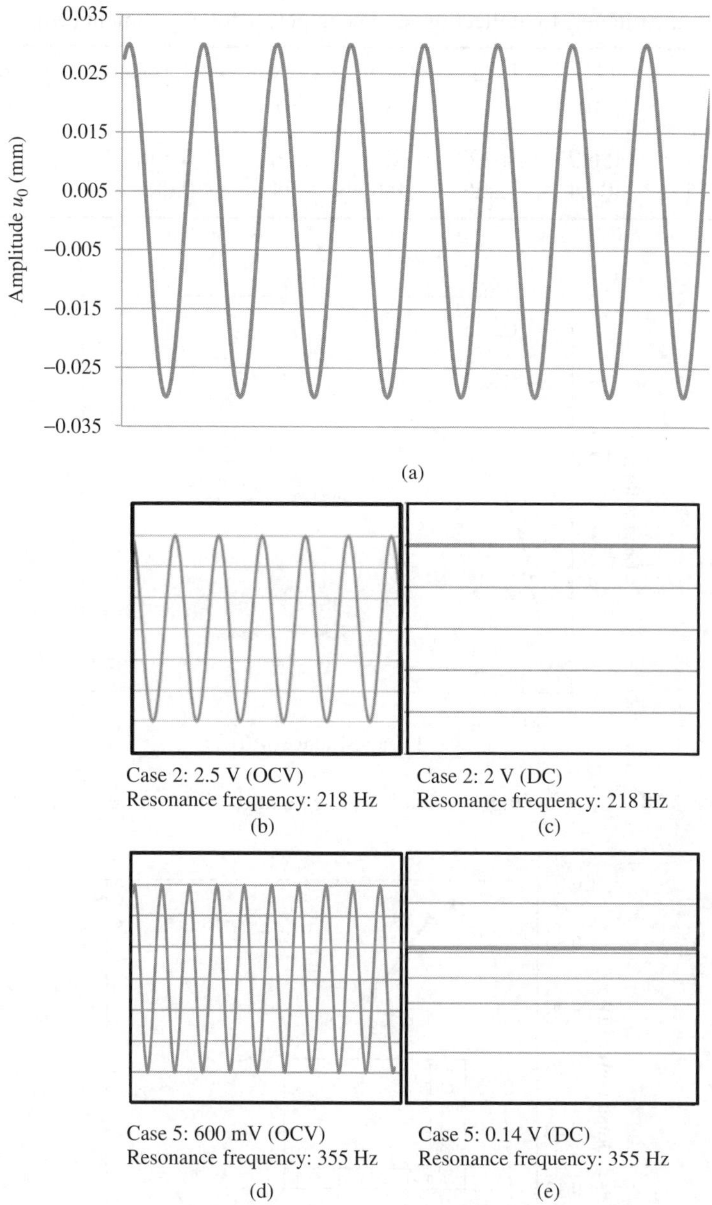

Figure 2.27 Experimental results: (a) vibration amplitude at the anchor side, (b) resonant frequency and harvested OCV signal of Case 2 (2.5 V under 218 Hz), (c) DC voltage 2 V across a 0.039 μF Mylar capacitor under 218 Hz of Case 2, (d) Case 5 (OCV 0.6 V under 355 Hz), and (e) DC voltage 0.14 V under 355 Hz of Case 5

Table 2.9 Peak voltages, tip deflection, and peak power for experiment [28]

Case No.	$\omega_{n(1)}$ (Hz)	U_z (μm)	U_z/u_0	OCV (V)	CCV (V)	Peak power (μW)	DC voltage across a capacitor (V)
Case 2	218	1456.2	44.127	2.5	2.16	0.556	2
Case 5	355	404.4	12.25	0.6	0.24	0.005	0.14

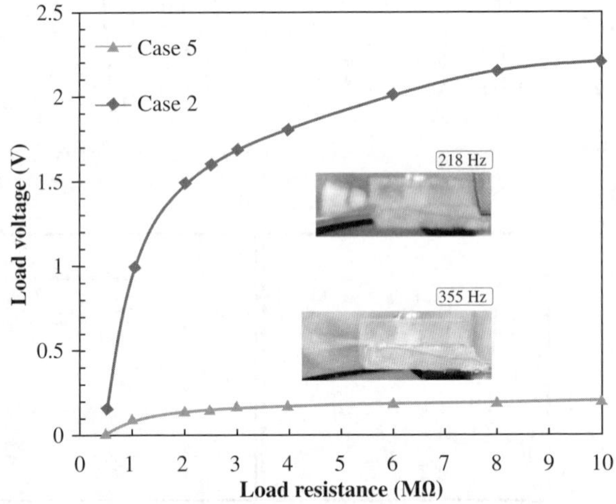

Figure 2.28 Load voltage vs. load resistance at the first mode resonance [28]

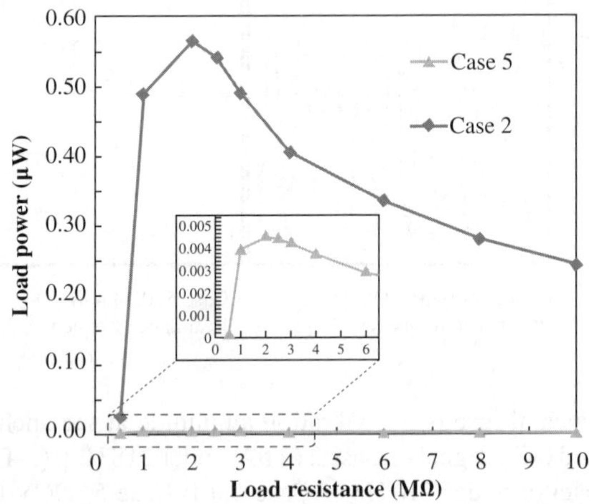

Figure 2.29 Load power vs. load resistance at the first mode resonance [28]

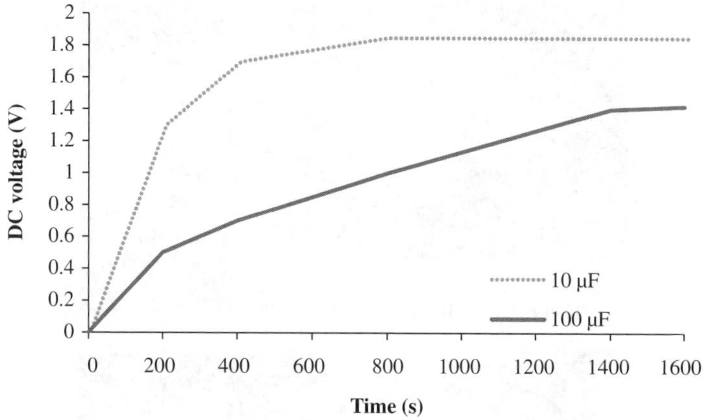

Figure 2.30 Charging results of Case 2 using 100 and 10 μF storage capacitors at 218 Hz

Case 2. Case 5 produced 0.005 μW from input vibrations of 164 m s^{-2} at 355 Hz. The output electric power was calculated according to peak power = $(CCV_{peak}/1.414)^2/R$ where R is resistance.

2.3.6 Application of ZnO/PET-Based Generator to Flash Signal LED Module

The experimental results of using this energy-harvesting system show that the charge performance of the storage capacitor is proportional to the peak power. The energy-harvesting system of Case 2 was tested at 218 Hz. The charging results of Case 2 (Table 2.2) are shown in Figure 2.30. The DC voltages across the storage capacitors (100 and 10 μF) after rectification were measured. The 100 μF storage capacitor required 2250 seconds to accumulate sufficient electrical energy of approximately 1.572 V to drive an external flash LED (light-emitting diode) module, as shown in Figure 2.31a, and the 10 μF storage capacitor required 330 seconds to store electrical energy (nearly 1.58 V).

In addition, the harvesting system was also driven using a domestic electric fan to scavenge wind power under non-resonant conditions. Figure 2.31b shows the experimental process of air flow affecting the energy harvesters. In an air flow (wind speed of 2.5 m s^{-1} and air flow of 0.56 m^3 s^{-1}), the energy harvesters regularly shed alternating vortices on either side because of the turbulence effect. The resulting pressure differential caused by the vortices forces the energy harvesters to move in an oscillating motion (a vibration frequency of 56 Hz and a tip deflection of 2125 μm were measured using a laser displacement meter). Following AC to DC rectification, the maximum DC output voltage across a 0.039 μF capacitor remained at 3.7 V. The maximum power of 1.247 μW was obtained with a load resistance of 2 MΩ. When the harvesting

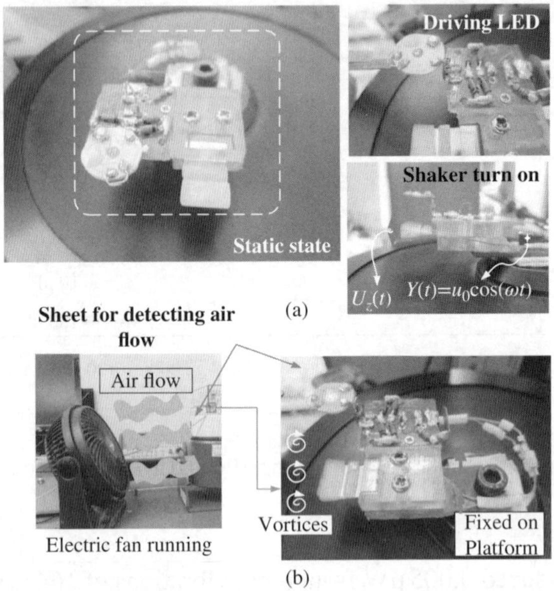

Figure 2.31 Vibration energy scavenging for self-powered flash signal LED module: (a) excited at 218 Hz by shaker and (b) excited at 56 Hz with tip deflection of 2125 μm using a domestic electric fan (wind speed: 150 m/min and air flow: 34 m^3/min)

system was running, the storage module (using a 10 μF capacitor) quickly accumulated sufficient electric potential after approximately 47 seconds and overcame the forward voltage of the LED module. Subsequently, the LED module was lit four times for a period of 2 s by the limited electrical energy output from a storage capacitor.

2.3.7 Design and Performance of a Broad Bandwidth Energy Harvesting System

Four single harvesters are assembled as an array in parallel cantilever structures, as shown in Figure 2.32. The homemade clamping apparatus is used to hold the four single harvesters. The broad bandwidth harvesting system was placed on a shaker with an adjustable vibration frequency. The shaker was controlled by the function generator with a power amplifier, and the amplitude of vibration was set at a fixed displacement to supply sufficient acceleration for the harvesting system. A laser displacement meter was used to measure the deflection in the free end of the cantilever plate. The induced electric potential signal from the harvester was measured using the oscilloscope (Figure 2.33).

To measure the resonant frequency, OCV, CCV, and load electric power, the vibration-induced energy-harvesting module with a bridge rectifier and a wave filter was prepared. To investigate the resonant mode and induced electric potential,

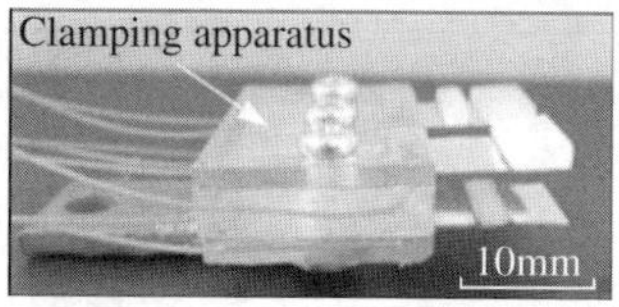

Figure 2.32 The four single harvesters are assembled as an array with a homemade clamping apparatus [29]

Figure 2.33 The photo of the measurement apparatuses [29]

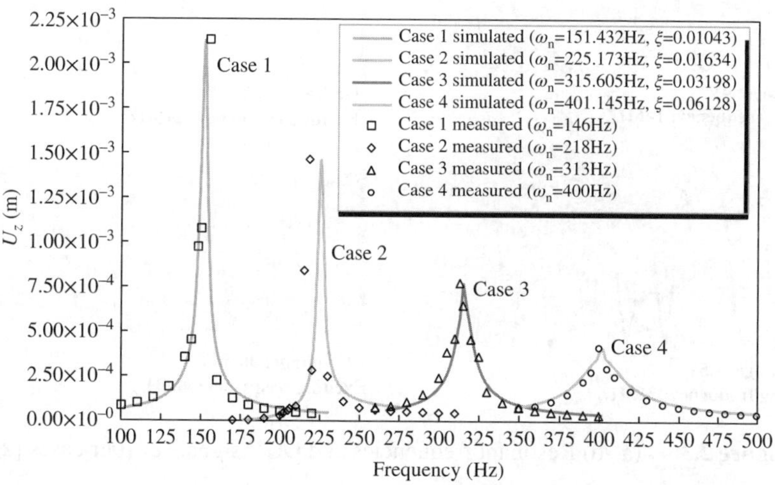

Figure 2.34 Deflection as a function of excited frequency of four cases [29]

Case 1 (Table 2.2) was directly excited at a fixed amplitude scanning of 100–230 Hz. Case 2, Case 3, and Case 4 were scanned from 175 to 315 Hz, 260 to 400 Hz, and 350 to 500 Hz, respectively. The tip deflections of the four cases are displayed in Figure 2.34, showing the simulated and experimental results. The resonant frequency of Case 1 is the minimum among the four cases, and Case 4 is the maximum case.

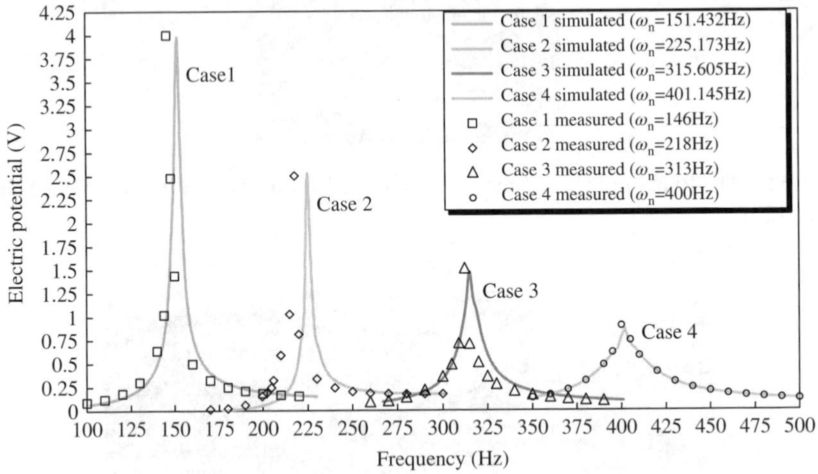

Figure 2.35 Electric potential as a function of excited frequency of four cases [29]

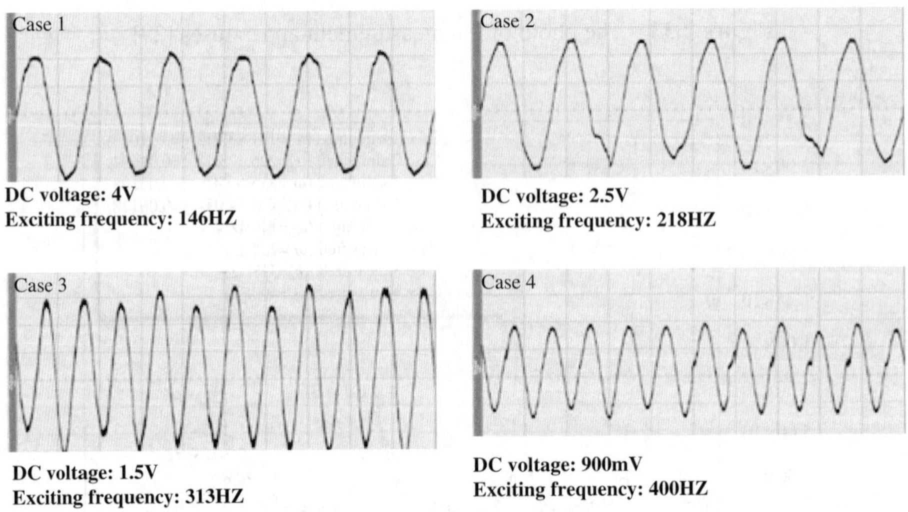

Figure 2.36 (a–d) Resonant frequencies and OCV signals of four cases [29]

Figure 2.35 shows the simulated and the experimental results of electric potential for the four cases. Case 1 has the largest tip deflection and a maximum OCV up to 4 V. The resonant frequency and OCV signal were recorded by the oscillograph, as shown in Figure 2.36.

Table 2.10 lists the experimental results of OCV, CCV, tip deflection, DC voltage across a capacitor, and peak power for the experiment. As described, the bridge rectifier circuit consists of four Schottky diodes with a low forward voltage drop, but

Table 2.10 Peak voltages, tip deflection, peak power, and DC voltages across a capacitor for experiment [28]

Case No.	$\omega_{n(1)}$ (Hz)	U_z (μm)	U_z/u_0	OCV (V)	CCV (V)	Peak power (μW)	DC voltage across a capacitor (V)
Case 1	146	2123.3	64.342	4	3.2	1.247	3.5
Case 2	218	1456.2	44.127	2.5	2.16	0.556	2
Case 3	313	769.4	23.31	1.5	1.08	0.138	1
Case 4	400	454.4	13.769	0.9	0.25	0.007	0.4

the rectifying bridge circuit must consume an electric potential bias of 442 mV. A 0.039 μF storage capacitor is chosen to be a wave filter because it can rapidly recharge. This allows for the largest DC voltage signal to output across this capacitor. The DC voltage of Case 1 across a wave filter after rectification is up to 3.5 V, 2 V for Case 2, 1 V for Case 3, and 0.4 V for Case 4.

The four single harvesters are connected to the different load resistances from 100 kΩ to 10 MΩ. Figure 2.37 shows the variation of load voltage (CCV) and load power with various load resistors for the four cases. The load voltage increases with the resistive load. The results show Case 1 of 3.2 V (Figure 2.37a), Case 2 of 2.16 V (Figure 2.37b), Case 3 of 1.08 V (Figure 2.37c) and Case 4 of 0.25 V at 10 MΩ (Figure 2.37d). The maximum electric power outputs of the four harvesters are 1.247 μW for Case 1, 0.556 μW for Case 2, 0.138 μW for Case 3, and 0.007 μW for Case 4. At a fixed amplitude of shaker vibration, the electric power has a peak value for each case when the load resistance is set at 2 MΩ. The electric potential and power are all proportional to the maximum vertical tip deflection; consequently, they are proportional to the inverse of the damping ratio, as described by Equations (2.19) and (2.22) (refer to Section 2.2.5). The mechanical damping ratio in Cases 3 and 4 therefore increases because of the lower additional mass of the lump structures on the cantilever plate.

The size of a broad bandwidth harvesting system composed of four connected-in-parallel harvesters as an array is 5 cm × 3 cm × 0.8 cm, and all four of the bridge rectifiers have a parallel connection (B_1–B_4). The four electrolytic capacitors (10 μF × 4), connected in parallel (C_1–C_4) as a rechargeable module, were investigated to determine the amount of electric energy that can be harvested at a fixed amplitude of shaker vibration, as shown in Figure 2.38. In addition, the storage gold capacitor (Panasonic 0.022 F/5.5 V) can also be used as a battery to store adequate electric energy stably; however, it requires additional charging time. The four electrolytic capacitors are used because the external flash LED module has a lower forward voltage bias of 1.5–1.55 V and consumes less electrical energy. Figure 2.39 shows an electrical connection of an AC to DC rectification module with a flash LED module. Promising application areas for this design concept of low frequency excitation with broad bandwidth are independent, self-powered devices powered by a varying ambient vibration.

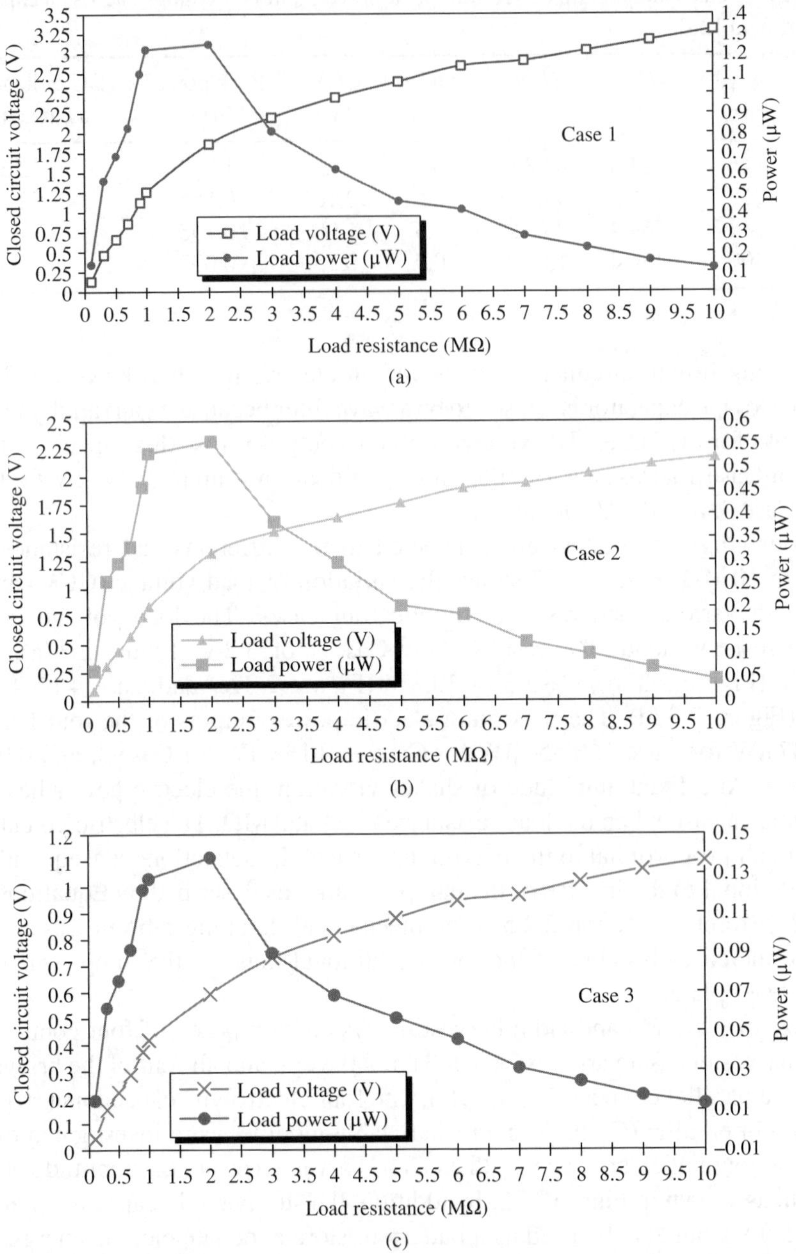

Figure 2.37 Load voltage and load power vs. load resistance of four cases: (a) Case 1 of 3.2 V, (b) Case 2 of 2.16 V, (c) Case 3 of 1.08 V, and (d) Case 4 of 0.25 V at 10 MΩ load [29]

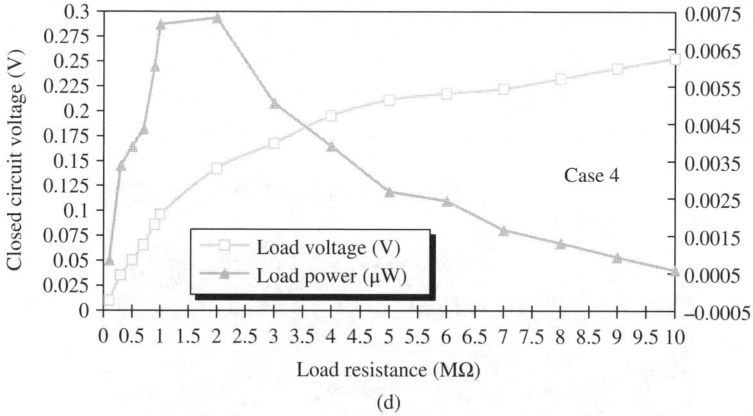

Figure 2.37 (*continued*)

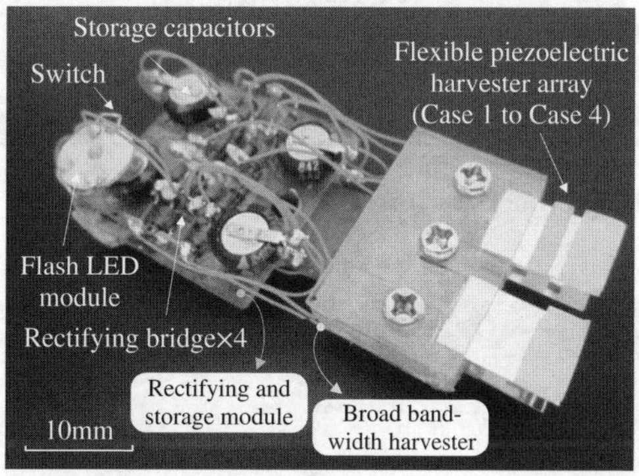

Figure 2.38 The broad bandwidth piezoelectric harvesting system with bridge rectifier circuit and storage module [29]

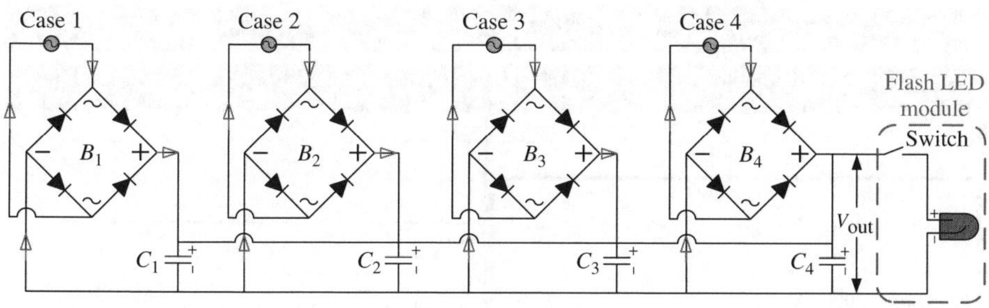

Figure 2.39 AC to DC rectification module with a flash LED module [29]

The system is excited by frequencies ranging from 100 to 400 Hz, as shown in Figure 2.40. First, the broad bandwidth system is directly tested by excitation at 146 Hz. (Case 1 is designed to resonate at 146 Hz.) The amplitude of shaker vibration is set at a fixed amplitude to supply sufficient acceleration for the harvesting system. The result shown in Figure 2.41a was recorded by an oscillograph. The figure shows

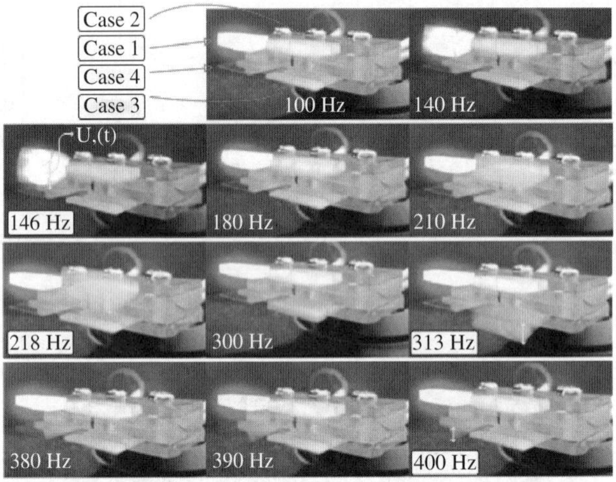

Figure 2.40 The working process of broad bandwidth system with exciting frequency from 100 to 400 Hz [29]

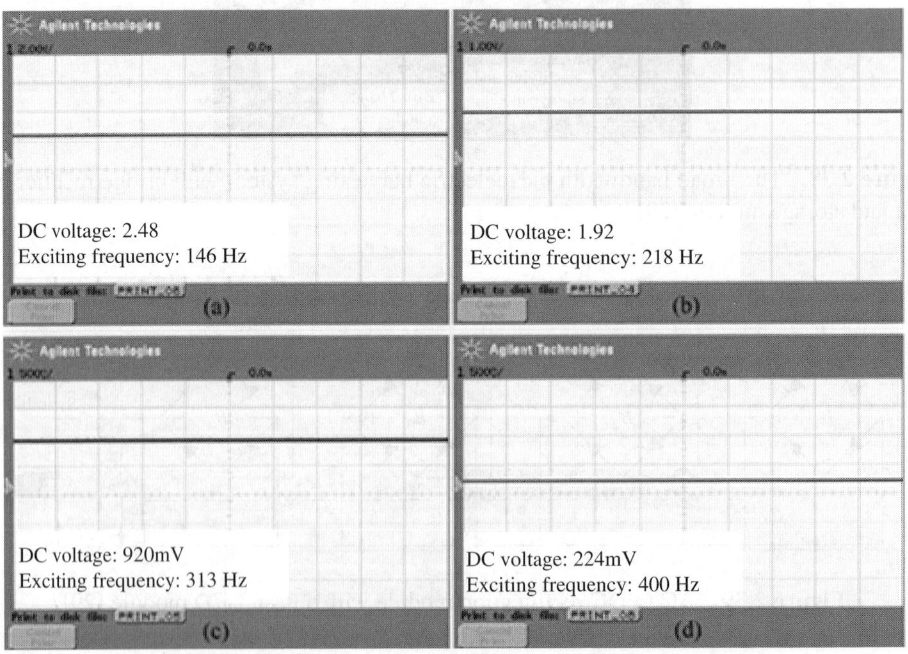

Figure 2.41 DC voltage across a storage module after 60 minutes: (a) harvesting system excited at 146 Hz, (b) 218 Hz, (c) 313 Hz, and (d) 400 Hz [29]

the DC voltage across a storage module after rectification reaches 2.48 V. By contrast, Figure 2.41b–d show the DC voltages for the other three exciting frequencies, 218, 313, and 400 Hz, as well as the DC voltage across a storage module after rectification reaches 1.92 V, 920 mV, and 224 mV, respectively. The charging results of the four cases are shown in Figure 2.42. The storage module can be charged quickly by a full-wave rectified electric potential within 10 minutes. The experimental results of the broad bandwidth system show that the electric power is proportional to the charging time. The charge efficiency of Case 1 is therefore higher than those of the other three cases. When the self-powered harvesting system is running, the storage module can accumulate sufficient electric potential quickly and overcome the forward voltage of the LED module. The LED flashes eight times for 4 s, consuming only approximately 1.1 μJ of electrical energy as shown in Figure 2.43.

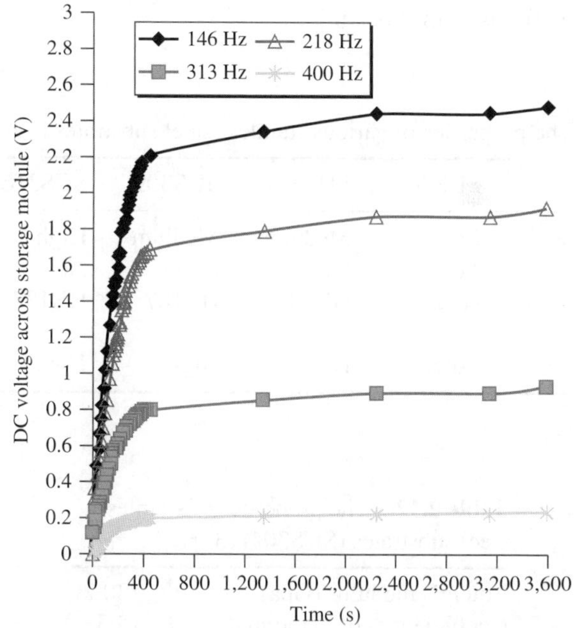

Figure 2.42 Charging results of DC voltage across a storage module when the harvesting system was excited at different operating frequencies [29]

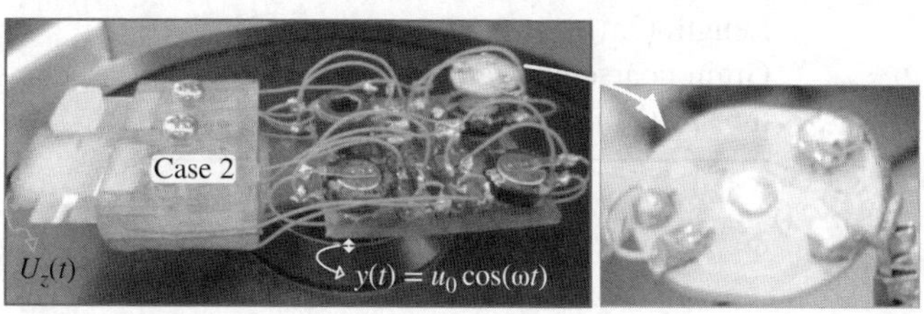

Figure 2.43 The flash LED module driven by the broad bandwidth energy harvester [29]

2.4 Fabrication and Performance of Flexible ZnO/SUS304-Based Piezoelectric Generators

2.4.1 Deposition of Piezoelectric ZnO Thin Films on Stainless Steel Substrates

We now discuss the use of a flexible stainless steel substrate as the bottom electrode for investigating the ZnO-based piezoelectric generator; the substrate is optimal for use because of its flexibility and conductivity. SUS304 has excellent properties such as heat tolerance, anti-corrosion, wear resistance, and mechanical strength. In general, five types of stainless steel substrates have been used in industries, as listed in Table 2.11 [39]. The non-magnetic property is a dominant restricting factor in exploiting microgenerators; the cost is also a considerable obstacle. SUS304 is suitable for use as a substrate in flexible energy generators. Table 2.12 lists the properties and dimensions of SUS304 used in this study.

Table 2.11 The properties of various stainless steel substrates [39]

Type	SUS201	SUS304	SUS309	SUS316	SUS430
Heat resistance	Low	Medium	Medium	High	High
Magnetism	Yes	No	No	No	Yes
Hardness (Brinell hardness)	HB241	HB187	HB187	HB187	HB183
Cost	Low	Low	High	High	Low

Table 2.12 The properties of stainless steel substrates (SUS304) [39]

Young's modulus (Gpa)	130.2
Coefficient of thermal expansion (ppm °C^{-1})	15.3
Thickness (μm)	45
Density (g m^{-3})	7.85
Breadth (cm)	2
Length (cm)	3
Grain coarsening temperature T_g (°C)	>1300

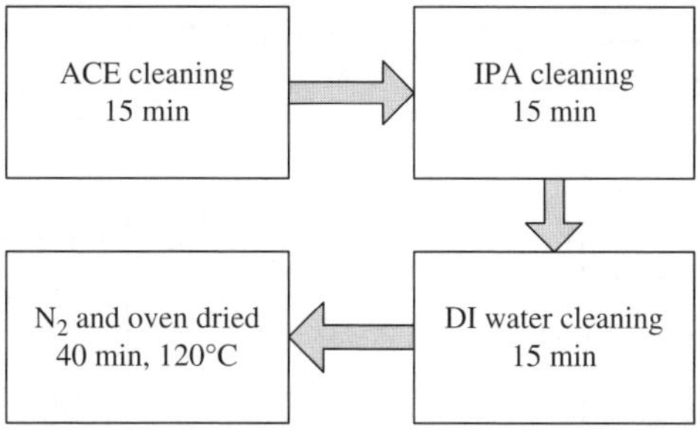

Figure 2.44 The standard spotless cleaning processes

This section introduces the experimental procedures, including the cleaning processes of SUS304, the thin film (electrodes and piezoelectric thin films) depositions using DC/RF sputtering, the optimized sputtering parameters regarding the piezoelectric thin films, the fabrication steps of single-sided and double-sided generators, the analysis of thin films, and the performance of piezoelectric generators.

SUS304 with a thickness of 45 μm was adopted as both flexible substrate and bottom electrode in this study. The substrates were diced into 20 mm × 30 mm pieces and cleaned by conducting the following standard spotless processes to remove the surface contaminants and enhance the adhesion of piezoelectric thin films on substrates. Figure 2.44 schematically depicts the standard spotless cleaning processes. The substrates are cleaned ultrasonically in acetone (ACE), iso-propanol (IPA), and de-ionized (DI) water for 15 minutes each. Finally, the substrates are dried using a nitrogen blow gun and baked for 40 minutes at 120 °C in an oven. After the substrates have been cleaned by the standard spotless cleaning processes, the piezoelectric thin films are deposited.

To deposit piezoelectric ZnO thin films, the RF sputtering system with a 5 cm (2 inch) Zn target is used, and the reactive RF magnetron sputtering method is adopted. The RF sputtering system is equipped with an RF source and a water-cooling pipeline inside to prevent meltdown of the target and demagnetization of the circular magnetic component. The high-purity oxygen (O_2) and argon (Ar) gases of 99.995% are provided as the sputtering gases. The distance between the target and the substrate holder is fixed at 45 mm. The schematic operation procedures are depicted in Figure 2.45.

Another DC sputtering system is adopted to deposit Ti and Pt thin films. This system is equipped with two 2 inch magnetron DC sources with a water-cooling pipeline inside to prevent meltdown of the target and demagnetization of the circular magnetic component. Ti and Pt targets of 99.995% are installed on the two DC guns. The Ar gas of 99.995% purity is provided as the sputtering gas.

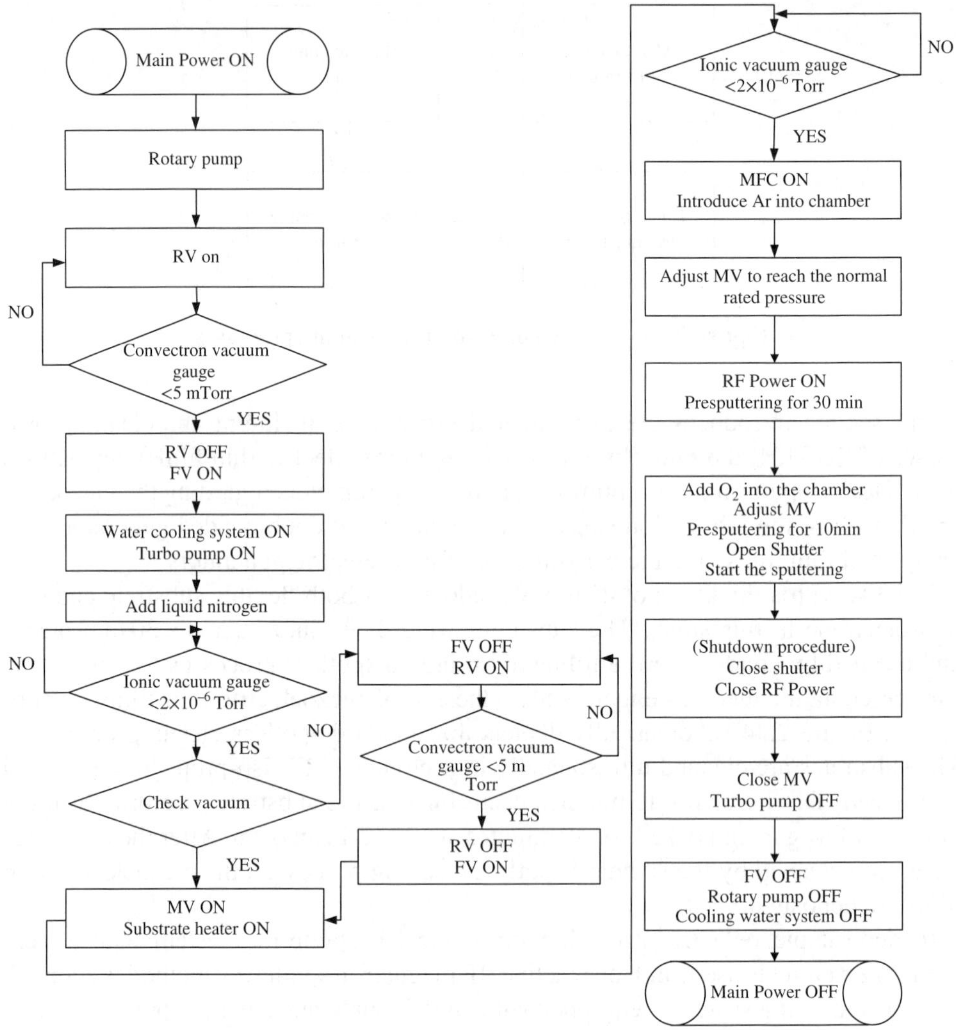

Figure 2.45 Operation procedures of the RF sputtering system

2.4.2 *Single-Sided ZnO/SUS304-Based Flexible Piezoelectric Generator*

The cleaned SUS304 substrates were adopted as flexible substrates and bottom electrodes. The ZnO piezoelectric thin film was deposited using RF magnetron sputtering with heating at various deposition parameters. Table 2.13 details the sputtering parameters to prepare *c*-axis-oriented piezoelectric ZnO thin films on SUS304 [40].

After the ZnO/SUS304 structure was formed, a copper (Cu) foil was attached to the ZnO/SUS304 as the top electrode of the piezoelectric generator because it has a lower resistivity and is economical. Finally, the electrical properties of the generator (Cu/ZnO/SUS304) were measured using a vibration testing system. To increase

Table 2.13 Deposition parameters of ZnO thin films on SUS304 [40]

Target	Zn (99.995%)				
Base pressure (Torr)	5×10^{-6}				
Substrate-to-target distance (mm)	45				
O_2 concentration ($O_2/O_2 + Ar$, %)	60				
Substrate temperature (°C)	RT	100	200	300	400
Sputtering pressure (mTorr)	3	9	15	21	27
RF power (W)	50	75	100	125	150

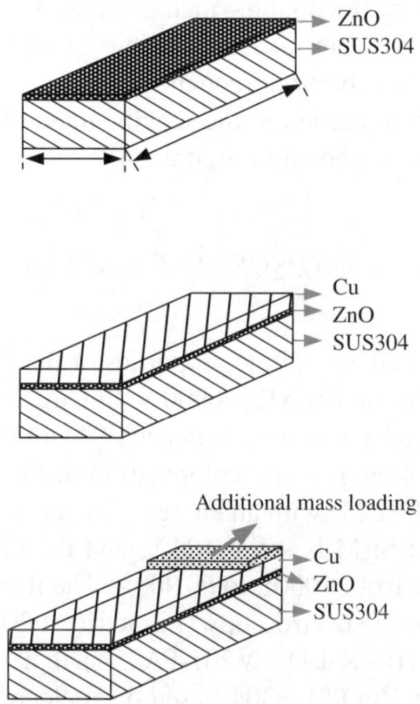

Figure 2.46 (a–c) The fabrication processes of single-sided ZnO/SUS340-based piezoelectric generator [39]

the amplitude of vibration of the generator, an additional mass on the cantilever was attached to the generators. Figure 2.46 depicts the fabrication processes of the device.

2.4.3 Double-Sided ZnO/SUS304-Based Flexible Piezoelectric Generator

After the substrates were cleaned by the standard spotless processes, titanium (Ti) and platinum (Pt) layers were deposited using a dual-gun DC sputtering system on the back side of the SUS304 as a protective layer before depositing the ZnO thin films. Because the back surface of the SUS304 exhibits a color transition, which is caused by heating at 300 °C during the deposition of the first ZnO layer (front side), the second ZnO layer (back side) does not adhere to the SUS304 unless the Pt/Ti layer is formed.

The front-side and back-side ZnO piezoelectric thin films (1st-ZnO and 2nd-ZnO) were deposited using RF magnetron sputtering with heating. Table 2.14 details the sputtering conditions in preparing first and second *c*-axis-oriented ZnO piezoelectric thin films. After the 1st-ZnO/SUS304/Ti/Pt/2nd-ZnO structure was formed, the Cu foils with low resistivity and low cost were attached to the ZnO surfaces as the top and bottom electrodes. The fabrication steps of a double-sided piezoelectric generator are depicted by Figure 2.47.

The electrical properties of the double-sided piezoelectric generators (Cu/1st-ZnO/SUS304/Ti/Pt/2nd-ZnO/Cu) were measured using a vibration-testing system. For a fixed value of amplitude, the vibration frequency is reduced with the added mass loading. To achieve the designed frequency, the specific mass loading must be attached to the double-sided generator as shown in Figure 2.48.

2.4.4 Characterization of ZnO/SUS304-Based Flexible Piezoelectric Generators

The preferred orientation and crystalline properties of the ZnO thin films were determined by XRD using a Shimadzu XRD-6000 with CuKα radiation. The diffraction angles of ZnO are revealed by scanning between 20° and 60° at a speed of 0.05° per second. The surface morphologies and composition of the thin films were observed by SEM (Philips XL40 FESEM) with an energy-dispersive spectrometer (EDS). The accelerated voltage of the SEM is set at 10 kV, and the magnification is controlled over a range of magnitude from 3000× to 60 000×. The thicknesses of the films were determined using reflectance spectroscopy (Filmetrics F20). Figure 2.49 depicts the vibration-testing system (B&K-LDS V201-PA25E) that was used to measure the electrical properties of the ZnO/SUS304-based piezoelectric generators. The system includes a laser displacement sensor (MEL M72L/4) for measuring the amplitude of vibration.

Table 2.14 Deposition parameters of double-sided ZnO thin films on SUS304 [39]

Target	Substrate-to-target distance (mm)	Base pressure (Torr)	Substrate temperature (°C)	RF power (W)	Sputtering pressure (mTorr)	O_2 concentration (O_2/O_2+Ar, %)
Zn (99.995%)	45	5×10^{-6}	300	75	9	60

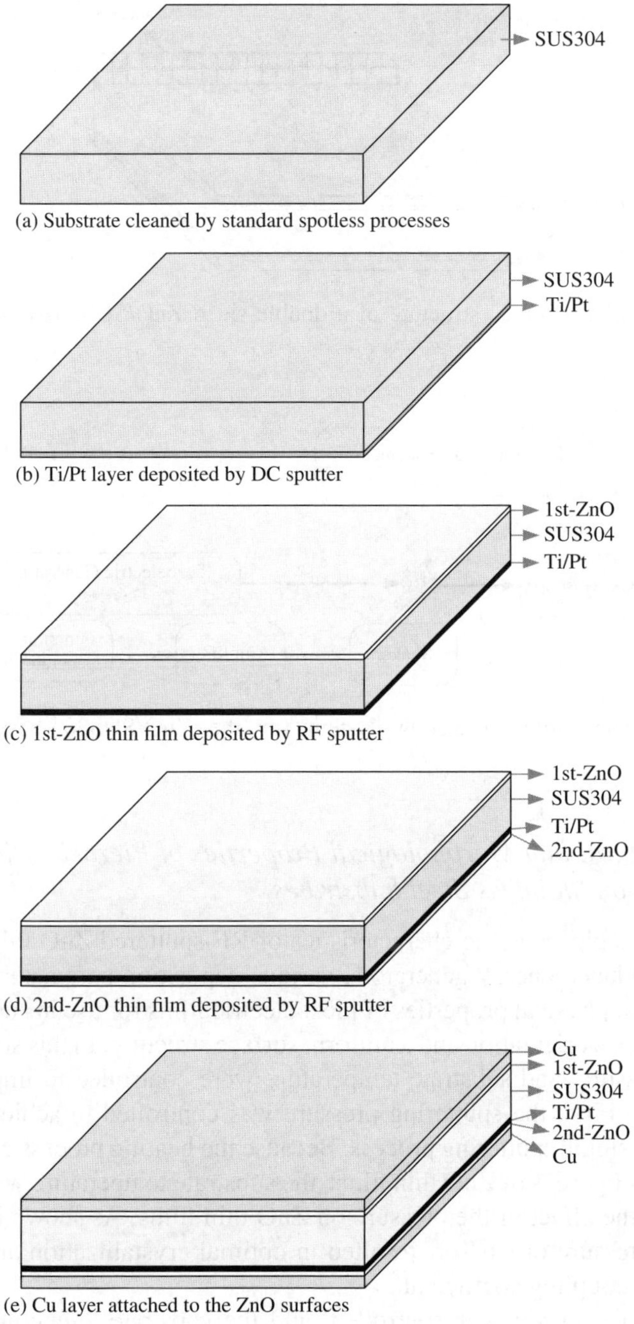

Figure 2.47 (a–e) The fabrication steps of double-sided ZnO/SUS304-based piezoelectric generator

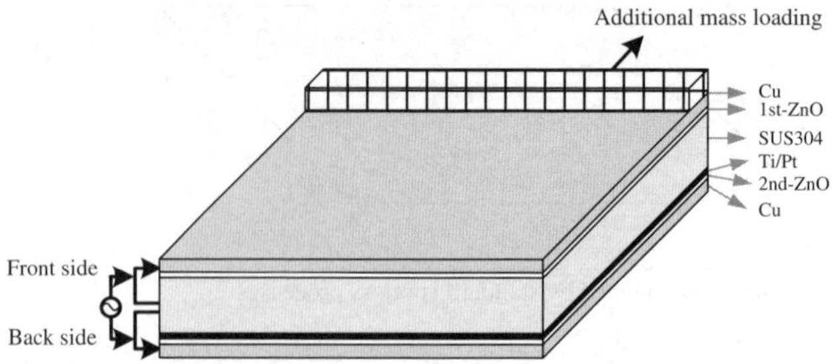

Figure 2.48 The schematic structure of a double-sided ZnO/SUS304-based piezoelectric generator

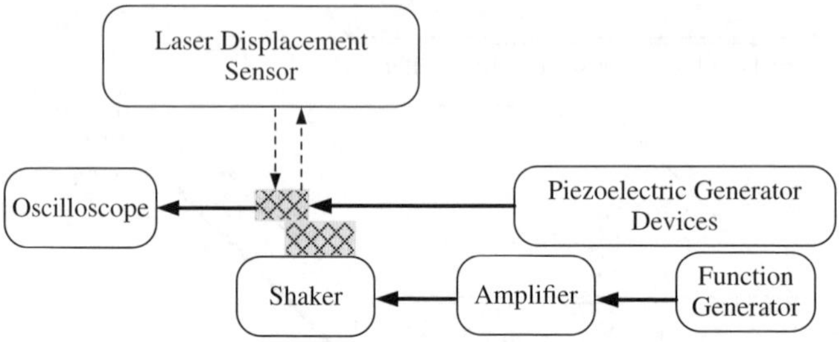

Figure 2.49 The vibration testing system for the ZnO/SUS304-based piezoelectric generators

2.4.5 *Structural and Morphological Properties of Piezoelectric ZnO Thin Films on Stainless Steel Substrates*

To optimize the piezoelectric characteristics of RF-sputtered ZnO thin films for use in vibration-induced energy generators, various deposition parameters were investigated. The ideal physical properties of piezoelectric films for use in such applications are a strong *c*-axis orientation and a uniform surface structure. In this study, RF power, sputtering pressure, and substrate temperature were controlled to improve the ZnO characteristics. First, the sputtering pressure was controlled to achieve the optimal ZnO layer in a single sputtering process. Because the heating process can improve the piezoelectric properties of ZnO thin films, the substrate temperature was set at 300 °C to investigate the effect of the pressure on ZnO thin films. As shown in Figure 2.50, a sputtering pressure of 9 mTorr resulted in optimal crystallization and a high electromechanical coupling coefficient.

Second, the RF power was controlled, and the substrate temperature was set at 300 °C to investigate the influence of RF power on ZnO thin films. The XRD patterns in Figure 2.51 show that ZnO thin films deposited at 75 W yield a sharp (002) peak with a small FWHM. The RF power was therefore fixed at 75 W to produce a highly textured structure.

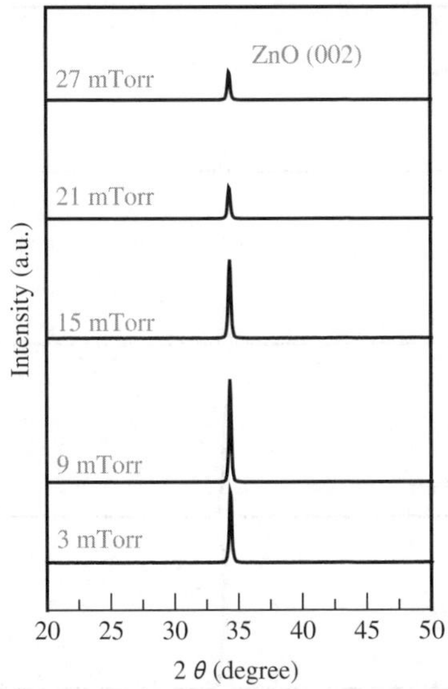

Figure 2.50 The $\theta-2\theta$ X-ray scans of the ZnO thin films deposited under various sputtering pressures. (Fixed sputtering parameters: substrate temperature of 300 °C, RF power of 75 W, O_2 concentration of 60%.)

Third, Figure 2.52 depicts the variation of the XRD patterns of the ZnO thin films deposited on the SUS304 at various substrate temperatures. A sharp (002) peak and narrow FWHM is obtained at substrate temperatures from room temperature (RT) to 300 °C. A stronger XRD intensity indicates that the ZnO thin film has a higher electromechanical coupling coefficient of piezoelectricity. The diffraction angle of ZnO thin films matches that of the standard ZnO films. The internal resistance of the ZnO thin films increased with the piezoelectricity however, as described by Table 2.15 [41].

To select the parameter of the substrate temperature between the piezoelectricity and electricity, four cases of generators were fabricated with ZnO thin films that were deposited at RT, 100, 200, and 300 °C, respectively. The output open voltages of four generators were measured using an electrical property measurement system combined with the vibration-testing system, as shown in Figure 2.49. Table 2.16 shows the OCV ($V_{\text{p-p}}$, peak-to-peak voltage). The OCVs in Generators 1–3 are lower than that in Generator 4. In this study, the substrate temperature was set at 300 °C to ensure the high OCV ($V_{\text{p-p}}$).

Finally, the optimal substrate temperature of 300 °C, RF power of 75 W, and sputtering pressure of 9 mTorr were adopted to deposit piezoelectric ZnO films. The resulting morphologies of ZnO films on the SUS304 appeared dense with smooth surfaces, uniform grain sizes, and strongly textured columnar structures. Hexagonal crystals such as ZnO are believed to exhibit piezoelectricity of unity with a (002)-preferred orientation.

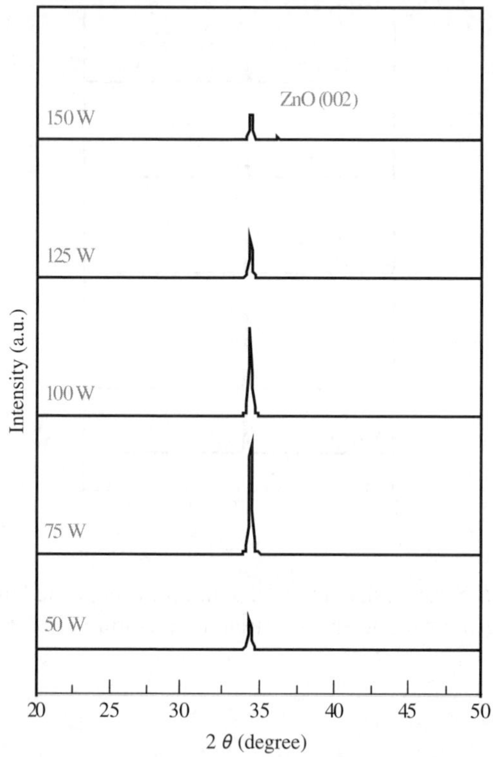

Figure 2.51 The θ–2θ X-ray scans of the ZnO thin films deposited under various RF powers. (Fixed sputtering parameters: substrate temperature of 300 °C, sputtering pressure of 9 mTorr, O_2 concentration of 60%.)

2.4.6 Analysis of Adhesion of ZnO Thin Films on Stainless Steel Substrates

In this investigation, a nano-indenter and a nano-scratch tester were applied to determine the mechanical properties of ZnO thin films on stainless steel substrates. Table 2.17 presents the conditions for measuring the mechanical properties of ZnO thin films. The OM images, as shown in Figure 2.53, indicate that the ZnO/SUS304 structure has several favorable properties such as high compressive strength, high fracture strength, and the absence of any breakdown phenomenon. Figure 2.54 presents the scratch depth of the ZnO thin film that was deposited on the SUS304 at various applied stresses. Point A in Figure 2.54 indicates that the ZnO/SUS304 structure endured 500 mN of applied stress without any change to the shape of the surface. Point B in Figure 2.54 shows both the end point and the start point, indicating that no extrusion phenomenon occurs. The ZnO/SUS304-based structure is therefore highly stable and exhibits long-term vibration in natural environments.

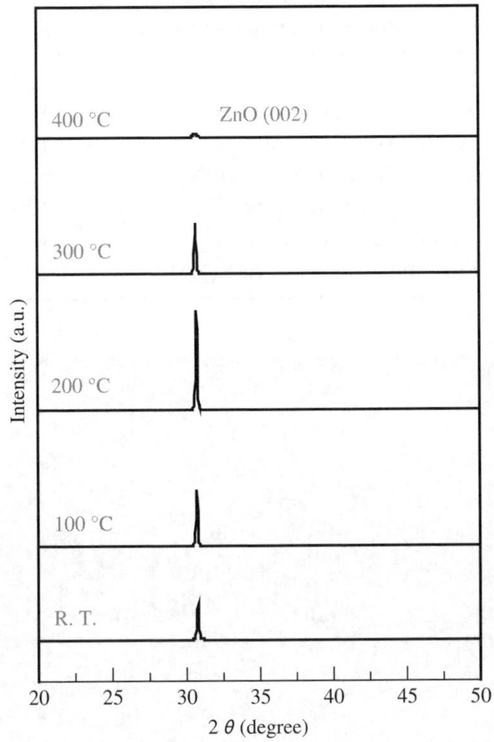

Figure 2.52 The θ–2θ X-ray scans of the ZnO thin films deposited under various substrate temperatures. (Fixed sputtering parameters: sputtering pressure of 9 mTorr, RF power of 75 W, O_2 concentration of 60%.)

Table 2.15 The internal resistance of ZnO thin films with various substrate temperatures

Substrate temperature	RT	100 °C	200 °C	300 °C	400 °C
R_{ZnO} (MΩ)	1.254	2.923	5.471	5.120	NA

Table 2.16 The output voltages of four generators [41]

Generator No.	1	2	3	4
Substrate temperature	RT	100 °C	200 °C	300 °C
$V_{p\text{-}p}$ (V)	1.215	1.525	2.687	7.796

Table 2.17 Parameters of the nano-indenter and nano-scratch test

Scratch length (μm)	500
Starting scratch load (mN)	1
Scratch velocity (nm s^{-1})	100
Maximum scratch load (mN)	520
Load applied during profiling (μN)	100
Allowable drift rate (nm s^{-1})	1
Profiling acquisition rate (Hz)	20

(a)

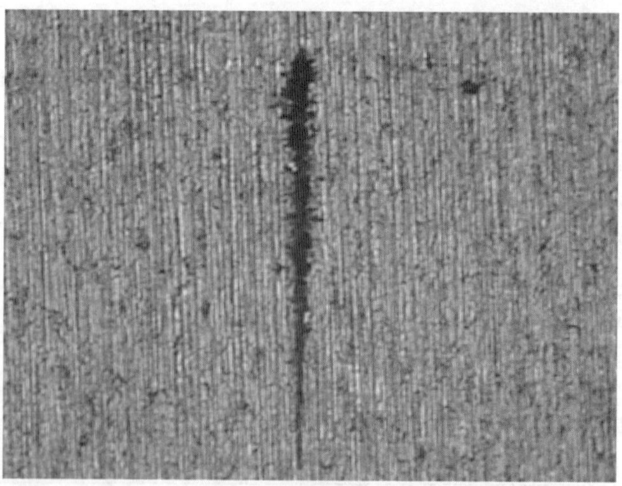

(b)

Figure 2.53 The optical microscopy (OM) drawing of the ZnO thin film on SUS304: (a) the reset condition and (b) after the nano-indenter and nano-scratch test [40]

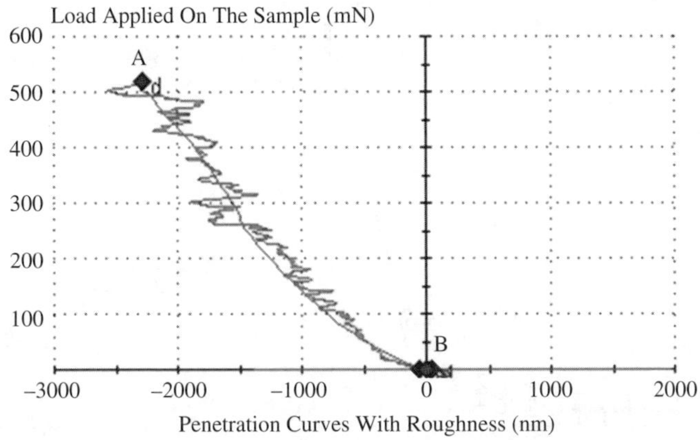

Figure 2.54 The indentation depth variations of the ZnO thin films deposited on the SUS304 with various extra stresses

2.4.7 Electrical Properties of Single-Sided ZnO/SUS304-Based Flexible Piezoelectric Generator

As reported, the excitation frequency and amplitude of vibration critically influence the output voltage of a generator. Figure 2.55 shows the variations of OCV ($V_{p\text{-}p}$) with excitation frequencies at a vibration amplitude of 1 mm. The maximum V_{oc} (open-circuit voltage) of 7.52 V was achieved at an excitation frequency of 75 Hz. Clearly, the ZnO/SUS304-based generator in this study is suitable for low-frequency application. Figure 2.56 shows a plot of the variations of OCV ($V_{p\text{-}p}$) with the amplitude of vibration. The maximum OCV of 10.5 V ($V_{p\text{-}p}$) is reached at a vibration amplitude of 1.19 mm at the vibration frequency of 75 Hz.

The piezoelectric generator can be characterized as high voltage with a low-current output. For applications in this study, the ZnO-based generators were connected to an energy storage or filtering capacitor with a load resistance of 5 MΩ and four Schottky diodes (1N5711) for rectification. The load resistance of 5 MΩ was matched by the internal resistance of the ZnO thin films. Figure 2.57 shows the output DC voltage of 4.64 V, which was rectified by the bridge rectifier circuit and filtered with a 33 nF capacitor at a vibration amplitude of 1.19 mm and a vibration frequency of 75 Hz. As 5 MΩ was loaded, the maximum output DC voltage of 2.24 V and a power output of 1.0 μW cm^{-2} were obtained, thereby demonstrating excellent performance.

To determine the optimal additional mass loading to obtain the higher-power output and lower-vibration frequency, six different mass loadings were added to the single-sided ZnO/SUS304-based generators: 0.05, 0.10, 0.15, 0.30, 0.50, and 0.57 g. The open-circuit output voltages ($V_{p\text{-}p}$) of the six generators were measured, as shown in Table 2.18 [42]. The output voltages in Generators 1, 2, 3, 4, and 6 are lower than

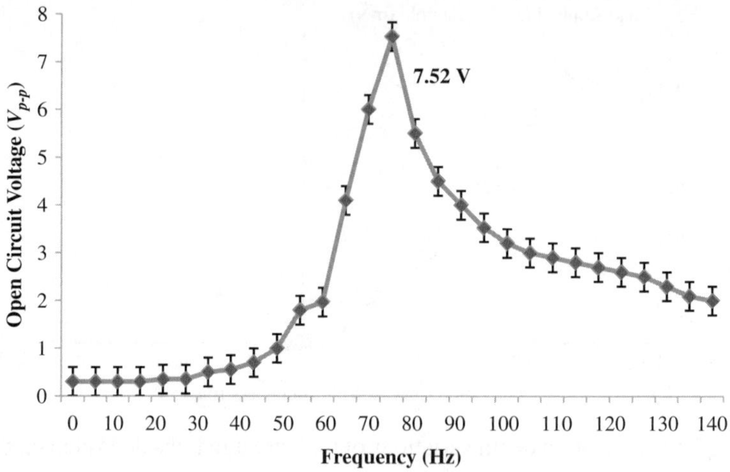

Figure 2.55 Open circuit voltages (OCV, V_{p-p}) with various excitation frequencies (amplitude of vibration: 1 mm)

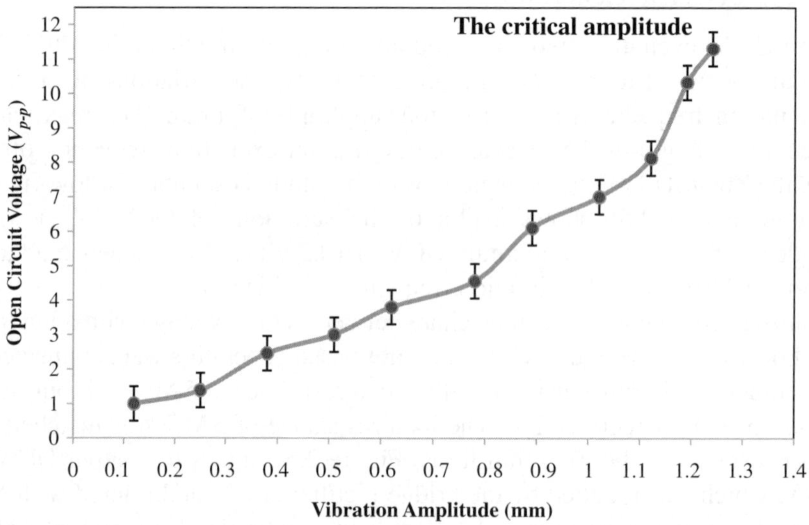

Figure 2.56 The variations of OCV (V_{p-p}) with various vibration amplitudes (vibration frequency: 75 Hz)

that of Generator 5. In this study, the additional mass loading of 0.5 g was adopted to optimize the output voltage. Finally, the maximum OCV (V_{p-p}) of 10.693 V for the single-sided ZnO/SUS304-based generators was reached at the vibration frequency of 82 Hz. Clearly, this single-sided ZnO/SUS304-based generator can be used in vibration-powered applications. Based on these results, a high-performance scheme for designing double-sided generators is warranted.

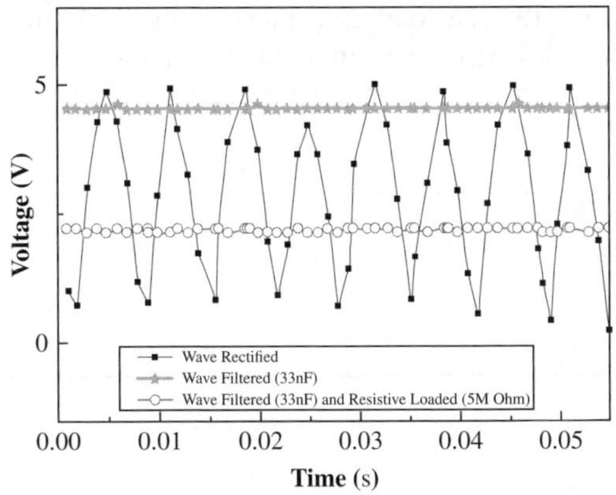

Figure 2.57 The wave forms of various output voltages of optimized piezoelectric generator after rectified, filtered, and resistive loaded

Table 2.18 The output voltages of six generators with various additional mass loading

Generator No.	Additional mass loading (g)	OCV ($V_{p\text{-}p}$)	Vibration frequency (Hz)
1	0.05	1.147	158
2	0.10	2.443	150
3	0.15	3.104	147
4	0.30	5.918	91
5	0.50	10.693	82
6	0.57	10.487	77

2.4.8 Characterization of Double-Sided ZnO/SUS304-Based Flexible Piezoelectric Generator: Analysis and Modification of Back Surface of SUS304

To increase the power output density, the second ZnO piezoelectric thin film should be deposited on the back of the 1st-ZnO/SUS304 directly to form a double-sided generator. However, the results in this study showed that the quality of the second ZnO thin film is poor. The back of the SUS304 exhibited a color transition after heating at 300 °C, and the surface did not favor the growth of the second ZnO layer. Therefore, the back surface before and after 300 °C heating was analyzed using an EDS. Table 2.19 shows how the oxygen content (O_2%) in the back surface of the SUS304 increased considerably after heating at 300 °C.

The results of roughness analyses in Figure 2.58 and Table 2.20 show that, after heating at 300 °C, the surface roughness of the SUS304 increases from 1.455 to 12.592 nm,

Table 2.19 The oxygen contents in the back surface of SUS304 before and after 300 °C heating [39]

Sample No.	O_2 (%) before 300 °C heating	O_2 (%) after 300 °C heating
1	0	14.08
2	4.39	11.73
3	2.42	10.94
4	0	4.89

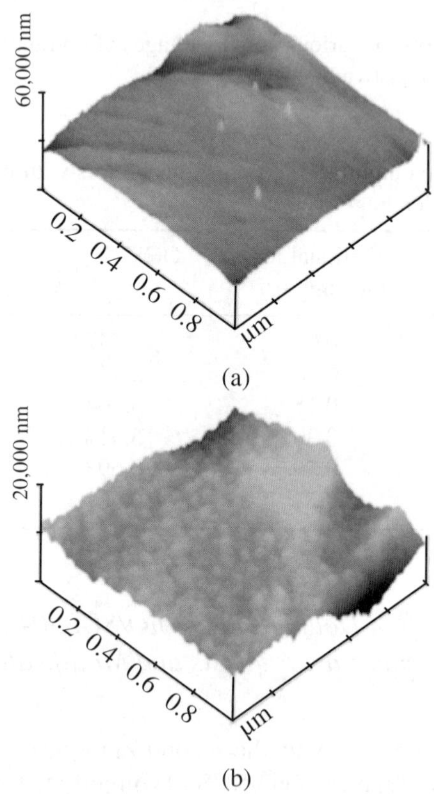

Figure 2.58 Atomic force microscopy (AFM) images of the back surface of SUS304: (a) before and (b) after 300 °C heating [42]

and the second ZnO piezoelectric thin film cannot be deposited on the rough back of the SUS304. To solve this problem, Ti and Pt layers were deposited on the back side of the SUS304 before the first ZnO thin film deposition to improve the adhesion between the second ZnO piezoelectric thin film and the SUS304. Figure 2.59 shows a strong

Table 2.20 The roughness analyses of the back surface of SUS304 before and after 300 °C heating [42]

	Before 300 °C heating	After 300 °C heating
Root mean square roughness R_q (nm)	1.455	12.592
Arithmetic average roughness R_a (nm)	1.075	8.583
Max roughness R_{max} (nm)	26.127	128.03

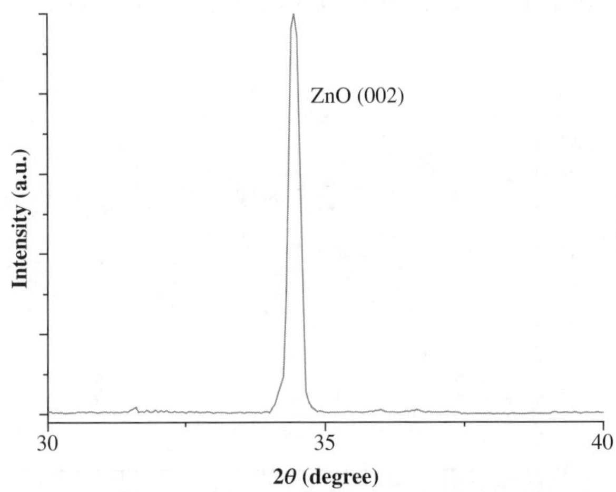

Figure 2.59 The θ–2θ X-ray scans of the 2nd-ZnO thin film

(002) peak of the second ZnO layer. In addition, the second ZnO layer exhibits a dense and smooth surface morphology with a highly textured structure, as shown in Figure 2.60. The second ZnO layer deposited on the back of SUS304/Ti/Pt therefore has excellent properties and is effective in double-sided piezoelectric generators.

2.4.9 Electrical Properties of Double-Sided ZnO/SUS304-Based Piezoelectric Generator

Based on previous investigations [43, 44], excitation frequency and amplitude of vibration critically affect the output voltage of a double-sided generator. Figure 2.61 shows a plot of the variation of OCV ($V_{p\text{-}p}$) of a double-sided ZnO/SUS304-based generator at the fixed 1 mm vibration amplitude and various excitation frequencies, in which OCV ($V_{p\text{-}p}$) reached a maximum of 18.2 V at an excitation frequency of 80 Hz.

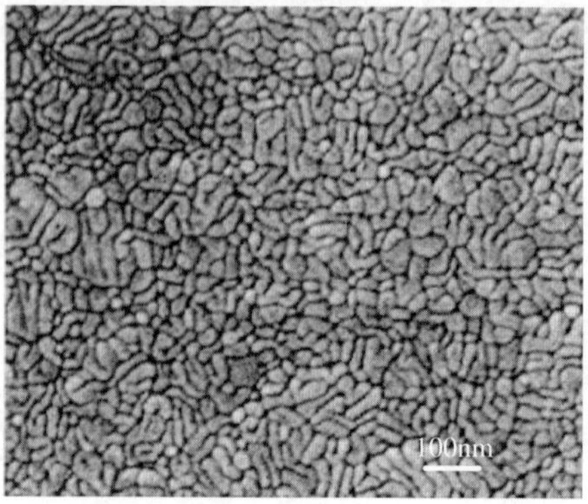

Figure 2.60 The surface morphologies of 2nd-ZnO thin film

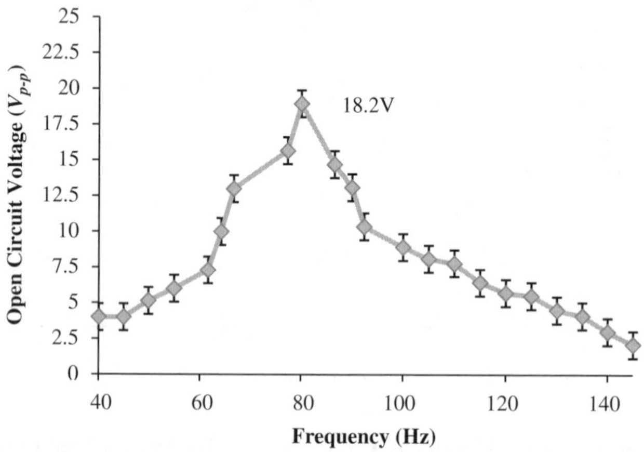

Figure 2.61 The open circuit voltages (V_{p-p}) under various excitation frequencies

Clearly, the double-sided generator used in this study is effective at low-frequency vibration. Figure 2.62 shows a plot of the variation of OCV (V_{p-p}) with the vibration amplitude. The stable value of OCV (V_{p-p}) of 18.2 V is reached at a vibration amplitude of 1 mm and a vibration frequency of 80 Hz. A higher OCV (V_{p-p}) of 20.6 V can be obtained at a higher vibration amplitude; however, the cantilever collapses within a few hours.

In practical applications, double-sided ZnO/SUS304-based generators are rectified using a bridge circuit with four Schottky diodes (1N5711) and filtered using a capacitor of 33 nF. To determine the maximum power extraction, a load resistance of 6 MΩ was used to match the internal resistance of the double-sided ZnO/SUS304-based generator. Figure 2.63 shows the output voltage rectified by the bridge circuit, the voltage that was filtered using a 33 nF capacitor, and the voltage with a resistor of

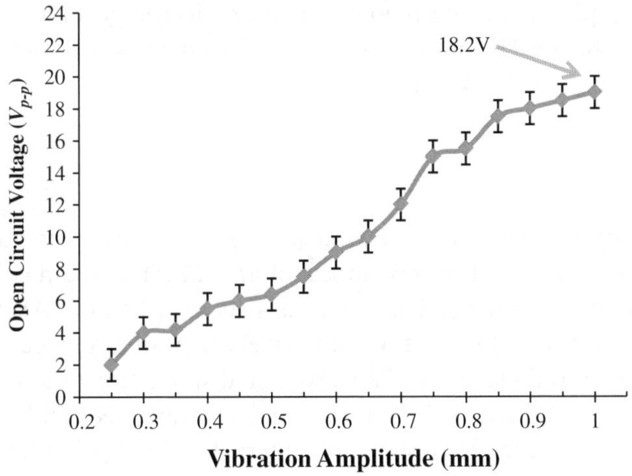

Figure 2.62 The open circuit voltages (V_{p-p}) under various vibration amplitudes

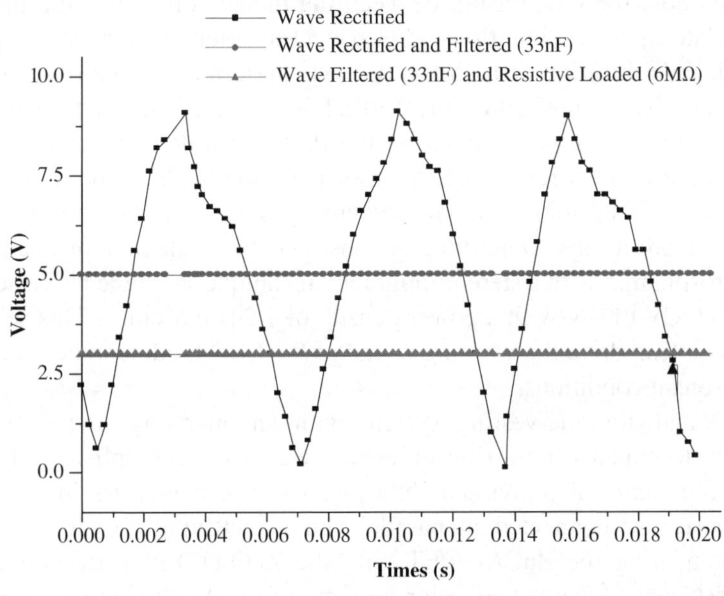

Figure 2.63 The wave forms of output voltages of optimized double-sided ZnO/SUS304-based piezoelectric generator after rectified, filtered, and resistive loaded

6 MΩ loaded, obtained using the optimal generator with vibration amplitude of 1 mm and vibration frequency of 80 Hz.

The maximum DC output voltage with 6 MΩ loading was 2.8 V, and the power output was 1.31 μW cm^{-2}. The variation of the power output ±0.001% was obtained after a continuous 24-hour-long test. To examine the maximum instant power output, a test with turbulent amplitude of vibration on the piezoelectric generator was conducted. The result shows that the DC output voltage with loading reached a maximum of 4.4 V and a power output of 3.23 μW cm^{-2}.

2.5 Summary

This chapter describes two types of flexible piezoelectric power generators using ZnO piezoelectric thin films. First, a new design of an Al/PET-based flexible energy harvester is proposed. The energy harvester consists of a flexible Al/PET-conductive substrate, piezoelectric ZnO thin film, selectively deposited UV-curable resin lump structures, and a Cu foil electrode. The design and simulation of a piezoelectric cantilever plate is described after using commercial software ANSYS FEA to determine the optimal thickness of the PET substrate, internal stress distribution, operation frequency, and electric potential. Because of the optimal thickness predicted by performing developed accurate, analytical, formulaic analysis, the one-way mechanical strain enhances the induced electric potential can be controlled within the piezoelectric ZnO layer. In addition, the relationship between the model solution of the piezoelectric cantilever plate equation, vibration-induced electric potential, and electric power was determined. ZnO thin film of high (002) *c*-axis-preferred orientation with excellent piezoelectricity was deposited on the Al/PET by using RF magnetron sputtering at room temperature. Al was sputtered on the PET substrate as the bottom electrode because of its low sheet resistance, superior adhesion to PET, and lattice constants matching that of ZnO thin film. The selectively deposited UV-curable resin lump structures as a proof mass were directly constructed on a flexible piezoelectric plate using electrospinning with a stereolithography technique. A single harvester achieves a maximum OCV of 4 V with a power density of 1.247 μW cm^{-2}. This self-powered storage system can drive the warning signal of the LED module under both resonant and non-resonant conditions.

A broad bandwidth harvesting system with an operating frequency range of 100–400 Hz to enhance powering efficiency was also accomplished. This system comprises four units of individual ZnO piezoelectric harvesters in the form of a cantilever structure connected in parallel and a rectifying circuit with a storage module. Comparing the ZnO/Al/PET with the ZnO/ITO/PET structures after the vibration test showed that the adhesion strength of the Al film between the ZnO film and the PET substrate was superior to that of the ITO film. This indicates that Al film provides excellent adhesion between the ZnO thin film and the PET substrate.

The *c*-axis-oriented ZnO thin films were also effectively deposited on flexible stainless steel substrates with particular sputtering parameters to develop another type of piezoelectric generator. The ZnO film not only had a uniform grain size and smooth surface but also a strong *c*-axis orientation and a columnar structure. The strong crystalline properties of thin film contributed substantially to the desired piezoelectricity of the piezoelectric generator. The ZnO/SUS304-based piezoelectric generator provides a power output of 1.0 μW cm^{-2}. In addition, Ti and Pt layers were deposited using a dual-gun DC sputtering system between the ZnO piezoelectric thin film and the back side of the SUS304 to promote the adhesion between the second ZnO piezoelectric thin film and the SUS304. The second ZnO thin film not only had a uniform grain size and a smooth surface but also a strong *c*-axis orientation and a columnar structure. The double-sided vibration-powered generator therefore provides a smooth and steady power output of 1.3 μW cm^{-2} and a maximum power output of 3.0 μW cm^{-2} under turbulent vibration.

References

1. Celso Penche and European (2004) Guide on How to Develop a Small Hydropower Plant, Small Hydropower Association (ESHA).
2. Hatziargyriou, N. and Zervos, A. (2001) Wind power development in Europe. *Proceedings of the IEEE*, **89** (12), 1765–1782.
3. Blunden, L.S. and Bahaj, A.S. (2007) Tidal energy resource assessment for tidal stream generators. *Proceedings of the Institution of Mechanical Engineers, Part A: Journal of Power and Energy*, **221** (2), 137–146.
4. O'Regan, B. and Gratzel, M. (1991) A low-cost, high-efficiency solar cell based on dye-sensitized colloidal TiO_2 films. *Nature*, **353**, 737–740.
5. Green, M.A., Emery, K., King, D.L. *et al.* (2007) Solar cell efficiency tables (Version 29). *Progress in Photovoltaics: Research and Applications*, **15** (1), 35–40.
6. Britt, J. and Ferekides, C. (1993) Thin-film CdS/CdTe solar cell with 15.8% efficiency. *Applied Physics Letters*, **62** (22), 2851–2852.
7. Stephen, N.G. (2006) On energy harvesting from ambient vibration. *Journal of Sound and Vibration*, **293** (1–2), 409–425.
8. Bouendeu, E., Greiner, A., Smith, P.J. and Korvink, J.G. (2011) A low-cost electromagnetic generator for vibration energy harvesting. *IEEE Sensors Journal*, **11** (1), 107–113.
9. Wang, Z.L. (2011) *Nanogenerators for Self-Powered Devices and Systems*, Georgia Institute of Technology, Atlanta, GA.
10. Wang, G.J., Yu, W.C., Lin, Y.H. and Yang, H. (2006) Modeling and fabrication of a piezoelectric vibration-induced micro power generator. *Journal of the Chinese Institute of Chemical Engineers*, **29** (4), 697–706.
11. Williams, C.B. and Yates, R.B. (1996) Analysis of a micro-electric generator for microsystems. *Sensors and Actuators A: Physical*, **52** (1–3), 8–11.
12. Feng, G.H. and Hung, J.C. (2008) Development of wide frequency range-operated micromachined piezoelectric generators based on figure-of-merit analysis. *Microsystem Technologies*, **14** (3), 419–425.
13. Watanabe, S. and Fujiu, T. (1996) PZT thin film actuator/sensor for atomic force microscope. Proceedings of 10th IEEE International Symposium on Applications of Ferroelectrics, vol. 1, pp. 199–204.
14. Jones, P.G., Beeby, S.P. and White, N.M. (2001) Towards a piezoelectric vibration-powered microgenerator. *IEE Proceedings: Science, Measurement & Technology*, **148** (2), 69–72.
15. Roundy, S., Leland, E.S., Baker, J. *et al.* (2005) Improving power output for vibration-based energy scavengers. *Energy Harvesting & Conservation*, **4** (1), 28–36.
16. Fang, H.B., Liu, J.Q., Xu, Z.Y. *et al.* (2006) Fabrication and performance of MEMS-based piezoelectric power generator for vibration energy harvesting. *Microelectronics Journal*, **37** (11), 1280–1284.
17. Minazara, E., Vasic, D., Costa, F. and Poulin, G. (2006) Piezoelectric diaphragm for vibration energy harvesting. *Ultrasonics*, **44** (1), e699–e703.

18. Kulkarni, S., Koukharenko, E., Tudor, J. *et al.* (2007) Fabrication and test of integrated micro-scale vibration based electromagnetic generator. IEEE Solid-State Sensors, Actuators and Microsystems Conference, Lyon, France, 1–14 June, pp. 879–882.
19. O'Donnell, T., Saha, C., Beeby, S. and Tudor, J. (2007) Scaling effects for electromagnetic vibrational power generators. *Microsystem Technologies*, **13** (11–12), 1637–1645.
20. Kulkarni, S., Koukharenko, E., Torah, R. *et al.* (2007) Design, fabrication and test of integrated micro-scale vibration-based electromagnetic generator. *Sensors and Actuators A: Physical*, **145–146**, 336–342.
21. Kim, S.I., Lee, D.H., Lee, Y.P. *et al.* (2008) Low frequency properties of micro power generator using a gold electroplated coil and magnet. *Current Applied Physics*, **8** (2), 138–141.
22. Ma, W., Zhu, R., Rufer, L. *et al.* (2007) An integrated floating-electrode electric microgenerator. *IEEE Journal of Microelectromechanical Systems*, **16** (1), 29–37.
23. Serre, C., Pérez-Rodríguez, A., Fondevilla, N. *et al.* (2007) Design and implementation of mechanical resonators for optimized inertial electromagnetic microgenerators. *Microsystem Technologies*, **14** (4–5), 653–658.
24. Giordano, C., Ingrosso, I., Todaro, M.T. *et al.* (2009) AlN on polysilicon piezoelectric cantilevers for sensors/actuators. *Microelectronic Engineering*, **86** (4), 1204–1207.
25. Chen, P.H. and Lin, S.C. (2007) Wind-powered piezo generators. The 33rd Annual Conference of the IEEE Industrial Electronics Society (IECON), Taipei, Taiwan, pp. 2163–2168.
26. Liu, J.Q., Fang, H.B., Xu, Z.Y. *et al.* (2008) A MEMS-based piezoelectric power generator array for vibration energy harvesting. *Microelectronics Journal*, **39** (5), 802–806.
27. Dutoit, N.E., Wardle, B.L. and Kim, S.G. (2005) Design considerations for MEMS-scale piezoelectric mechanical vibration energy harvesters. *Integrated Ferroelectrics*, vol. **71**(1), Taylor & Francis, pp. 121–160.
28. Liu, Z.H. (2012) Design and fabrication of flexible piezoelectric harvesters based on ZnO thin films and PVDF nanofibers. PhD thesis, National Sun Yat-Sen University, Kaohsiung, Taiwan.
29. Pan, C.T., Liu, Z.H. and Chen, Y.C. (2012) Study of broad bandwidth vibrational energy harvesting system with optimum thickness of PET substrate. *Current Applied Physics*, **12** (3), 684–696.
30. Yang, R., Qin, Y., Dai, L. and Wang, Z.L. (2009) Power generation with laterally packaged piezoelectric fine wires. *Nature Nanotechnology*, **4** (1), 34–39.
31. Bien, D.C. S., Mitchell, N.S. J. and Gamble, H.S. (2006) *Micro-Machined Passive Valves: Fabrication Techniques, Characterisation and Their Application*. Springer, pp. 741–800.
32. Lu, F., Lee, H.P. and Lim, S.P. (2004) Modeling and analysis of micro piezoelectric power generators for micro-electromechanical-systems applications. *Smart Materials and Structures*, **13** (1), 57–63.
33. Ugural, A.C. (1999) *Stresses in Plates and Shells*, 2nd edn, McGraw-Hill, New York.
34. Detwiler, D.T., Shen, M.-H.H. and Venkayya, V.B. (1995) Finite element analysis of laminated composite structures containing distributed piezoelectric actuators and sensors. *Finite Elements in Analysis and Design*, **20** (2), 87–100.
35. Hwang, W.S. and Park, H.C. (1993) Finite element modeling of piezoelectric sensors and actuators. *American Institute of Aeronautics and Astronautics*, **31** (5), 930–937.
36. Sohn, J.W., Choi, S.B. and Lee, D.Y. (2005) An investigation on piezoelectric energy harvesting for MEMS power sources. *Journal of Mechanical Engineering Science*, **219** (4), 429–436.
37. Chen, S.N., Wang, G.J. and Chien, M.C. (2006) Analytical modeling of piezoelectric vibration-induced micro power generator. *Mechatronics*, **16** (7), 379–387.
38. Changand, M.Y. and Kuo, J.J. (2004) *The Study of Deposited ITO Films On Flexible Substrates by R.F. Sputtering Technology At Room Temperature*, Taiwan Photonics Society.
39. Chang, W.T., Chen, Y.C., Kao, K.S. *et al.* (2012) Design and fabrication of a double-sided piezoelectric transducer for harvesting vibration power. *Thin Solid Films*, **529**, 39–44.

40. Chang, W.T., Chen, Y.C., Lin, R.C. *et al.* (2011) Wind-power generators based on ZnO piezoelectric thin films on stainless steel substrates. *Current Applied Physics*, **11** (1), S333–S338.
41. Huang, Y.-C. (2009) Investigation of ZnO thin films deposited on stainless steel substrates for piezoelectric transducers application. MSc thesis, National Sun Yat-Sen University, Kaohsiung, Taiwan.
42. Chu, Y.-H. (2010) Study of double-sided ZnO piezoelectric transducer. MSc thesis, National Sun Yat-Sen University, Kaohsiung, Taiwan.
43. Taha, L.Y., Majlis, B.Y., and Al Ali, A. (2006) Modelling and analysis of a transformer based MEMS piezoelectric vibration type microgenerator. IEEE International Conference on Semiconductor Electronics, ICSE, pp. 294–298.
44. Mehraeen, S., Jagannathan, S. and Corzine, K.A. (2010) Energy harvesting from vibration with alternate scavenging circuitry and tapered cantilever beam. *IEEE Transactions on Industrial Electronics*, **57** (3), 820–830.

3

Design and Fabrication of Vibration-Induced Electromagnetic Microgenerators

3.1 Introduction

To develop energy devices that are more compactly renewable, MEMS technologies may be used. Using MEMS (microelectromechanical systems) to fabricate a microgenerator, the energy source can be retrieved by an electromechanical mechanism [1] used in sensor systems and networks for automotive, robotics, medical electronics, wearable computers, and household applications. In such applications, a self-powered device can be used for renewable energies obtained from the ambient environment, reduce pollutants on Earth, and provide an independent power source for a microsystem without requiring conventional solutions such as power supplies or batteries [2].

Many achievements in self-powered devices have been reported. In 2006, Beeby *et al.* [3] reviewed state-of-the-art self-powered microsystems and discussed MEMS technology developments in the context of microfabrication and design capabilities. These efforts have resulted in a wide variety of compact sensors and actuators. In 2007, Arnold [4] summarized the power density (PD) (W cm^{-3}) of some state-of-the-art generators, indicating that a PD exceeding 2 mW cm^{-3} is possible for a compact magnetic power generation system.

Some researchers have used a commercially available finite-element code to examine the response of a resonating spring-mass system [5–10]. For example, in

Design and Fabrication of Self-Powered Micro-Harvesters: Rotating and Vibrating Micro-Power Systems, First Edition.
C. T. Pan, Y. M. Hwang, Liwei Lin and Ying-Chung Chen.

2001, El-hami *et al.* [5] employed MATLAB® to predict the power output of an electromagnetic generator and used the computer-aided design (CAD) software package VF-OPERA to analyze the magnetic field distribution. Ching *et al.* [6], Shearwood and Yates [7], Williams *et al.* [8, 9], Beeby *et al.* [10], and Buren and Troster [11] used ANSYS to model the mechanical motion of a spring-mass system on an electromagnetic device with a magnet and a spring, and have investigated the results of voltage outputs within various governing vibration parameters.

El-hami *et al.* [5] used stainless steel sheets to manufacture a cantilever beam vibration spring, which is shown in Figure 3.1. The micromagnetic generator is composed of a vibration spring, a copper inducer coil, and a permanent magnet (PM) of neodymium-iron-boron (NdFeB). The volume and weight of this microgenerator are 240 mm^3 and 500 mg, respectively. This microgenerator can operate at a resonant vibration frequency of 322 Hz and an amplitude of 25 μm; its mean-square-root voltage and power output are 12.8 mV and 53 μW, respectively, with a loading resistor of 0.28 Ω.

Amirtharajah and Chandrakasan [12] developed a tubular electro-magnetic microgenerator composed of a cylindrical spiral spring, a mass with an induction coil, and a PM, as shown in Figure 3.2. The vibration mass of this device weighs approximately 500 mg. An induced voltage of 0.18 V and a power output of 400 μW were obtained using an induction coil of 10 Ω under a vibration of amplitude 20 mm and frequency of *c*. 100 Hz.

Li and colleagues [13] used laser-forming technology to manufacture a copper spiral microspring constituting a vibration-type electro-magnetic microgenerator with a volume of 1 cm^3, which is shown in Figure 3.3. This device could generate a voltage of 2–4 V and a power output of 10 μW at a frequency of 200 Hz and an amplitude of less than 200 μm.

Pan *et al.* [14] used deep reactive ion etching (DRIE) technology and a silicon wafer to develop a spiral microspring on which a Ferro-platinum thin film is sputtered

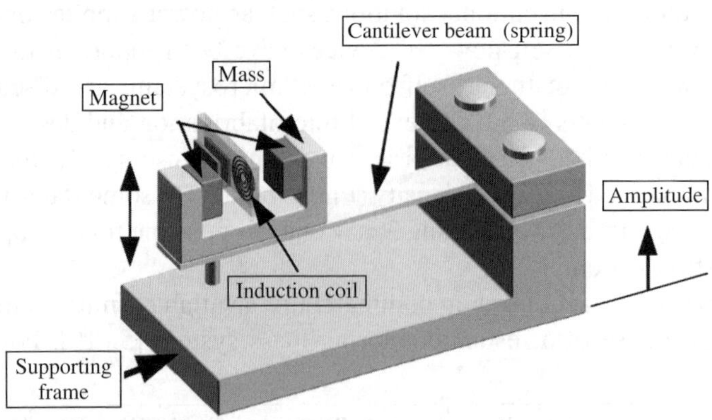

Figure 3.1 Schematic of a stainless steel canti-lever beam microgenerator

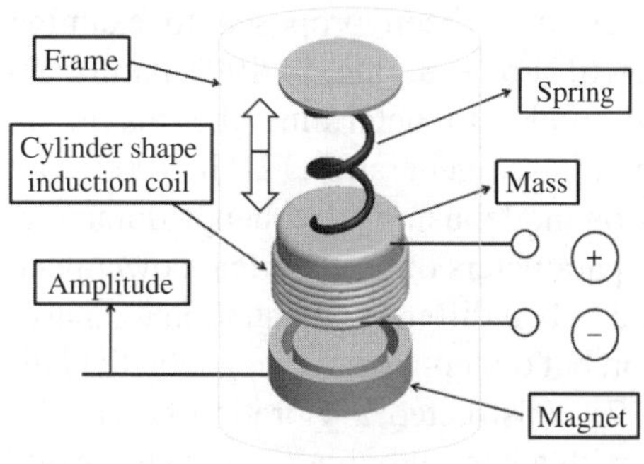

Figure 3.2 Schematic of an electromagnetic column-shape microgenerator

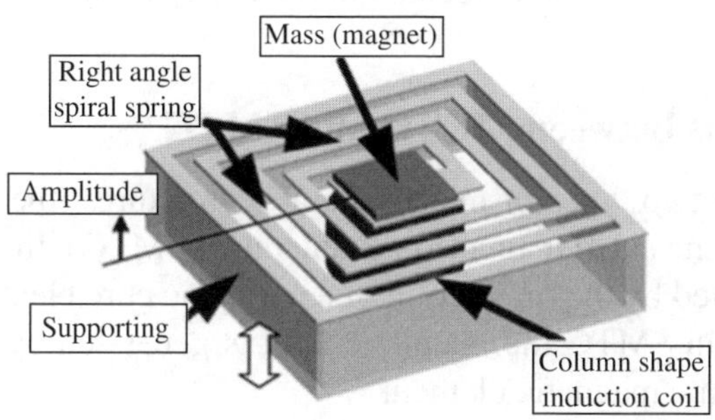

Figure 3.3 Schematic of copper spiral spring microgenerator

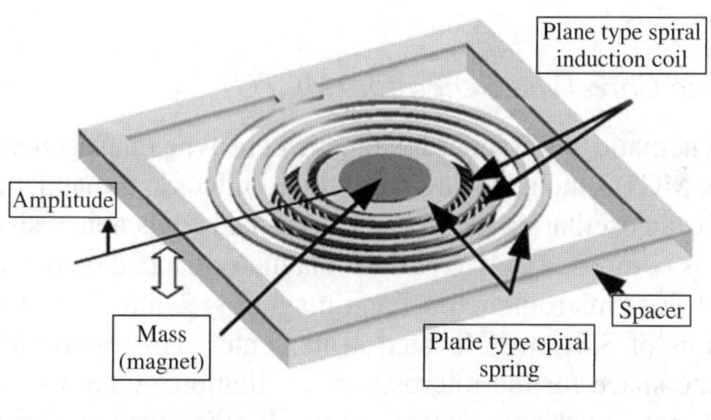

Figure 3.4 Schematic of silicon spiral microspring generator

at its center. By combining a plane-type spiral induction coil, a thin-film vibration electromagnetic microgenerator was developed as shown in Figure 3.4. The volume and thickness are less than 0.34 cm^3 and 0.34 mm, respectively. An induction voltage of 40 mV and a power output of 100 μW were obtained at a resonant frequency of 60 Hz and an amplitude of 0.03 mm.

Some analytical models have been proposed to examine the vibration system movement [12, 13, 15, 16]. For example, in 1998 Amirtharajah and Chandrakasan developed a linearized model to determine the maximum power output of an electromagnetic vibration-based generator [12]. In 2000, Li *et al.* [13, 15] proposed a simple power generation model using a frequency domain approach to examine the effects of the governing parameters of an electrical power generation system. In 1998, Williams *et al.* [16] used only a differential equation to analyze a vibration system in a microelectric generator, but did not discuss magnetic field distributions. The authors of the current study [17] constructed a vibration-based electromagnetic generator consisting of a magnet with a central cylindrical cavity inside and a movable helical low-temperature co-fired ceramic (LTCC) microspring with a multilayer silver induction coil. A simple analytical model was developed to analyze the mechanical motion of the spring-mass system located in the cavity of the magnet.

3.2 Comparisons between MCTG and SMTG

Generally, there are two types of microgenerator structures: the magnetic core-type generator (MCTG) and sided magnet-type generator (SMTG). In an MCTG, the magnetic field is generated by the interaction of a magnetic core element and a magnet in the center hole; in an SMTG the magnetic field is generated by the microinduction coil on the sides of the magnetic element.

Table 3.1 lists the differences between the elements of a MCTG and a SMTG. Properties such as volume and the magnetic field generation source, the assembly concept and the inductor structure for MCTG and SMTG types also differ slightly.

3.2.1 Magnetic Core-Type Generator (MCTG)

The assembly schematic of a standard magnetic core-type microgenerator is shown in Figure 3.5. This MCTG includes an LTCC microinductor, a magnetic core element, a magnet module, an oscillation mass, and an attached adjustable resistor in which the magnet module is composed of two NdFeB magnets attached on the upper and lower surfaces of the LTCC microinductor. Each magnet is 2 mm thick, and a cylindrical hole with a radius of 8.6 mm is located at its center. This cylindrical hole accommodates adequate space for the microspring oscillation. Accordingly, the magnetic elements provide not only the magnetic field but also the supporting frame for mechanical oscillation; 2-mm-thick NdFeB magnets are the commonly used specification. Because NdFeB magnets are commercially available, they present an advantage for cost-reduced mass-production.

A circular hole with a radius of 1.3 mm is at the center of the LTCC microinductor to create a vacuum leak inside, which results in the difficulty in the elements' fixtures. This problem can be resolved by using a microinductor with the same pattern to fix the elements; this is at the expense of increasing the manufacturing cost, however.

Table 3.1 Elements of types of microgenerator: MCTG and SMTG

Specification	Micro generator types	
	MCTG type	SMTG type
Magnet No.	2	2
Total magnet thickness (mm)	4	4
Magnet hole radius (mm)	8.6	–
Total micro inductor thickness (μm)	240	180
Induction coil cross-section area (width × thickness) (μm)	90 × 10	88 × 6.2
Resistor inside induction coil (Ω)	0.3	0.49
Outside resistor (Ω)	0.31	0.51
Inductor hole radius (mm)	1.3	–
Spacer thickness (mm)	–	3 (2 × 1.5)
Spacer hole radius (mm)	–	8.6
Magnetic core height (mm)	4	–
Magnetic core outer diameter (mm)	1.1	–
Vibration mass (mg)	27.6	16
Total volume (mm^3)	424	718

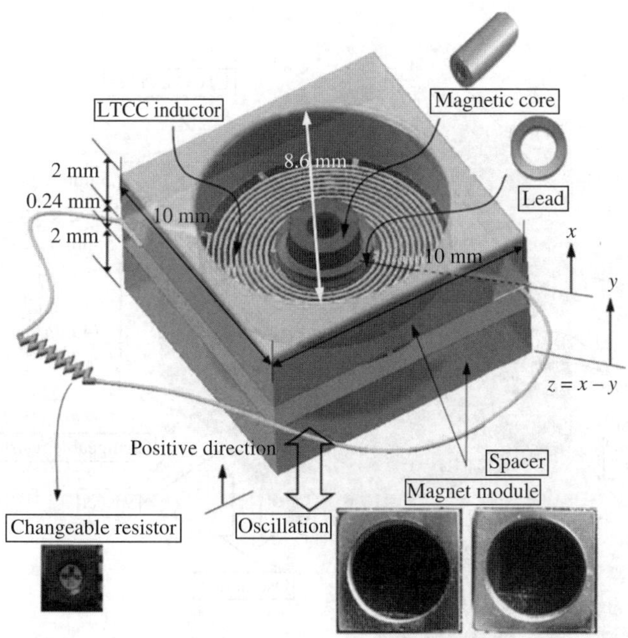

Figure 3.5 Assembly of a MCTG oscillation-type microgenerator [18]

The tubular magnetic core element has an inner and outer radii of 0.5 mm and 1.1 mm, respectively. It passes through the central hole of the microinducer and is located at central hole of the PM. Finally, a lead ring 27.6 mg in weight is used as the oscillation mass and is attached to the surface of the central ring structure of the inducer.

3.2.2 Sided Magnet-Type Generator (SMTG)

Figure 3.6 shows the assembly schematic of an SMTG which includes an LTCC inducer, a magnet module, two plastic spacers, an oscillation mass, and an attached changeable resistor.

3.3 Analysis of Electromagnetic Vibration-Induced Microgenerators

Analytical models of a microgenerator involve vibration-induced coupling issues of mechanical kinematics and electromagnetism. Some analytical derivations are developed to determine the relationship between these coupling issues. First, a formula is developed for vibrating kinematics. Based on Biot-Savart's law, the distribution of the electromagnetic spatial field is then established. Finally, according to Faraday's law, the behavior of electric inductance is predicted. The notation used in this model is summarized in Table 3.2.

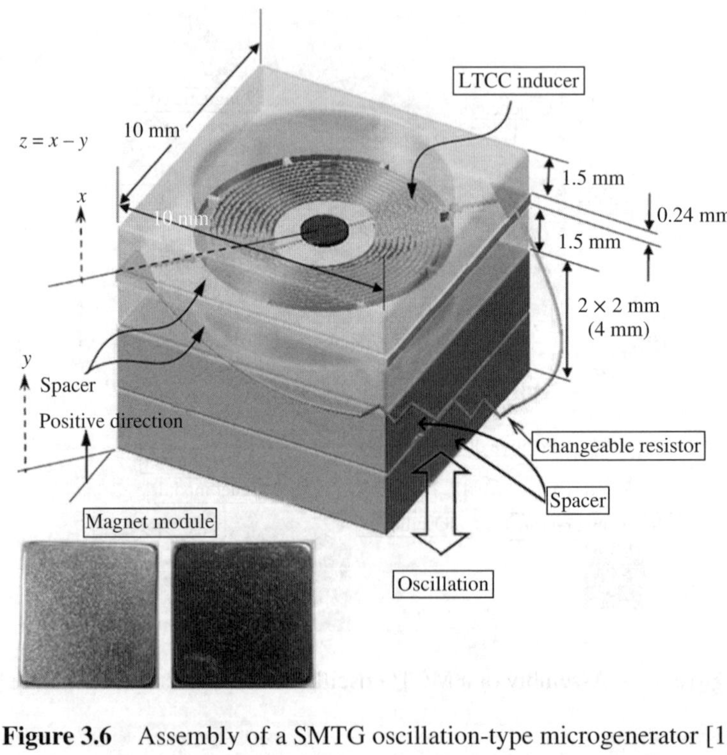

Figure 3.6 Assembly of a SMTG oscillation-type microgenerator [18]

Table 3.2 List of symbols, names, and units [18]

Symbol	Name	Unit
A	Cross-section area of microspring	μm^2
$\vec{a}_D$	Radial unit vector	–
$\vec{a}_\theta$	Tangent unit vector	–
B	Magnetic field	T
B_e	Effective magnetic field	T
$\dot{B}_e$	Time-varying effective magnetic field	T s^{-1}
b	Loop radius of microspring	mm
F_s	Springing load force	N
I	Electric current	A
J	Second polar moment of microspring	μm^4
k	Spring of stiffness	N m^{-1}
k_e	Equivalent spring of stiffness	N m^{-1}
r	Turn radius of coil	–
T_s	Springing torsion	N m
t	Time	s
U	Springing strain energy	J
Y	Maximum amplitude	mm
y	Input excitation	mm
z	Vibration amplitude	mm
$\dot{z}$	Vibration speed	mm s^{-1}
Φ	Magnetic flux	WB
δ	Including angle	rad
ε (or EMF)	Electromotive force	V
ε_i	Coil electromotive force	V
ε_t	Total electromotive force	V
ϕ	Phase angle	rad
ω	Input frequency	rad s^{-1}
ω_n	Resonant frequency	rad s^{-1}

3.3.1 Design of Electromagnetic Vibration-Induced Microgenerators

Figure 3.7a illustrates the configuration of a vibration-based electromagnetic micropower generator and its dimensions. Figure 3.7b shows a cross-sectional diagram, revealing the integration of the helical ceramic microspring and helical silver microcoil used to construct an inducer. Figure 3.7c shows a diagram of a single-layer inducer and design of the double supporting beam. The device shown in Figure 3.7a consists of two plastic spacers, a NdFeB PM and an LTCC inducer. The spacers are glued on the inducer to provide mechanical support and form a cylindrical cavity that is 8.6 mm in diameter and 1.5 mm deep, thereby providing sufficient space for vibration of the microspring. Each spacer is 1.5 mm thick. The NdFeB magnet has

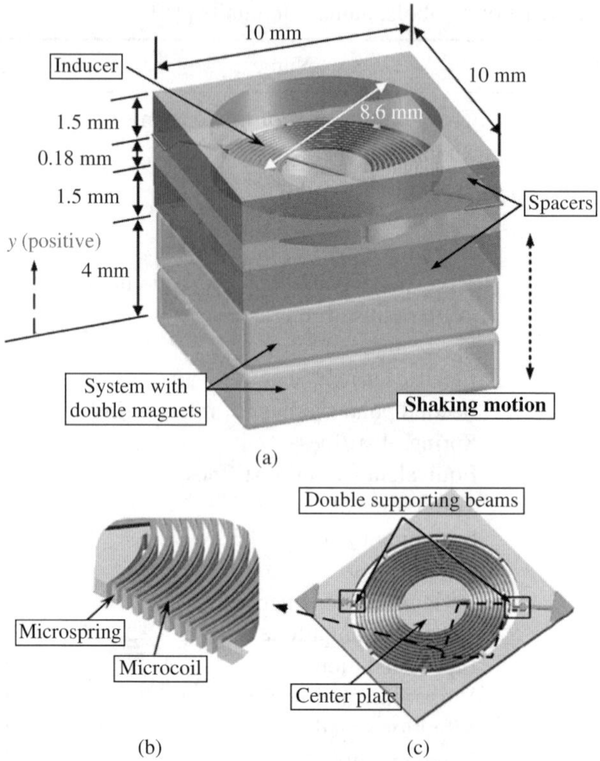

Figure 3.7 Design of vibration-based electromagnetic power generator: (a) configuration and the dimensions, (b) diagram of the cross-section, showing the integrated microspring and microcoil, and (c) diagram of single-layer inducer and design of the double supporting beam [18]

a residual induction (B_r) of 1.46 T. A system with double magnets has also been designed. The double magnet system has a total thickness of 4 mm, whereas the system with a single magnet is 2 mm thick. The entire volume of the meso-generator with double magnets is 10 mm × 10 mm × 7.18 mm (718 mm^3). The oscillating mass of the center plate for the vibration is 16 mg. The mentioned cavity is of sufficient depth to accommodate the maximum vibration amplitude of the microspring.

3.3.2 *Analysis Mode of the Microvibration Structure*

Figure 3.8 shows a schematic of a vibration-induced generator; Figure 3.8a depicts the operating kinematics, Figure 3.8b the supporting beam of the microspring and its geometric parameters, and Figure 3.8c configuration of the spiral microspring. The vibration system is composed of a seismic mass *m*, a spring stiffness *k*, and a damping coefficient *c*. When the rigid housing is moved with an input excitation *y*, the inertia

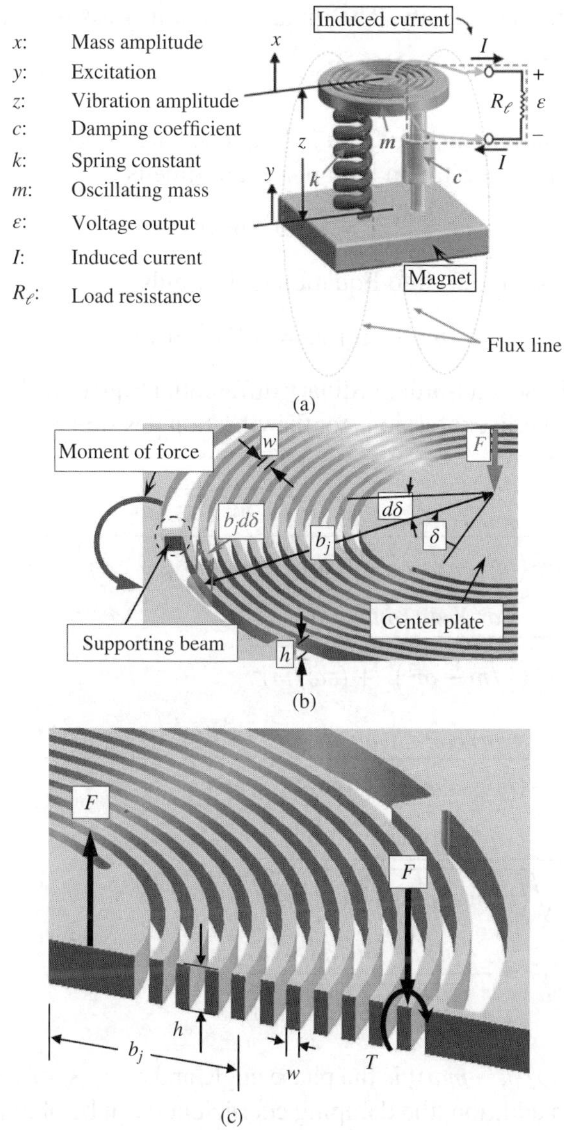

Figure 3.8 Illustration of the meso-generator: (a) in operating kinematics, (b) supporting beam of the microspring and its geometric parameters, and (c) configuration of the spiral microspring [18]

mass has a relative motion z to the housing. The magnetic movement relative to the induction coil causes variations in magnetic fields and can then drive the conversion of mechanical energy to electric energy. The kinematics of the generator is described as

$$m\ddot{z} + c\dot{z} + kz = -m\ddot{y} \tag{3.1}$$

where $\dot{z}$ is the vibration velocity. The input excitation is assumed to be a sinusoidal function expressed as

$$y = Y \sin \omega t \tag{3.2}$$

where Y is the excitation amplitude, ω is frequency, and t is time. The second-order differential of the input excitation with respect to time is

$$\ddot{y} = -Y\omega^2 \sin \omega t \tag{3.3}$$

Substituting Equation (3.3) into Equation (3.1) yields

$$m\ddot{z} + c\dot{z} + kz = mY\omega^2 \sin \omega t \tag{3.4}$$

The solution to the second-order ordinary differential Equation (3.4) can be obtained analytically. The general expressions for the vibration position z and vibration velocity $\dot{z}$ in the steady state are:

$$\begin{aligned} z &= \frac{-mYc\omega^3 \cos \omega t}{m^2\left({\omega^2}_{\mathrm{n}} - \omega^2\right)^2 + (c\omega)^2} + \frac{m^3Y\omega^2\left(\omega_{\mathrm{n}} - \omega^2\right) \sin \omega t}{m^2\left({\omega^2}_{\mathrm{n}} - \omega^2\right)^2 + (c\omega)^2} \\ &= \frac{\omega^2 Y \sin(\omega t - \varphi)}{\sqrt{\left(k/m - \omega^2\right)^2 + (c\omega/m)^2}} \end{aligned} \tag{3.5}$$

$$\begin{aligned} \dot{z} &= \frac{mYc\omega^4 \sin \omega t}{m^2\left({\omega^2}_{\mathrm{n}} - \omega^2\right)^2 + (c\omega)^2} + \frac{m^3Y\omega^3\left(\omega_{\mathrm{n}} - \omega^2\right) \cos \omega t}{m^2\left({\omega^2}_{\mathrm{n}} - \omega^2\right)^2 + (c\omega)^2} \\ &= \frac{\omega^3 Y \cos(\omega t - \varphi)}{\sqrt{\left(k/m - \omega^2\right)^2 + (c\omega/m)^2}} \end{aligned} \tag{3.6}$$

$$\ddot{z} = \frac{d\dot{z}}{dt} = \frac{-mY\omega^4 \sin(\omega t - \varphi)}{\sqrt{\left(k - m\omega^2\right)^2 + (c\omega)^2}} \tag{3.7}$$

where $\varphi = \tan^{-1}[c\omega/(k - m\omega)]$ is the phase angle and $\omega_{\mathrm{n}} = \sqrt{k/m}$ is the natural resonant frequency. In addition, the damping coefficient c can be obtained by measuring the vibration position z and velocity $\dot{z}$.

The springing strain energy U for one loop microspring-vibration system is composed of two terms from the spring torsion T_{s} and the applied force F_{s}, defined:

$$U = \frac{{T_{\mathrm{s}}}^2 \pi b}{4GJ} + \frac{{F_{\mathrm{s}}}^2 \pi b}{4AG} \tag{3.8}$$

$$T_{\mathrm{s}} = F_{\mathrm{s}} b/4 \tag{3.9}$$

$$J = wh\left(w^2 + h^2\right)/12 \tag{3.10}$$

$$A = wh \tag{3.11}$$

where G, J, and A are the silicon shear module, second polar moment of the microspring, and cross-section area of the microspring, respectively. The variables w and h are the width and the thickness of the microspring, respectively, and b is the radius of the microspring. Castigliano's theorem [18] indicates that the maximum displacement d is the partial derivative of springing strain energy U of Equation (3.8) with respect to the applied force F_s; that is,

$$d = \frac{\partial U}{\partial F_s} = F_s \left(\frac{\pi b^3}{32GJ} + \frac{\pi b}{2AG} \right) \tag{3.12}$$

Hooke's law states that the spring stiffness k is the ratio of the applied force F_s to the maximum displacement d, as given by using

$$k = \frac{F_s}{d} = \frac{32GJA}{\pi b^3 A + 16\pi bJ} \tag{3.13}$$

Substituting $J = wh(w^2 + h^2)/12$ and $A = wh$ into Equation (3.13), the spring stiffness of a microspring with one single loop can be expressed:

$$k = \frac{8Gwh\left(w^2 + h^2\right)}{4\pi b\left(w^2 + h^2\right) + 3\pi b^3} \tag{3.14}$$

For an n-loop spring vibration system in the device, the equivalent spring stiffness k_e can be obtained using

$$k_e = \left(\sum \frac{1}{k_j}\right)^{-1} = \left[\frac{2Gwh\left(w^2 + h^2\right)}{\pi\left(w^2 + h^2\right)\sum b_j + 0.75\pi \sum b_j^3}\right] \tag{3.15}$$

where b_j is the radius of the jth loop microspring.

3.3.3 Analysis Mode of Magnetic Field

The behavior of a spatial magnetic field $\vec{B}$ is described according to the relative position vector $\vec{r}$ from the magnet to the induction coil and the tangent vector $\vec{r}_T$ at the induction coil position in the spherical coordinate system, as shown in Figure 3.9a. Based on the image shown in Figure 3.9a, the geometric parameters of θ and $|\vec{r}|$ can be expressed as $\theta = \tan^{-1}[(|\vec{p}|)/(1.5 - z)]$ and $|\vec{r}| = \sqrt{|\vec{p}|^2 + (1.5 - z)^2}$. Symbol $\vec{p}$ represents the projected radial vector of the induction coil. The spacer thickness is 1.5 mm. When the magnet moves away from the induction coil, the magnetic field density around the induction coil becomes low. The magnitude of $\vec{B}$ decays with the cubic of the distance $|\vec{r}|$, and is expressed regarding vacuum permeability μ_0, relative

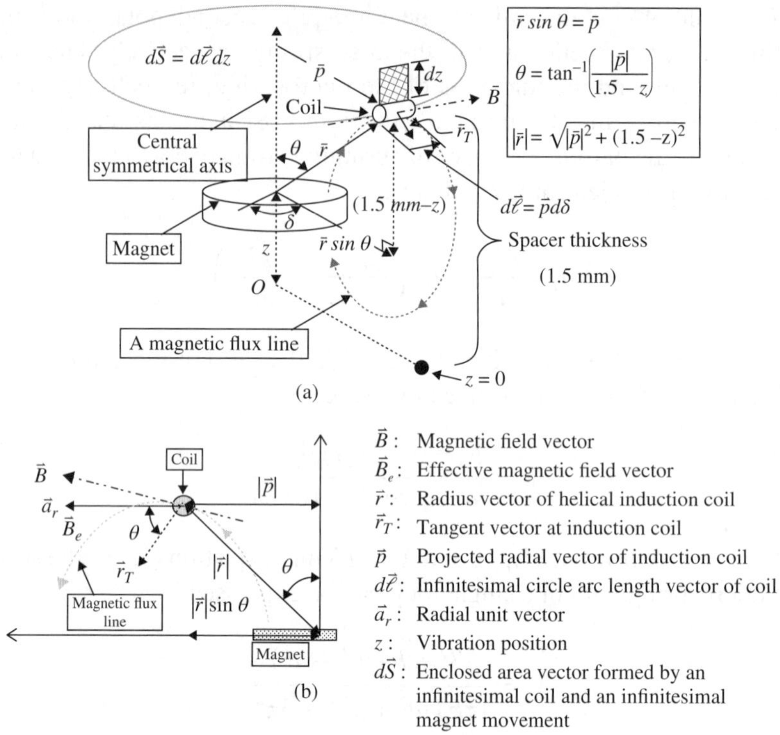

Figure 3.9 Schematic of (a) a three-dimensional magnetic field in the spherical coordination system (R, δ, θ) and (b) effective magnetic field [18]

permeability μ_r, and the magnetic dipole of magnet m_H. From Biot-Savart's law [19], we obtain

$$\overrightarrow{B} = \frac{\mu_0 \mu_r m_H}{4\pi |\overrightarrow{r}|^3} \left(2 \cos\theta \overrightarrow{r} + \sin\theta \overrightarrow{r}_T\right) \tag{3.16}$$

$$\cos\theta = (1.5 - z) / |\overrightarrow{r}| = (1.5 - z) / \left(|\overrightarrow{p}|^2 + (1.5 - z)^2\right)^{1/2} \tag{3.17}$$

$$\sin\theta = |\overrightarrow{p}| / |\overrightarrow{r}| = |\overrightarrow{p}| / \left(|\overrightarrow{p}|^2 + (1.5 - z)^2\right)^{1/2} \tag{3.18}$$

where θ is the inclination angle from the induction coil to the central symmetrical axis. Faraday's law of electromagnetism can predict the induced voltage output. Because symmetrical $\overrightarrow{B}$ is evidently independent of the circumferential angle δ, a value of $-\pi/2$ to $\pi/2$ of δ is assumed (shown in Figure 3.9a) to simplify the analysis of ε. According to this law, the inducible ε electromotive force (EMF) is equal to the rate of the change of the flux while the magnetic field line is passing through an enclosed area vector ($d\overrightarrow{S} = d\overrightarrow{l}\, dz$) formed by an induction coil of an infinitesimal circle arc length ($d\overrightarrow{l} = \overrightarrow{p}\, d\delta$) and a magnetic displacement dz. The inducible ε (EMF) is obtained using

$$\varepsilon = -\frac{d\Phi}{dt} = -\frac{d}{dt}\int_S \vec{B}\,d\vec{S} = -\left(\int_S \frac{d\vec{B}}{dt}d\vec{S} + \int_S \vec{B}\frac{d\vec{S}}{dt}\right)$$
$$= \int_{-\delta}^{\delta}\left(z\dot{\vec{B}}\right)d\vec{l} + \int_{-\delta}^{\delta}\left(\dot{z}\vec{B}\right)d\vec{l} = \int_{-\pi/2}^{\pi/2}\left(z\dot{\vec{B}}\right)d\vec{l} + \int_{-\pi/2}^{\pi/2}\left(\dot{z}\vec{B}\right)d\vec{l} \tag{3.19}$$

where Φ is the magnetic flux and $\vec{S}$ is the enclosed area vector.

Based on terms $z\dot{\vec{B}}_e$ and $\dot{z}\vec{B}_e$ in Equation (3.19), only the component of $\vec{B}$ perpendicular to the vibration direction of the magnet contributes to the EMF. The effective magnetic field $\vec{B}_e$ shown in Figure 3.9b can be derived as

$$\vec{B}_e = \frac{\mu_0\mu_r m_H}{4\pi|\vec{r}|^3}\left(2\cos\theta\sin\theta\,\vec{a}_r + \sin\theta\cos\theta\,\vec{a}_r\right) = \frac{3\mu_0\mu_r m_H}{4\pi|\vec{r}|^3}\cos\theta\sin\theta\,\vec{a}_r \tag{3.20}$$

or

$$\vec{B}_e = \frac{3\mu_0\mu_r m_H|\vec{p}|(z-1.5)}{4\pi\left[|\vec{p}|^2 + (1.5-z)^2\right]^{5/2}}\vec{a}_r \tag{3.21}$$

where $\vec{a}_r$ is the radial unit vector. The effective magnetic field rate $\dot{\vec{B}}_e$ that is perpendicular to the vibration direction of the magnet can be obtained using

$$\dot{\vec{B}}_e = \frac{d\vec{B}_e}{dt} = \frac{3\mu_0\mu_r m_H|\vec{r}|}{4\pi|\vec{r}|^4}\left[|\vec{r}|\left(1-2\sin^2\theta\right)\dot{\theta} - 3|\dot{\vec{r}}|\sin\theta\cos\theta\right]\vec{a}_r \tag{3.22}$$

or

$$\dot{\vec{B}}_e = \frac{3\mu_0\mu_r m_H|\vec{p}|\dot{z}}{4\pi}\left\{\frac{\left[|\vec{p}|^2+(1.5-z)^2\right]^{1/2} - 5(z-1.5)^2/\left[|\vec{p}|^2+(1.5-z)^2\right]^{3/2}}{\left[|\vec{p}|^2+(1.5-z)^2\right]^3}\right\}\vec{a}_r \tag{3.23}$$

By substituting Equations (3.20) and (3.22) into Equation (3.19), the EMF for a single-turn induction coil can be rewritten as

$$\varepsilon = \int_{-\pi/2}^{\pi/2} z\dot{\vec{B}}_e\vec{p}\,d\delta + \int_{-\pi/2}^{\pi/2}\dot{z}\vec{B}_e\vec{p}\,d\delta = 2\pi|\vec{p}|\left(z\dot{\vec{B}}_e + \dot{z}\vec{B}_e\right) \tag{3.24}$$

The variable ε_i represents the EMFs on an induction coil of the ith turn. For a multi-turn induction coil, the EMF for the ith turn is derived by substituting Equations (3.21) and (3.23) into Equation (3.24) to yield:

$$\varepsilon_i = \frac{3\mu_0\mu_r m_H|\vec{p}_i|^2}{4\pi}\left\{\frac{\dot{z}_i(z_i-1.5)}{\left[|\vec{p}_i|^2+(1.5-z_i)^2\right]^{5/2}} + \frac{z_i\dot{z}_i\left\{\left[|\vec{p}_i|^2+(1.5-z_i)^2\right]^{1/2}-5(z_i-1.5)^2/\left[|\vec{p}_i|^2+(1.5-z_i)^2\right]^{3/2}\right\}}{\left[|\vec{p}_i|^2+(1.5-z_i)^2\right]^3}\right\} \tag{3.25}$$

Based on Equation (3.25), the term of ε_i can be predicted by m_H, μ_0, μ_r, $|\vec{p}|$, z (Equation (3.5)), and $\dot{z}$ (Equation (3.6)). The EMF on all of the induction coils is the sum of $\varepsilon_1 + \varepsilon_2 + \varepsilon_3 + \ldots \cdots + \varepsilon_n$ in the device, and can be written:

$$\varepsilon_t = \frac{3\mu_0\mu_r m_H}{4\pi}\sum_{i=1}^{n}|\vec{p}_i|^2\left\{\frac{\dot{z}_i(z_i-1.5)}{\left[|\vec{p}_i|^2+(1.5-z_i)^2\right]^{5/2}} + \frac{z_i\dot{z}_i\left\{\left[|\vec{p}_i|^2+(1.5-z_i)^2\right]^{1/2}-5(z_i-1.5)^2/\left[|\vec{p}_i|^2+(1.5-z_i)^2\right]^{3/2}\right\}}{\left[|\vec{p}_i|^2+(1.5-z_i)^2\right]^3}\right\} \tag{3.26}$$

If an electric resistance R is linked with the induction coil, the induced current can be obtained using

$$I = \frac{\varepsilon_t}{R} \tag{3.27}$$

The external resistance R is much larger than that inside the copper induction coil. For a sinusoidal EMF, the average power generation is expressed:

$$\text{Power} = (I)_{\text{rms}}(\varepsilon_t)_{\text{rms}} \tag{3.28}$$

where the subscript rms denotes the root-mean-square value.

3.3.4 Evaluation of Various Parameters of Power Output

According to Equation 3.15, the equivalent spring stiffness k_e decreases non-linearly with the sum of the loop numbers. As shown in Figure 3.10, the curve of k_e exhibits a negative slope. Using a device with 12 loops as an example, the value of k_e is $5.05\,\text{N m}^{-1}$, in which the dimensions are 100 μm in width and 100 μm in interspace, the innermost loop radius is 1.7 mm and the minimum feature size is 40 μm in depth.

The magnetic field $|\vec{B}|$ varies non-linearly with the position when the PM moves up and down. The simulated result based on Equation (3.16) is shown in Figure 3.11. The permeability μ_0 and relative permeability μ_r of copper induction coils are $4\pi \times 10^{-7}\,\text{H m}^{-1}$ and 0.9999, respectively. The graph indicates that the maximum value of $|\vec{B}|$ is 0.314 T at the innermost turn of $|\vec{p}| = 0.8$ mm when the vibration position z is 1.5 mm. The direction of z is illustrated in Figure 3.7. The positive vibration direction points toward the induction coil, meaning that the distance between the PM and induction coil is reduced. A higher value of $|\vec{B}|$ was found when

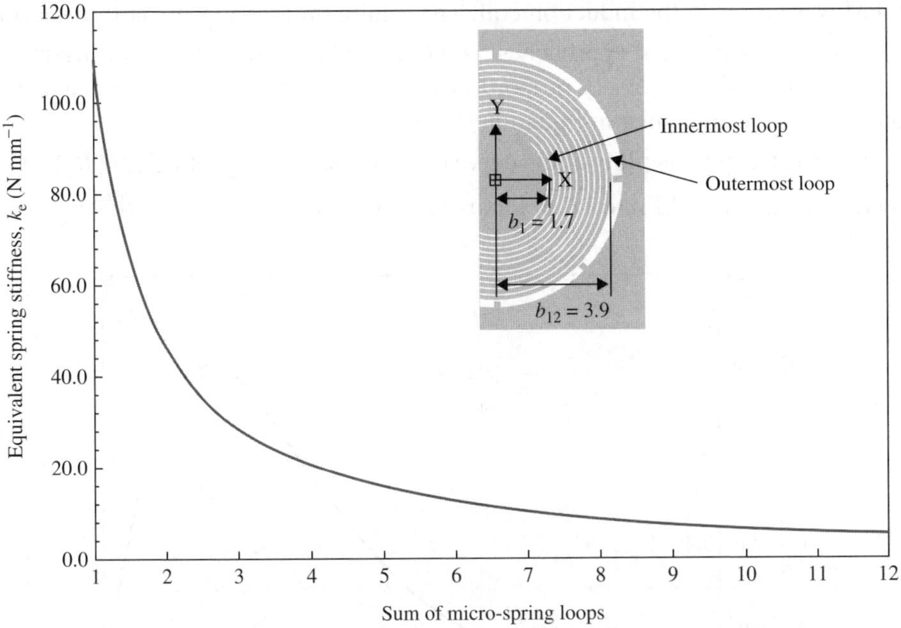

Figure 3.10 Graph of equivalent spring stiffness as a function of loop number of the microspring, microspring radius $b_1 = 1.7\,\text{mm}$, $b_{J+1} = b_j + 200\,\mu\text{m}$, microspring width $w = 100\,\mu\text{m}$, and microspring thickness $h = 40\,\mu\text{m}$ [18]

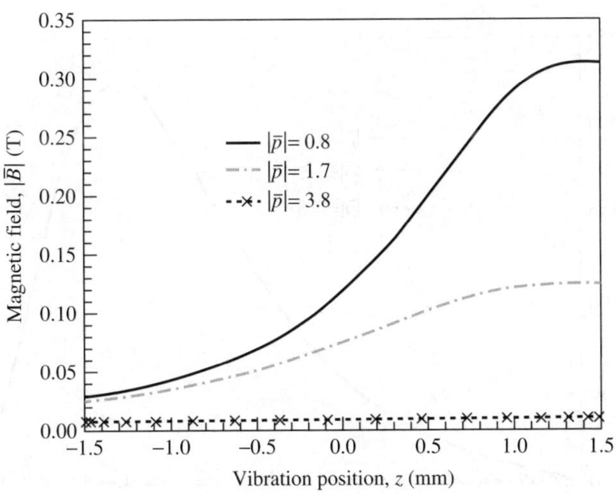

Figure 3.11 Analytical magnetic field on each induction coil of three unequal radii against vibration position [18]

the PM was closer to the induction coil. The minimum $|\vec{B}|$ appears at the outermost turn of $|\vec{B}| = 3.8$ mm and approaches a constant value of 0.01 T, demonstrating that the magnetic field is inversely proportional to the distance between the magnet and induction coil, as indicated in Equation (3.16).

Figure 3.12a,b depicts plots of the curve of the effective magnetic field $|\vec{B}_e|$ and the effective magnetic field rate $|\dot{\vec{B}}_e|$ versus vibration position z, respectively, on each

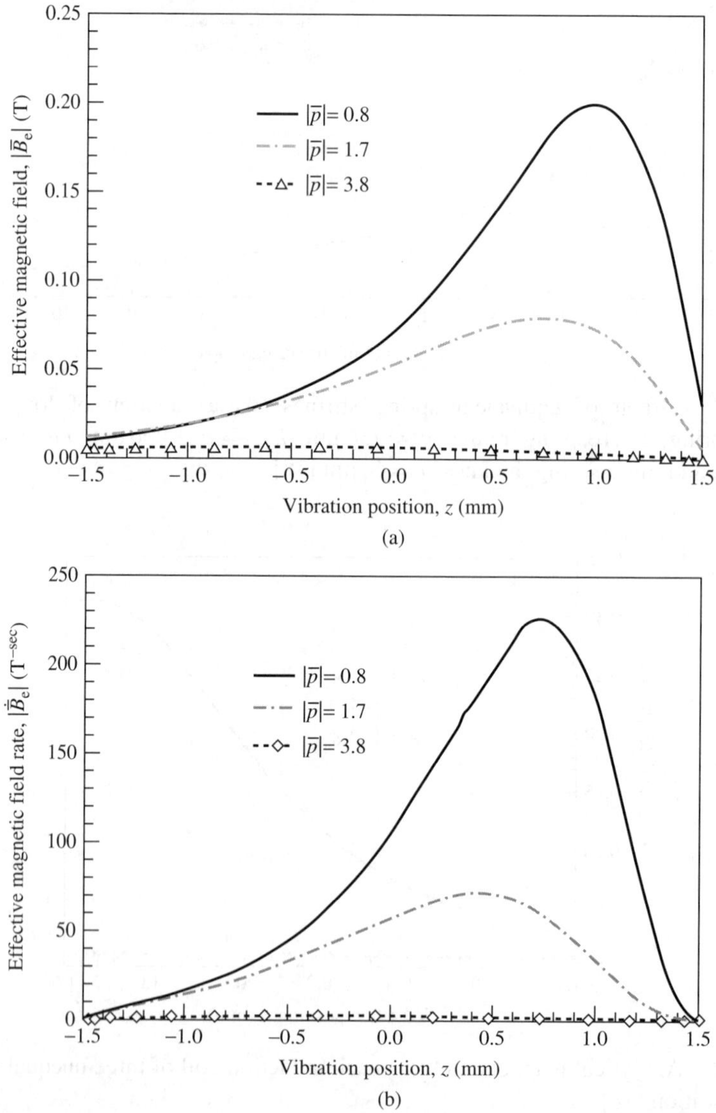

Figure 3.12 (a) Analytical effective magnetic field and (b) analytical effective magnetic field rate on each induction coil of three unequal radii against vibration position [18]

copper induction coil of three unequal radii $|\vec{p}|$. In Figure 3.12a, the magnetic field curve of $|\vec{B}_e|$ is obtained from Equation (3.21). A maximum value was observed at each curve. The maximum value of $|\vec{B}_e|$ for $|\vec{p}| = 0.8$ mm of approximately 0.201 T occurred at $z = 0.96$ mm, which can be obtained by setting the differential of Equation (3.21) to zero. The positions for the maximum magnetic field vary with the induction coil because of the non-linear magnetic flux lines (Figure 3.7). For instance, the $|\vec{B}_e|$ for $|\vec{p}| = 0.8$ mm reduces to a minimum beyond $z = 0.96$ mm which is equal to $y = -0.54$ mm for a net distance of 0.54 mm and θ of $\pi/4$. This net difference between y and z is caused by the existence of vibrating-damping effects. In addition, a smaller $|\vec{B}_e|$ is found at z of -1.5 and 1.5 mm, respectively. The value of $|\vec{B}_e|$ is close to a constant of 0.005 T for $|\vec{p}|$ equal to 3.8 mm.

In Figure 3.12b, according to Equation (3.23) the largest value of $|\dot{\vec{B}}_e|$ is generated when the PM is at the positive vibration position z closest to the induction coil. The appearance of curves shows positive and then negative slopes with respect to z, in which the reason for the curvatures is as the $|\vec{B}_e|$ case. For example, the maximum $|\dot{\vec{B}}_e|$ at z of 0.72 mm is 227.8 T s^{-1} for $|\vec{p}|$ equal to 0.8 mm. Values close to zero are found at z of -1.5 mm and z of 1.5 mm for each turn of the induction coil, respectively.

Based on Equations (3.5), (3.21), and (3.23), the relationship between $|\vec{B}_e|$ and $|\dot{\vec{B}}_e|$ for $|\vec{p}|$ equal to 0.8 mm, with respect to vibration position z, is obtained as shown in Figure 3.13. Both curves of $|\vec{B}_e|$ and $|\dot{\vec{B}}_e|$ show a positive slope at a smaller value of z, which increases gradually according to z, and show an intersection at z of 0.8 mm corresponding to $|\vec{B}_e|$ of 0.2 T and $|\dot{\vec{B}}_e|$ of 0.2 kT s^{-1}. Based on

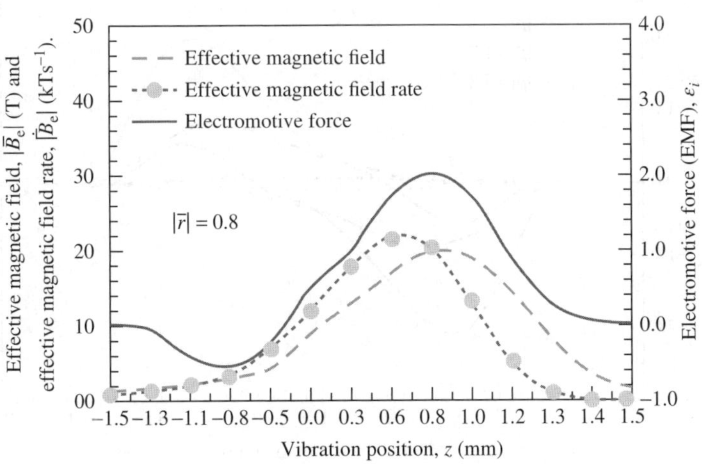

Figure 3.13 Comparisons between effective magnetic field, effective magnetic field rate and EMF with respect to vibration position [18]

Equation (3.24), Figure 3.13 shows that the maximum value of EMF ε_i of the ith turn induction coil of approximately 2.02 mV is obtained at the intersection ($z = 0.8$ mm). The value of $\varepsilon_i = 2\pi\vec{r}(z\dot{\vec{B}}_{\mathrm{e}} + \dot{z}\vec{B}_{\mathrm{e}})$ is zero because $|\dot{\vec{B}}_{\mathrm{e}}| = 0$ and $\dot{z} = 0$ at the vibration position z of 1.5 mm.

Figure 3.14 shows the analytical value of the EMF ε_i from Equation (3.25) on each induction coil as a function of input excitation y ($y = z - 1.5$). The curves show that the highest value of ε_i is 2.02 mV for $|\vec{p}|$ of 0.8 mm at y of −0.7 mm ($z = 0.8$ mm). This result agrees with that shown in Figure 3.13. The graph also indicates the variation of ε_i on each induction coil. Furthermore, for a larger $|\vec{p}|$, the generator exhibits a smaller ε_i. When y reaches −1.5 or 1.5 mm, the value of ε_i on each induction coil is approximately zero because $|\dot{\vec{B}}_{\mathrm{e}}| = 0$ and $\dot{z} = 0$.

3.4 Analytical Results and Discussion

If the microspring is composed of two sets of springs (one clockwise or cw and the other counterclockwise or ccw), then the spring constant of this combined structure can be expressed as

$$k_{\mathrm{comb}} = \left(k_{\mathrm{eq}}\right)_{\mathrm{cw}} + \left(k_{\mathrm{eq}}\right)_{\mathrm{ccw}} \tag{3.29}$$

According to Equation (3.29), the relationship between the spring constant and the loop number is shown in Figure 3.15. Clearly, the spring constant decreases with increasing the loop number.

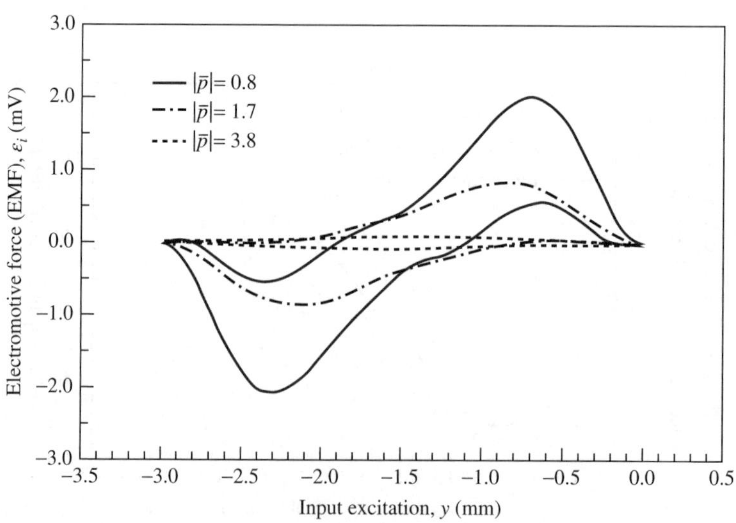

Figure 3.14 Analytical EMF of each induction coil of three unequal radii against input excitation; $Y = 1$ mm, $\omega_{\mathrm{n}} = 120\pi$ rad s^{-1}, and $y = z - 1.5$ [18]

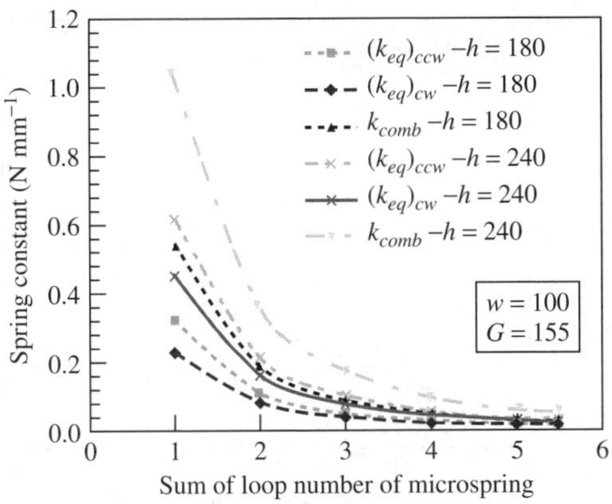

Figure 3.15 Relationship between the spring constant and the loop number [18]

For example, a combined spring with two loops is designed. The inner and outer radii of the spiral are 1.7 and 3.9 mm, respectively. The thickness, width, and gap of the elastic beam are 240, 90, and 400 μm, respectively. The combined spring constant (k_{comb}) is 252 N m^{-1}. Under the same dimensional conditions, for a spiral number of 5.5 k_{comb} becomes 38.5 N m^{-1}. Accordingly, $(k_{\text{eq}})_{\text{cw}}$, $(k_{\text{eq}})_{\text{ccw}}$, and k_{comb} have the same tendency. The spring constant increases as the spiral radius b and loop number decrease. For example, $(k_{\text{eq}})_{\text{cw}} = 298$ N m^{-1} for $b = 1.9$ mm is smaller than $(k_{\text{eq}})_{\text{ccw}} = 417$ N m^{-1} for $b = 1.7$ mm.

After rearranging Equation (3.20), the resonance frequency ω_{n} can be expressed

$$\omega_{\text{n}} = \sqrt{\frac{k}{m}} = \sqrt{\frac{\left(k_{\text{eq}}\right)_{\text{cw}} + \left(k_{\text{eq}}\right)_{\text{ccw}}}{m}} \tag{3.30}$$

or

$$\omega_{\text{n}} = \left\{ \frac{1}{\pi m} \left[\frac{2Gwh\left(w^2+h^2\right)}{\left(w^2+h^2\right)\sum_{i=1}^{q\ell} b_{2i-1} + 0.75\sum_{i=1}^{q\ell}\left(b_{2i-1}\right)^3} + \frac{2Gwh\left(w^2+h^2\right)}{\left(w^2+h^2\right)\sum_{i=1}^{q\ell} b_{2i+1} + 0.75\sum_{i=1}^{q\ell}\left(b_{2i+1}\right)^3} \right] \right\}^{1/2} \tag{3.31}$$

According to Equation (3.31), the effects of spring beam width w and thickness h on the resonance frequency ω_{n} are shown in Figure 3.16. It is known that, as the width w

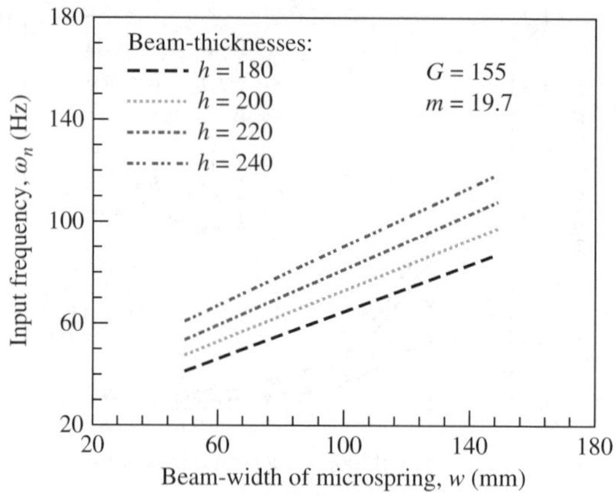

Figure 3.16 Effects of spring beam width w and thickness h on the resonance frequency ω_n [18]

increases from 50 to 150 μm and 180 μm < h < 240 μm, ω_n increases with w and h. This result indicates that the resonance frequency of the spring structure is affected by the spring beam width and thickness.

3.4.1 Analysis of Bending Stress within the Supporting Beam of the Spiral Microspring

During system oscillation, the spiral microspring supports the oscillation structure. The bending moment on the surface of the supporting beam, which results in a bending stress, is shown in Figure 3.17. The differential bending moment dM results from a force F and a differential length dl, that is,

$$dM = F\,dl = F\,(b\,d\alpha) \tag{3.32}$$

Integrating Equation (3.32) along one loop yields the total bending moment on one loop of the spiral spring.

A microspring is a structure of many loops. Thus, the total bending moments on the entire spiral spring are

$$(M)_{\mathrm{cw}} = \int d\left(M_{\mathrm{cw}}\right) = \sum_{i=1}^{i} b_{2i-1} \int_{-\pi/2}^{\pi/2} F\,d\beta \text{ (clockwise)} \tag{3.33}$$

and

$$(M)_{\mathrm{ccw}} = \int d\left(M_{\mathrm{ccw}}\right) = \sum_{i=1}^{q} b_{2i+1} \int_{-\pi/2}^{\pi/2} F\,d\beta \text{ (counterclockwise)} \tag{3.34}$$

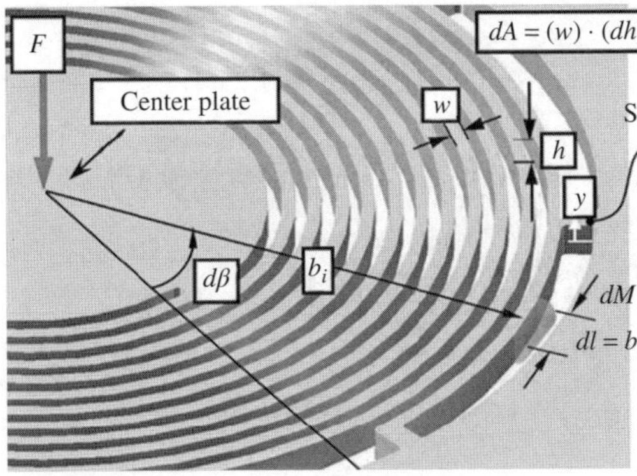

Figure 3.17 Schematic of force and bending moment on the supporting beam [18]

where F is the force resulting from acceleration and mass ($F = m\ddot{x} = m(\ddot{z} + \ddot{y})$). The maximum moments occurring at the maximum amplitude are

$$\left(M_{\text{cw}}\right)_{\text{max}} = m\sum_{i=1}^{q} b_{2i-1} \int_{-\pi/2}^{\pi/2} (\ddot{x})_{\text{max}} d\beta = m\sum_{i=1}^{q} b_{2i-1} \int_{-\pi/2}^{\pi/2} (\ddot{z} + \ddot{y})_{\text{max}} d\beta \tag{3.35}$$

and

$$\left(M_{\text{ccw}}\right)_{\text{max}} = m\sum_{i=1}^{q} b_{2i+1} \int_{-\pi/2}^{\pi/2} (\ddot{x})_{\text{max}} d\beta = m\sum_{i=1}^{q} b_{2i+1} \int_{-\pi/2}^{\pi/2} (\ddot{z} + \ddot{y})_{\text{max}} d\beta \tag{3.36}$$

where acceleration $\ddot{z}$ and $\ddot{y}$ can be obtained from Equations (3.7) and (3.3), respectively.

Figure 3.18 shows the configuration of the supporting beam at a bending moment. The supporting beam bending upwards denotes positive moment, as shown in the figure. Cross-section a–a is parallel to cross-section c–c before bending, whereas sections a–a and c–c form an angle $d\theta$ after the bending moment.

If the curvature radius of the neutral axis is ρ, the curve length du for the neutral axis with angle $d\theta$ is

$$du = \rho \, d\theta \tag{3.37}$$

and

$$ds = (h/2) \, d\theta \tag{3.38}$$

The strain ε is defined as the ratio of the displacement ds to the original length du, that is,

$$\varepsilon = -\frac{ds}{du} \tag{3.39}$$

Substituting Equation (3.38) into Equation (3.39) yields

$$\varepsilon = -\frac{h}{2\rho} \tag{3.40}$$

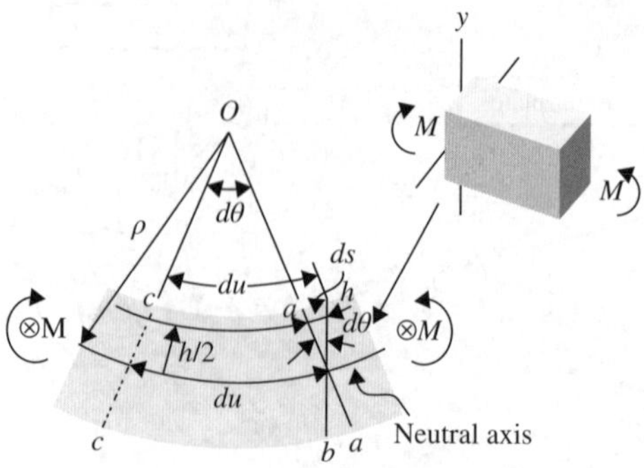

Figure 3.18 Configuration of the supporting beam under a bending moment [18]

According to the material mechanics definition, stress is the product of elastic modulus and strain; that is, $\sigma = E\varepsilon$ and the stress σ on the cross-section of the supporting beam can be determined via

$$\sigma = -\frac{Eh}{2\rho} \tag{3.41}$$

where the minus sign denotes a compressive stress. Based on the equilibrium force, the imposed moment must be equal to that resulting from the internal stress, that is,

$$M = \int (h/2)\,\sigma\, dA = \frac{E}{4\rho}\int (h)^2 dA \tag{3.42}$$

where dA ($= w\,dh$) is the infinitesimal cross-sectional area of the beam. Combining Equations (3.41) and (3.42) yields

$$\sigma = -\frac{Mh}{2I_b} \tag{3.43}$$

where I_b is the cross-sectional area moment inertia, defined

$$I_b = \frac{1}{4}\int h^2 dA \tag{3.44}$$

Equation (3.44) indicates that the bending stress σ is proportional to the bending moment M, and that the maximum stress σ_{max} occurs on the surface of the beam (with the maximum value of h). Combining Equations (3.43) and (3.41) yields the maximum stress, defined:

$$\sigma_{max} = \frac{\left(M_{cw}\right)_{max} h}{2I_b} = \frac{mh}{2I_b}\sum_{i=1}^{q} b_{2i-1}\int_{-\pi/2}^{\pi/2} (\ddot{z}+\ddot{y})_{max}\, d\beta \tag{3.45}$$

or

$$\sigma_{max} = \frac{\left(M_{ccw}\right)_{max} h}{2I_b} = \frac{mh}{2I_b}\sum_{i=1}^{q} b_{2i+1}\int_{-\pi/2}^{\pi/2} (\ddot{z}+\ddot{y})_{max}\, d\beta \tag{3.46}$$

where $\ddot{y}$ and $\ddot{z}$ are functions of the maximum amplitude Y, frequency ω and time t. Based on Equation (3.46), aside from the oscillation parameters $\ddot{y}$, $\ddot{z}$, and m, the maximum bending stress is also affected by the spiral microspring geometric parameters such as thickness h, width w, spiral radius b and loop number j. If the maximum stress is smaller than the material strength, the spring structure would be safe. The maximum stress can therefore be used as one of the criteria of the optimal structural design.

The effects of the beam width w and the maximum amplitude Y on the maximum bending stress σ_{max} are shown in Figure 3.19. The beam thickness is $h = 240\ \mu m$.

3.4.2 *Finite Element Models for Magnetic Density Distribution*

The commercial finite element software Ansoft Maxwell is used to analyze the magnetic field distribution inside a MCTG. An analytical model is proposed to analyze the magnetic field distribution inside a SMTG.

As shown in Figure 3.7, a space exists between the magnetic core element and the PM in a MCTG. The spiral microinduction coils oscillate inside the space, where an effective magnetic field exists. Finite element simulations can be used to observe the magnetic field distribution in the space. When the effective magnetic field is orthogonal to the movement of the induction coils, a larger induced EMF can be obtained. Ansoft Maxwell is commonly used finite-element software, which has strong pre- and post-processing abilities. The magnetic field distribution can be easily illustrated. To understand the effects of the magnetic thickness and magnetic core materials on the magnetic field, NdFeB and nickel-zinc-iron (NiZn) are used as the core materials and the magnetic thicknesses are 4 and 5 mm. NdFeB is a type of hard magnetic material. Compared to other PMs, it exhibits magnetic properties that are more favorable, such as the maximum magnetic energy product. In addition, NdFeB offers high magnetic density even under a small size or shape. NiZn is a type of soft magnetic material.

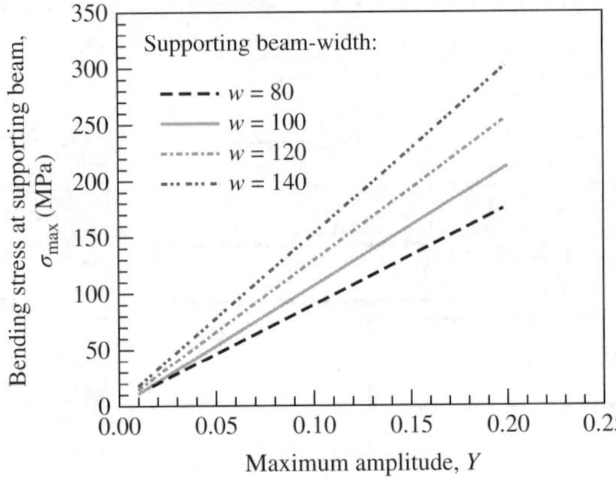

Figure 3.19 Effects of maximum amplitude on the maximum bending stress [18]

Because NiZn has the properties of low assembly cost and low eddy current loss, it is widely used and a potential magnetic material. The residual induction B_r densities for NdFeB and NiZn are 1.46 and 0.17 T, respectively. Table 3.3 summarizes the other magnetic properties.

Figures 3.20–3.23 show the simulation results with different magnetic thicknesses and core materials. Figures 3.20 and 3.21 show the results without using magnetic core materials and using magnets with thicknesses of 5 mm and 4 mm. In the figures, the maximum magnetic field densities are $B = 0.83$ and 0.82 T for Figures 3.20 and 3.21, respectively. The maximum magnetic field occurs near the upper and lower surfaces

Table 3.3 Physical magnetic properties of NdFeB and NiZn (measuring temperature is 25 °C) [18]

	Residual induction B_r (T)	Maximum energy product $B_{H,max}$ (kJ m^{-3})	Coercive force H_{cb} (kA m^{-1})	Intrinsic coercive force H_{cj} (kA m^{-1})
NdFeB	1.457	417.2	955.5	1009
NiZn	0.17	26.26	0.05	0.06

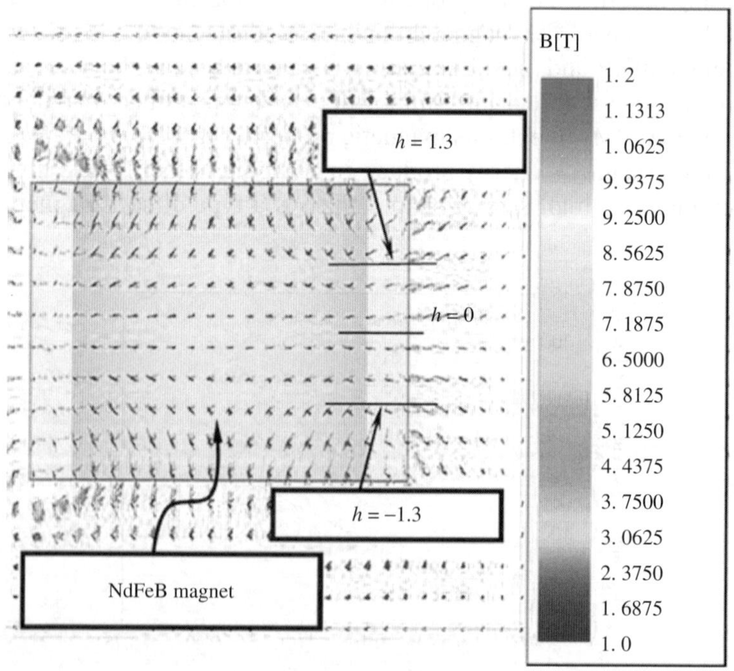

Figure 3.20 Simulation results for magnet thickness of 5 mm and without magnetic core materials [18]

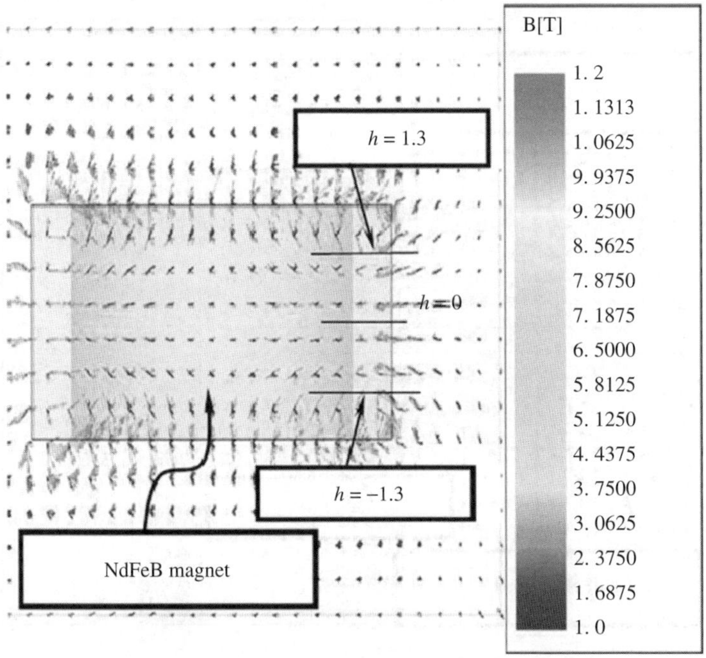

Figure 3.21 Simulation results for magnet thickness of 4 mm and without magnetic core materials [18]

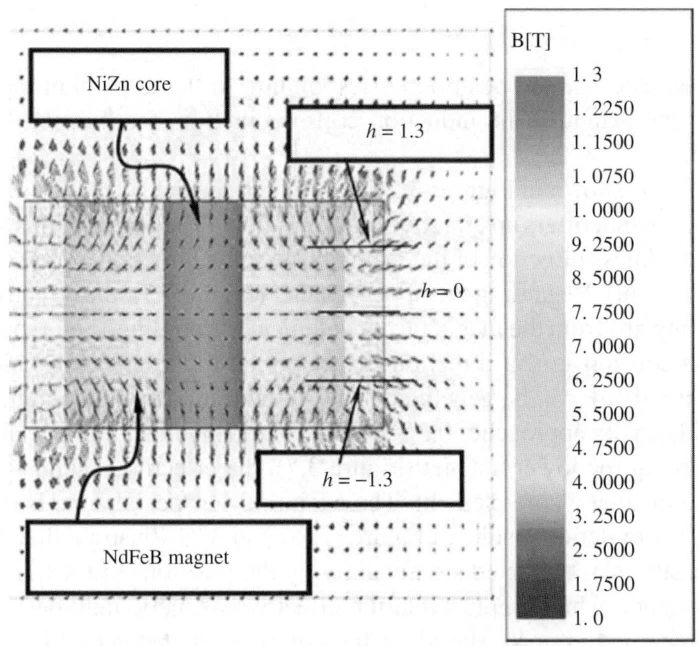

Figure 3.22 Simulation results for NiZn magnetic core material with magnet thickness of 4 mm [18]

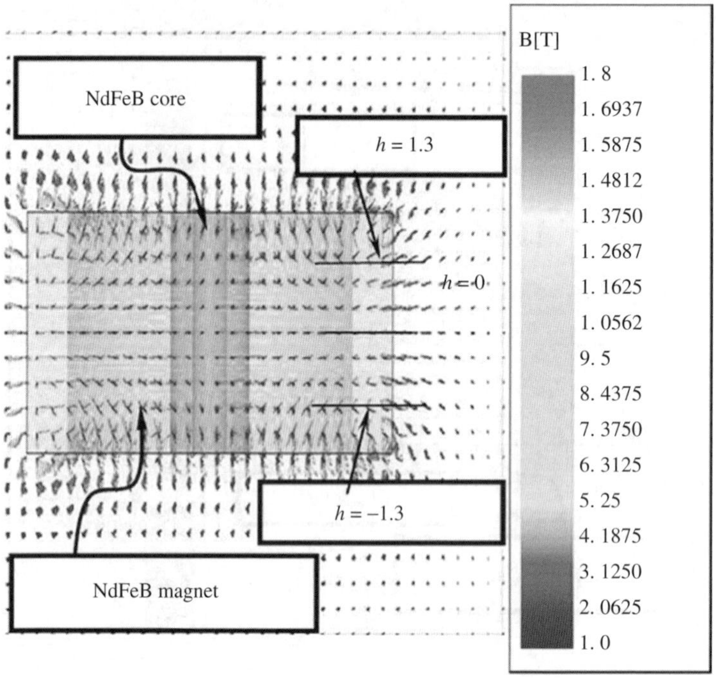

Figure 3.23 Simulation results for NdFeB magnetic core material with magnet thickness of 4 mm [18]

of the magnet. Because the position corresponding to the maximum magnetic field is not inside the extent of the induction coils oscillation, the induced EMF cannot be increased.

Nevertheless, as shown in Figures 3.20 and 3.21, a special zone is inside the central cylindrical space that offers an effective magnetic field. The extent of this special zone is along the thickness direction of the magnet (the same direction as that of the induction coil movement). Regarding the upper magnet (Figure 3.7), the effective magnetic field points outward from the magnetic core element. Accordingly, it is speculated that the induction coil movement direction is orthogonal to the effective field direction, and a maximum EMF can be generated. The simulation results also indicate that the effective field density approaches 0.3 T from $z = 0$ to 1.3 mm in the special zone. Conversely, regarding the lower magnet (Figure 3.7), the effective magnetic field points inward to the magnetic core element. The extent of the special zone is from $z = 0$ to −1.3 mm. The simulation results in Figures 3.20 and 3.21 illustrate that the effective magnetic density is not affected significantly by the positions of the thickness direction of the magnet, which implies that the effective magnetic field density variation rate can be regarded as zero. Based on the comparisons between Figures 3.20 and 3.21, the effects of the magnetic thickness on the magnetic density are quite small.

Figures 3.22 and 3.23 show the results of using a 4-mm-thick magnet and magnetic core materials of NiZn and NdFeB, respectively. The simulation results in Figures 3.22 and 3.23 illustrate that the patterns of the magnetic field are quite similar to those in Figures 3.20 and 3.21, and the maximum effective magnetic field densities are 0.35 T for the NiZn magnetic core and 0.46 T for the NdFeB magnetic core. These results imply that the NdFeB magnetic core is more effective than the NiZn magnetic core at increasing the effective magnetic field density.

3.4.3 Power Output Evaluation

Figure 3.24 shows the electron-shifting physical phenomenon as a conductor of an infinitesimal length dl moving with a velocity $\dot{z}$ within a magnetic field B. As the conductor moves, the magnetic field forces the electron to move to one end of the conductor. Conversely, the opposite end of the conductor has positive electricity. At that moment, there is a voltage drop between the two ends and there is also a Coulomb attraction force between them. This separation situation continues until the magnetic force and electric force counteract with each other and a balance situation is achieved. In the procedure, the infinitesimal length dl with a movement dz constitutes an infinitesimal area variation dS. The magnetic flux Φ passing through this area dS is

$$\Phi = \oint_S B\, dS \tag{3.47}$$

According to Faraday's law of electromagnetic induction, the EMF is the magnetic flux variation rate $\dot{\Phi}$ passing through a conductor with an area variation dS, that is,

$$\text{EMF} = -\dot{\Phi} = -\frac{d\Phi}{dt} = -\frac{d}{dt}\left(\int_S B\, dS\right) \tag{3.48}$$

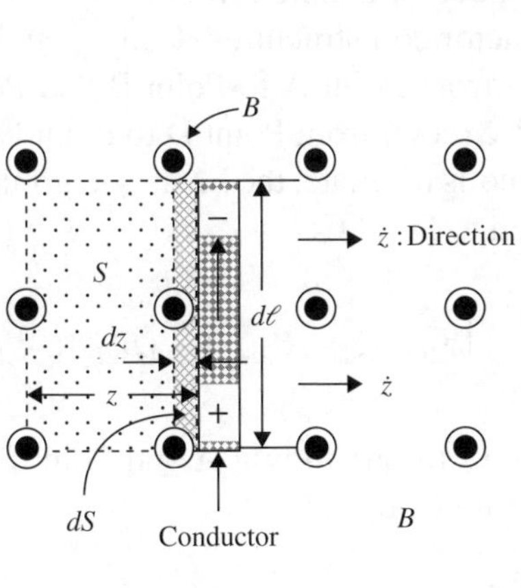

Figure 3.24 Electron shifting physical phenomenon as a conductor moving within a magnetic field [18]

By applying the differentiable product rule, Equation (3.48) can be rewritten as

$$\text{EMF} = -\int_S \frac{dB}{dt} S + \oint_S B \frac{dS}{dt} \tag{3.49}$$

where the term dB/dt is the magnetic flux variation rate $\dot{B}$. The term dS/dt denotes the infinitesimal area variation rate $\dot{S}$, and is equal to the product of the infinitesimal length dl and moving velocity $\dot{z}$. Total area S is the product of dl and moving distance z. Equation (3.49) can therefore be rewritten as

$$\text{EMF} = -\int_S \dot{B}\, z\, dl + \oint_\ell \dot{z}\, B\, dl \tag{3.50}$$

Based on Equation (3.50), the magnitude of EMF is affected not only by the magnetic field B, the magnetic field rate $\dot{B}$, but also by the oscillation amplitude z, oscillation speed $\dot{z}$, and the conductor length l. According to the vector product in the second term, the EMF achieves the maximum because B and $\dot{z}$ are orthogonal to each other.

If the spiral coil shape is circular with a constant radius b ($dl = b\, d\beta$), the EMF for a single loop microconductor is

$$\text{EMF} = -\int_0^{2\pi} \left(\dot{B}z\right)\ (b\, d\beta) + \int_0^{2\pi} (\dot{z}B)\ (b\, d\beta) \tag{3.51}$$

For a conductor coil with a multiloop structure, the EMF becomes

$$\text{EMF} = \sum_{i=1}^{ql} \int_0^{2\pi} \left[b_i\left(\dot{z}_i B_i\right) - \left(\dot{B}_i z_i b_i\right)\right] d\beta \tag{3.52}$$

where ql denotes the loop number of the spiral spring. Figure 3.25 shows the top view of an LTCC microconductor adopted from Figures 3.1 and 3.2, which indicate two groups of spiral conductor coil structures (Coil_{ABC} and Coil_{DEF}). The first group Coil_{ABC} is clockwise (cw) from Point A to Point B and Point C. The second group Coil_{DEF} is counterclockwise (ccw) from Point D to Point E and Point F.

As the microconduction coils oscillate, the EMF_{cw} in group Coil_{ABC} generates along Point A to Point B and then to Point C:

$$\text{EMF}_{\text{cw}} = \sum_{i=1}^{ql} \int_0^{2\pi} \left[b_{2i-1}\left(\dot{z}_{2i-1} B_{2i-1}\right) - \left(\dot{B}_{2i-1} z_{2i-1} b_{2i-1}\right)\right] d\beta \tag{3.53}$$

The induction current flows through Point A and Point B, and then to Point C. By contrast, the EMF_{ccw} in group Coil_{DEF} is

$$\text{EMF}_{\text{ccw}} = \sum_{i=1}^{ql} \int_0^{2\pi} \left[b_{2i+1}\left(\dot{z}_{2i+1} B_{2i+1}\right) - \left(\dot{B}_{2i+1} z_{2i+1} b_{2i+1}\right)\right] d\beta \tag{3.54}$$

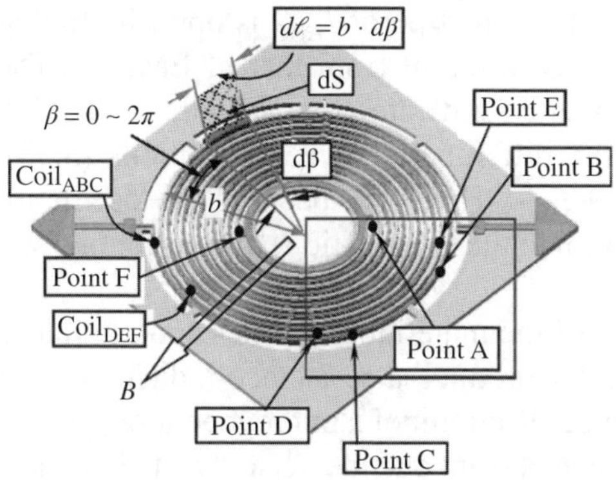

Figure 3.25 Configuration of two groups of spiral conduction coils within a spiral spring [18]

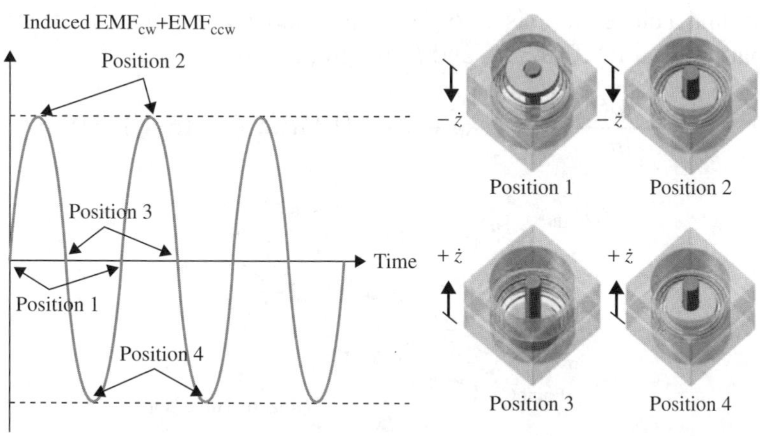

Figure 3.26 Relation of induced voltage to oscillation position and speed [18]

The induction current I_{DEF} in group Coil_{DEF} flows through Point D and Point E, and then to Point F. The total voltage of the EMF is obtained by adding Equations (3.53) and (3.54) ($\text{EMF} = \text{EMF}_{cw} + \text{EMF}_{ccw}$), and can be measured using an oscilloscope.

Figure 3.26 shows the relationship of the total induced voltage and the coil movement direction $\dot{z}$. As the path moves from Position 1 to Position 2 and then to Position 3, the induction voltage in the microinduction coils is positive. However, as the path moves from Position 4 to Position 1 and then to Position 2 (the lowest position to the highest position during the oscillation), the induced voltage is negative. The maximum induced voltage occurs at Position 2. The results are consistent with those obtained according to the previously stated magnetic field variation rate (Figure 3.22). These results also verify the validity of using the model with the magnetic field variation rate.

Figure 3.27 depicts the current flow direction in the two coil groups, Coil_{ABC} and Coil_{DEF}. The coils move from the top position to the bottom position (negative velocity). In the figure, the direction of I_{ABC} is opposite to that of I_{DEF}, because the voltage drop of EMF_{cw} is different from that of EMF_{ccw}. The EMF is affected by magnetic field B, magnetic field rate $\dot{B}$, amplitude z, oscillation speed $\dot{z}$, and coil length l. Comparing the difference between Equations (3.53) and (3.54), the total current is $I_{total} = I_{ABC} + I_{DEF}$. However, because the oscillation velocity is positive (from the bottom position to the top position), the current flow directions are opposite to those in Figure 3.27.

Because of the effect of the material properties and internal structure, the microinduction coils intrinsically produce a resistance-inductance (RL) effect. Figure 3.28 shows the equivalent electric circuit of a microgenerator system, where R_l denotes the changeable resistor attached from outside. R_c and L_c denote the internal resistance and inductance of the microinduction coils, respectively. As it is attached, a current output can be obtained.

According to Kirchhoff's voltage law, the sum of the voltage that drops across every element around a closed loop is zero. Thus, at the node in Figure 3.15, the sum of the voltage that drops at EMF, R_c, R_l, and L_c is zero, that is,

$$L_c \frac{d\,(I_{total})}{dt} + I_{total}\,(R_c + R_l) - \left(EMF_{cw} + EMF_{ccw}\right) = 0 \tag{3.55}$$

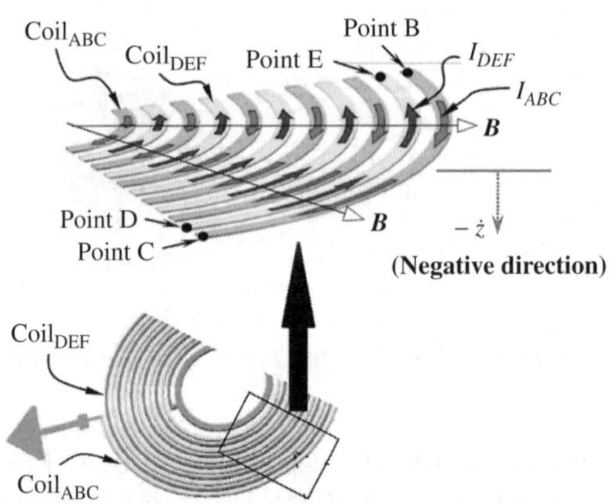

Figure 3.27 Schematic of current direction and induction coil oscillation [18]

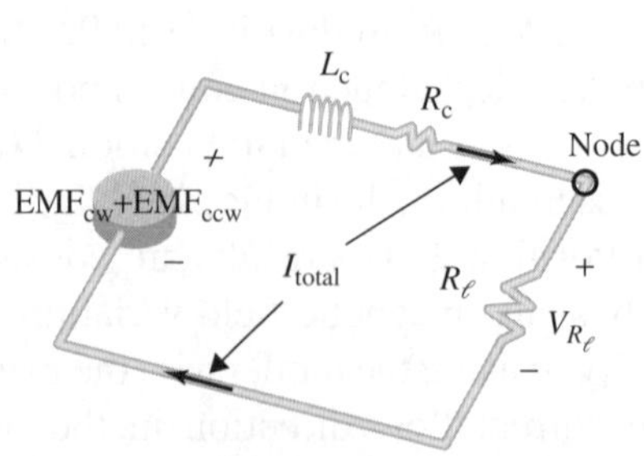

Figure 3.28 Equivalent electric circuit of a microgenerator [18]

The solution of the output current I in Equation (3.55) can be expressed as

$$I_{\text{total}} = \frac{\left(\text{EMF}_{\text{cw}} + \text{EMF}_{\text{ccw}}\right)\left(1 - \exp\left[-\left(R_{\text{l}} + R_{\text{c}}\right)t/L_{\text{c}}\right]\right)}{\left(R_{\text{l}} + R_{\text{c}}\right)} \tag{3.56}$$

The flow direction of the current is determined according to Fleming's right-hand rule, as shown in Figure 3.29, in which the directions of the forefinger, thumb, and middle finger denote the directions of the magnetic field, inductor movement, and current flow, respectively. They are orthogonal to one another. Because a current I_{total} flows through the circuit and a voltmeter is connected to the both ends of the changeable resistor R_{l}, an induced voltage $V_{R_{\text{l}}}$ can be measured, the value of which is

$$V_{R_{\text{l}}} = I_{\text{total}} R_{\text{l}} \tag{3.57}$$

Substituting Equation (3.56) into Equation (3.57) and rearranging yields

$$V_{R_{\text{l}}} = \frac{\left(\text{EMF}_{\text{cw}} + \text{EMF}_{\text{ccw}}\right)\left(1 - \exp\left[-\left(R_{\text{l}} + R_{\text{c}}\right)t/L_{\text{c}}\right]\right)}{\left(1 + R_{\text{c}}/R_{\text{l}}\right)} \tag{3.58}$$

Based on Equation (3.58), the voltage is a function of time. When time t is large, the term $\exp[-(R_{\text{l}} + R_{\text{c}})t/L_{\text{c}}]$ approaches zero and the steady state of the voltage response is

$$V_{R_{\text{l}}} = \frac{\text{EMF}_{\text{cw}} + \text{EMF}_{\text{ccw}}}{\left(1 + R_{\text{c}}/R_{\text{l}}\right)} \tag{3.59}$$

Because $R_{\text{l}} >> R_{\text{c}}$, the maximum value of $V_{R_{\text{l}}}$ can be obtained for the microgenerator.

If the microinduction coils are implemented into the multilayer parallel structure with the same node, and the resistance of each layer is assumed to be the same for R, then the resistor for each layer could be regarded as a parallel circuit and the total equivalent resistance R_{c} could be obtained using

$$R_{\text{c}} = \left(\frac{1}{R_1} + \frac{1}{R_2} + \ldots + \frac{1}{R_{\text{ml}}}\right)^{-1} = \left(\sum_{i=1}^{\text{ml}} \frac{1}{R_i}\right)^{-1} \tag{3.60}$$

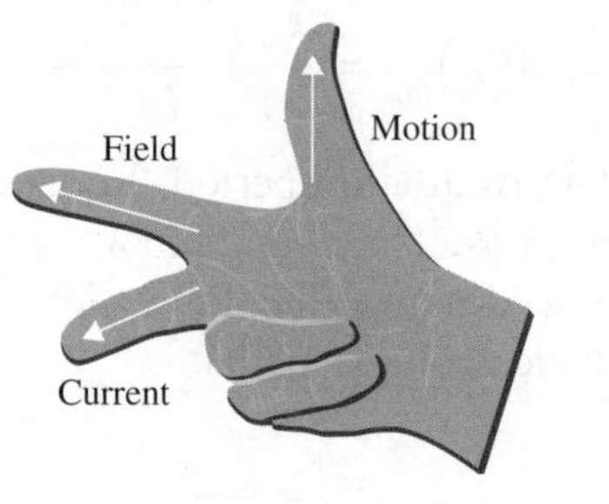

Figure 3.29 Fleming's right-hand rule

where the subscript ml denotes the number of microinduction coil layers. As the current flows through the circuit, an electromagnetic coupling action occurs, resulting in an inductance effect. Two sources result in this current-induced inductance effect. The first is called 'intergap inductance' and the second is 'interlayer inductance'. Intergap inductance L_S can be regarded as a series circuit, and the equivalent intergap resistance $(L_S)_{eq}$ can be calculated via

$$(L_S)_{eq} = (L_S)_1 + (L_S)_2 + \ldots \cdots + (L_S)_{ql} = \sum_{i=1}^{ql} (L_S)_i \tag{3.61}$$

The other inductance effect occurs at the interlayers. The interlayer inductance L_L can be regarded as a parallel circuit and the equivalent interlayer resistance $(L_L)_{eq}$ be calculated via

$$(L_L)_{eq} = \left[\left(\frac{1}{(L_L)_1}\right) + \left(\frac{1}{(L_L)_2}\right) + \ldots + \left(\frac{1}{(L_L)_{ml}}\right)\right]^{-1} = \left[\sum_{j=1}^{ml} \frac{1}{(L_L)_j}\right]^{-1} \tag{3.62}$$

where ml denotes the number of layers of the microinduction coils. Using electric circuit formulas, the whole equivalent inductance L_c can be obtained by incorporating Equations (3.61) and (3.62):

$$L_c = 2\pi \sum_{i=1}^{ql} \left[\sum_{j=1}^{ml} \frac{\mu_0\left((r_i)_j\right)^2}{\left(2\pi(r_i)_j + 0.25w_c h_c\right)}\right] \tag{3.63}$$

where μ_0, w_c, and h_c denote the magnetic-conductance coefficient ($4\pi \times 10^{-7}$ H m^{-1}), and the width and thickness of the induction coils, respectively.

Based on electric circuit theory, in an electric system the average power output P_{avg} is a function of the attached resistor R_l and the root-mean-square voltage output is $(V_{R_l})_{rms}$, as expressed below:

$$P_{avg} = \frac{\left[\left(V_{R_l}\right)_{rms}\right]^2}{R_l} \tag{3.64}$$

$$\left(V_{R_l}\right)_{rms} = \frac{\left(\int_T V_{R_l}{}^2 dt\right)^{1/2}}{T_P} \tag{3.65}$$

where T_P is the steady-state oscillation period. Equation (3.67) indicates that P_{avg} decreases with an increase in R_l. The maximum V_{R_l} occurs for $R_l >> R_c$; however, P_{avg} is not necessarily the maximum value. PD is defined as the ratio of the average power output P_{avg} to the device volume V, that is,

$$\mathrm{PD} = \frac{P_{avg}}{V} \tag{3.66}$$

Clearly, a microgenerator can have a relatively high PD. The characteristics of a microgenerator can also be evaluated according to volumetric power density per frequency (VPDF) or mass power density per frequency (MPDF). VPDF is defined as the ratio of the average power output P_{avg} to the product of device volume V and frequency ω, that is,

$$\mathrm{VPDF} = \frac{P_{avg}}{\omega V} \tag{3.67}$$

MPDF is defined as the ratio of the average power output P_{avg} to the product of oscillation mass m and frequency ω, that is,

$$\mathrm{MPDF} = \frac{P_{avg}}{\omega m} \tag{3.68}$$

Clearly, a microgenerator can have a relatively high VPDF or MPDF if the volume or mass is quite small.

3.5 Fabrication of Microcoil for Microgenerator

3.5.1 Microspring and Induction Coil

Figure 3.30a illustrates the key processes of the LTCC inducer. In Step 1, a printer (Hotshine Corp. 2000 model manufactured in Taiwan) prints silver paste onto the green sheet by using a screen-stencil with a coil pattern. The pattern with a 10-μm-thick emulsion is fabricated using standard photolithography techniques. After the printing procedure is complete, both the silver paste and the PET tape on each green sheet is oven-dried at 80 °C for 30 min. In Step 2, a carbon-dioxide (CO_2) laser driller (Sumitomo GSD 1400TW model) is employed to fabricate an interspace between coils. A helical microspring of 5.5 unequal radius loops and a center circle plate are formed by laser cutting. Each layer is fabricated on green tape by repeating these steps. In Step 3, the green tape is reversed with the PET facing upward. A rubber roller (Hotshine 200FA model) is used to strip the PET tape from the green sheet and a vacuum holder catches the residual ceramic film.

In Step 4, the vacuum holder stacks layers of film with an alignment accuracy of *c*. 18 μm, which places a limit on the total number of layers. For example, in the six-layer film stack, a stacking error of 108 μm is acceptable because it is smaller than the unfired microspring interspace of 116 μm. The vacuum packing in Step 5 protects the green structure and eliminates remaining bubbles. The films are laminated using an isotropic high press of 20.6 MPa in a pressing machine (ABB Corp. IL102010-SS model in USA) and liquid water is used as a medium. The films can be pre-adhered to each other after being maintained at 75 °C for 30 min. Using silver paste, two electrodes are painted over the two terminations of each coil and dried at 100 °C for 10 min.

In Step 6, the green material is hardened in a high-temperature furnace (UF Corp. K4U model manufactured in Taiwan). LTCC multilayer structures consist of different

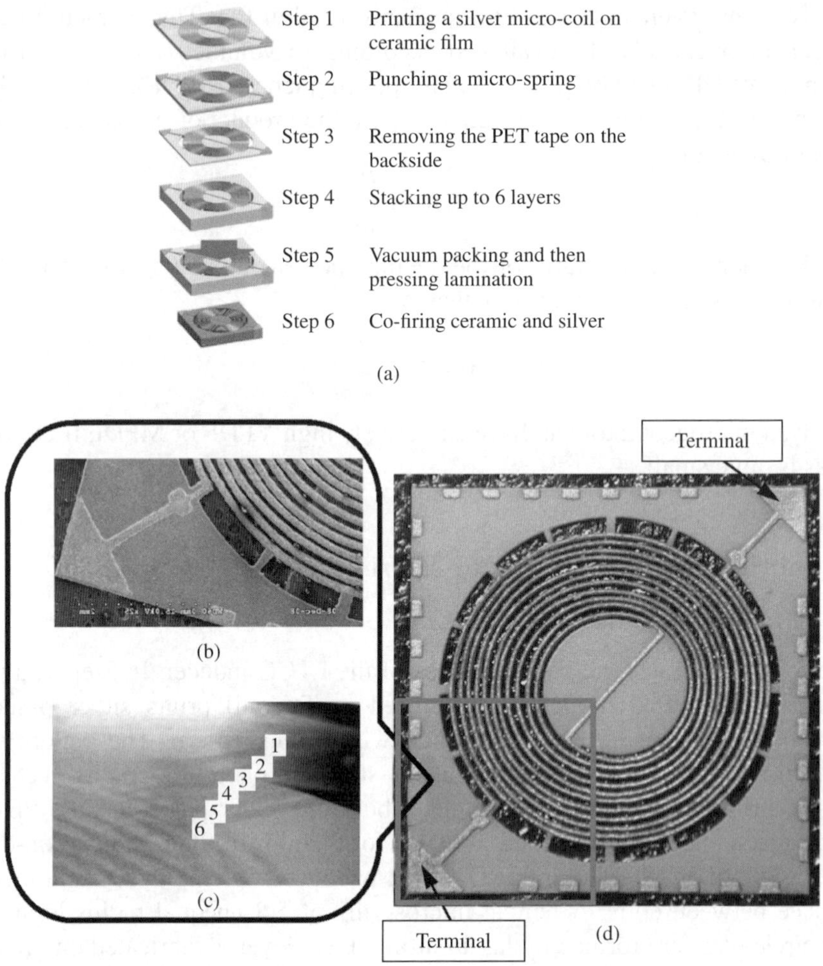

Figure 3.30 A sintered inducer: (a) illustration of LTCC fabrication, (b) observation of the six-layer coils via X-ray fluoroscopy method, (c) SEM image of the supporting beams and its terminal, and (d) photograph of appearance [18]

materials such as ceramic and silver, which are co-fired in one step. Sintering promoters play an essential role in the adhesion strength between the silver coil and the ceramic microspring. The LTCC co-firing fabrication allows for processing in ambient air without chemical oxidization. The heating profile is divided into two stages. The first is heating at 350 °C for 2 h. At this stage, the organic binder in the ceramic slurry is burnt away, leaving behind voids in the structure. The second stage involves sintering the materials and allows for the formation of covalent bonds. The temperature is ramped at 0.16 °C s^{-1} from room temperature and heats at 850 °C for 30 min. During the sintering process, the voids left by the released binder are filled with ceramic and glass and the entire volume of the inducer becomes dense. The resulting shrinkage values of the in-plane and in-thickness dimensions are 0.16 and 33.3%, respectively. The difference in the shrinkage values can be attributed to the anisotropic surface area. After this sintering step, an inducer with a helical-multilayer structure consisting of a ceramic microspring and silver microcoil is constructed.

The six-layer coils in the ceramic structure as seen using an X-ray fluoroscopy method are shown in Figure 3.30b. Figure 3.30c shows a scanning electron microscope (SEM) image of the microspring. The process of multilayer LTCC fabrication enables the inducer to be constructed. Coils in the different layers are connected at the terminals. Figure 3.30d shows a photograph of the sintered inducer. The sintered microspring has a beam thickness of 180 μm, a beam width of 100 μm, and a beam interspace of 100 μm. The resulting structure has a 1.7 mm innermost helical loop radius, and a 3.9 mm outermost helical loop radius. The cross-sectional area of sintered microcoil is 90 μm (line width) × 10 μm (line thickness). Table 3.4 lists all of the material properties and geometric parameters.

3.5.2 *Microspring and Magnet*

A microspring with a helical beam and a center plate was fabricated. The key techniques were DRIE, co-sputtering, and reactive ion etching (RIE), as schematically illustrated in Figure 3.31a. In Step 1, a silicon wafer is selected to fabricate the microspring structure using DRIE. In Step 2, a ferro-platinum (FePt) PM membrane is co-sputtered on the center plate. For the PM used in this study, a magnetic dipole m_H of 6.13×10^{-3} A m^2 can be achieved after annealing. In Steps 3–6, a silicon nitride (Si_3N_4) layer and etching mask are used to remove the back side of the wafer via RIE (refer to [20] for additional details). Figure 3.31b shows an SEM photo of the microspring structure.

Table 3.4 Copper and permanent magnet material properties and geometric parameters [18]

Symbol	Name	Constant	Unit
c	Damping coefficient	3×10^{-8}	kg m s
G	Silicon shear modulus	66.7	MPa
h	Thickness of microspring	40	μm
m	Seismic mass	8.35×10^{-7}	kg
m_H	Magnetic dipole	6.13×10^{-3}	A m^2
R	Copper resistance	16	Ω
w	Width of microspring	100	μm
μ_0	Copper permeability	$4\pi \times 10^{-7}$	H m^{-1}
μ_r	Copper relative permeability	0.9999	–
σ_c	Copper conductivity	5.8×10^{-7}	Ω m^{-1}

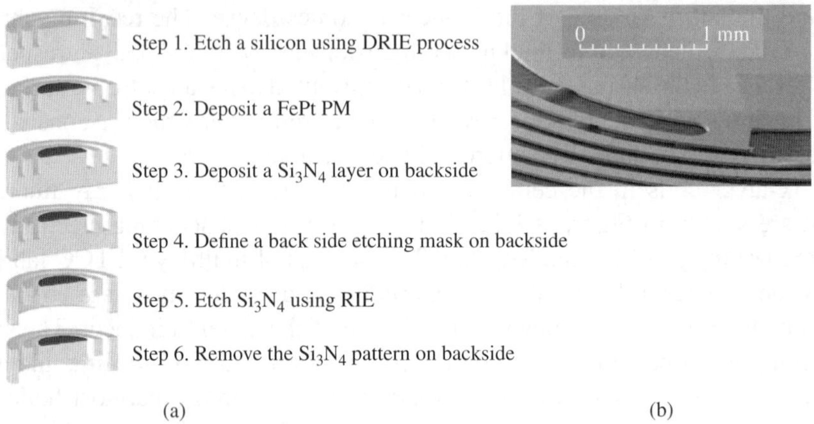

Figure 3.31 (a) Fabrication processes for microspring and FePt by DRIE method. (b) SEM photo of the microspring [18]

According to Equation (3.15), the non-linear equivalent spring stiffness k_e decreases with the loop number. As shown in Figure 3.13, the curve of k_e exhibits a negative slope. Considering a device with 12 loops, for example, the value of k_e is 5.05 N m^{-1}, in which the dimensions are 100 μm in width and 100 μm in interspace, and the innermost loop radius is 1.7 mm and the minimum feature size is 40 μm in depth. The value of k_e for the single loop spring is 108 N m^{-1}.

The material of the inducer before fabrication consists of green sheets (Heraeus HL2000) and silver paste (Heraeus TC0308). The dimensions of each green sheet are 150 mm × 150 mm in surface area and a thickness of 95 μm. The ceramic slurry is mixed with the aluminum oxide (Al_2O_3) powder, the glass powder, and binder powder in proportion. The soft, pliable, and abraded characteristics of green tape are beneficial to laser cutting. Laser cutting involves fabricating various shapes and special structures for MEMS. Furthermore, by using the silver material with favorable conductivity and a multilayer structure, the LTCC technique allows for inducer fabrication that benefits the current output induced in the energy conversion processes.

3.6 Tests and Experiments

3.6.1 Measurement System

After fabricating the meso-generator, several measurements were made to characterize the performance of the device. Figure 3.32 shows the block diagram of the measurement set-up and the photograph of the experimental instruments. As shown in the graph, a sinusoidal waveform of input amplitude and input frequency were derived from an arbitrary waveform generator (Agilent 33220A model). A power amplifier (Data Physic 100 W model) was used to magnify the signal from the

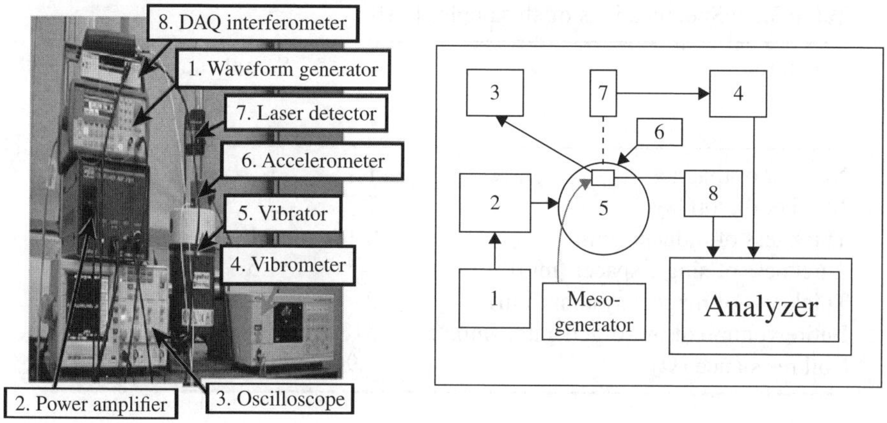

(a) (b)

Figure 3.32 Measurement set-up: (a) block diagram and (b) photograph of the experimental instruments [18]

waveform generator, allowing it to control a shaker (Data Physic PA100E) that provides the mechanical vibration input. An accelerometer (PCB 357B61 model) was attached to the top platform of the shaker to measure the vibration acceleration. Table 3.4 lists the material properties and geometric parameters used in analysis.

The meso-generator was fixed on the top of the accelerometer. An interferometer of data acquisition (DAQ) was used between the accelerometer and the analyzer to collect vibration-acceleration data. The shaking motion passes through the housing of the meso-generator, into the microinducer. A laser displacement vibrometer (Keyence LC-2401 model) with a laser detector was used to measure the mechanical vibration displacement. Finally, a variable resistor was connected directly to the terminals of the inducer. When the induced current passes through the variable resistor during vibration, a digital storage oscilloscope (GW Instek GDS-2062 model) measures both the induced voltage output and the induced current output from the meso-generator. In addition, the viscous damping coefficient can be measured by removing the variable resistor.

3.6.2 *Measurement Results and Discussion*

After measurement, a spring constant of $k = 29.4\,\mathrm{N\,m^{-1}}$ for 0.58 g ($g = 9.8\,\mathrm{m\,s^{-2}}$) of input vibration acceleration was obtained for the system at natural vibration frequency $\omega_n = 69$ Hz and maximum amplitude of $Y = 0.03$ mm. In addition to the suitable operating parameters in the system, to investigate the effects of number of coil layers and number of magnets on voltage and current output, four samples were made. Only one magnet was designed in Samples 1, 2, and 3, whereas two magnets were designed in Sample 4. In Samples 1, 2, 3, and 4, there were 1, 2, 6, and 6 layers of induction coil,

Table 3.5 Specifications of all samples [18]

Structures	Sample number			
	1	2	3	4
Number of magnet	1	1	1	2
Number of coil layer	1	2	6	6
Thickness of inducer (mm)	0.18	0.18	0.18	0.18
Thickness of single spacer (mm)	1.5	1.5	1.5	1.5
Thickness of magnet system (mm)	2	2	2	4
Entire volume of meso-generator (mm^3)	518	518	518	718
Coil resistance (Ω)	2.96	1.45	0.49	0.49

respectively. The specifications and the measured coil resistances for the four samples are listed in Table 3.5.

Figure 3.33a,b show the induced voltage output ε and induced current output I versus time, respectively. As shown in Figure 3.33a, a maximum voltage output in excess of 37 mV was found for Sample 1. The voltage output curves for Samples 1, 2, and 3 are similar, because the layered coils are all connected at common terminals. However, the maximum voltage in Sample 4 is 44.5 mV, which is larger than those in Samples 1–3. In comparing Samples 3 and 4, the voltage output is more sensitive to a change in the number of magnets than the number of coil layers. The voltage output for Sample 4 (2-magnet system) is *c*. 1.2 times that of Samples 1–3 (1-magnet systems). This demonstrates that increasing the voltage output involves increasing the number of magnets.

In Figure 3.33b, the induced current of Sample 3 is shown to be larger than that of Sample 1. The maximum induced current of 69 mA corresponding to Sample 3 is *c*. 5.95 times greater than the 11.6 mA value of Sample 1. Furthermore, the maximum current output of Sample 4 of 83.1 mA is *c*. 1.2 times greater than the output of Sample 3. This result reveals the special property of high-current and low-voltage output in this generator, which is similar to that found in Table 3.1 [8]. We also found that the increasing rate of induced current is proportional to both the number of coil layers and the number of magnets. The slope extracted from a plot of induced current versus time for Sample 4 is higher than the slope extracted from Sample 3 data, indicating that the influence of the number of magnets on the current output is larger than the number of coil layers. The maximum root-mean-square value of the measured current output induced in Sample 4 is $I_{rms} = 56.4$ mA.

In addition, Figure 3.34a and b depict the maximum average electrical power output versus number of coil layers and number of magnets, respectively. Figure 3.34a shows that the average power output in a multilayer coil system is higher than that in a single coil layer system, with the maximum P_{avg} increasing in proportion to the number of layers. For example, the maximum P_{avg} of 0.67 mW for Sample 3 (which

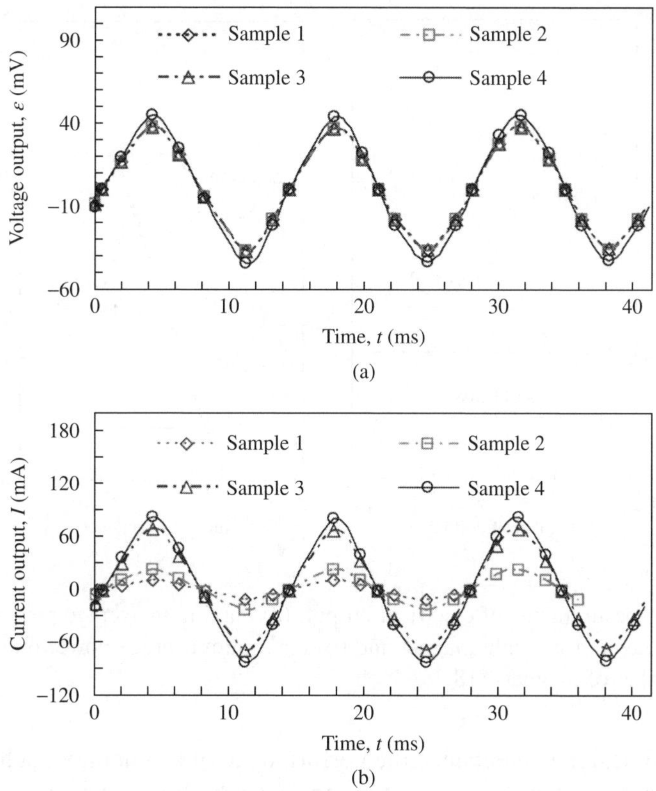

Figure 3.33 Results of electricity generation: (a) voltage output curve versus time and (b) current output curve versus time [18]

has six layers) is approximately six times larger than the maximum P_{avg} of 0.11 mW for Sample 1 (which has one layer). Sample 2 has a maximum P_{avg} of 0.22 mW, twice as large as that of Sample 1. Figure 3.34b indicates an increasing value of maximum average power output with the number of magnets. The double magnet system in Sample 4 yields a larger value of maximum P_{avg} (1.56 mW) than the single magnet system in Sample 3 (0.67 mW). Our results indicate that the number of magnets is more important than the number of coil layers, as was the case for the current output.

For such vibration-induced microsources, the energy-to-size ratio is a crucial criterion for determining when to evaluate the generator performance. The PD is a commonly cited criterion related to device volume, and is defined as the power output divided by the entire volume of the generator. Another parameter is the normalized power density (NPD), which is derived from the power output per square of input vibration acceleration. For example, Sample 4 has a higher PD of 2.17 $\mu W\ mm^{-3}$ than that of Sample 3, which is smaller only than the value of 2.2 $\mu W\ mm^{-3}$ [18]. Moreover, Sample 4 has a higher NPD of 6.46 $\mu W\ mm^{-3}\ g^{-2}$ ($g = 9.8\ m\ s^{-2}$) than that of Sample 3. Overall, results indicate that multiple magnets (high magnetic field) and a

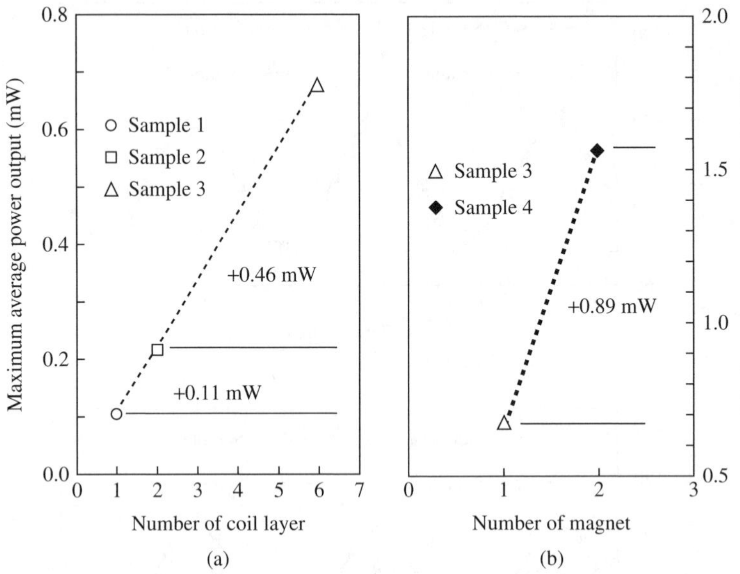

Figure 3.34 Measurements of electrical output: (a) maximum average power output versus number of coil layers for single magnet and (b) maximum average power output of six-layer coil versus number of magnets [18]

multilayer coil design can enhance the performance of a generator such as Sample 4. Measurements show that the meso-generator has a high power output because of the high current levels. Although the voltage output is sometimes as low as 44.5 mV, it can be increased to more reasonable levels by optimizing circuitry in the microsystem. Generators wired in series would have a much larger EMF than a single device.

3.6.3 Comparison between Measured Results and Analytical Values

Figure 3.35 shows the relationship of vibration amplitude z versus time. The orientation of vibration amplitude is shown in Figure 3.8a. This experiment is used to ensure that the distance swept out by the vibration is within the spacer thickness. Results confirm that the vibration ranges from vibration amplitude of $z = 1.47$ mm to $z = 1.53$ mm. The maximum amplitude of the coil is close to the maximum amplitude of the shaker, $Y = 0.03$ mm, because of the suitable spring constant. Because the oscillating mass is able to more completely sweep through the allowable space with a larger vibration acceleration and higher velocity, this implies that spring behavior in a generator increases the power output. In the same figure, the predicted vibration amplitude follows the same trend as measured result. The difference is 1.5%, possibly because of the predicted spring constant and damping coefficient.

In Figure 3.36a,b, the peak points of the analytical and experimental values match favorably. However, the waveforms of the simulation results show different slopes at

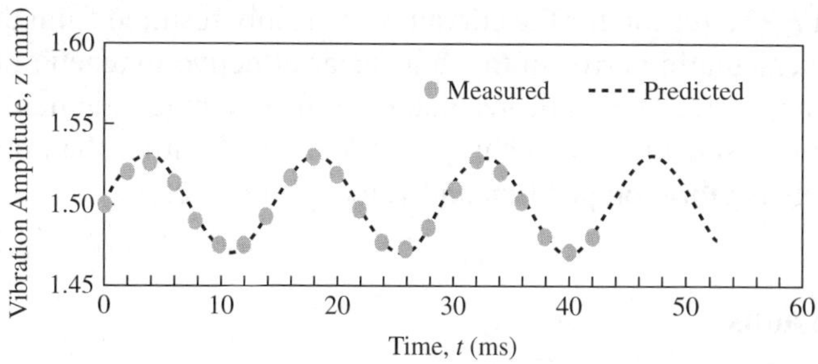

Figure 3.35 Experimental and predicted vibration amplitude versus time [18]

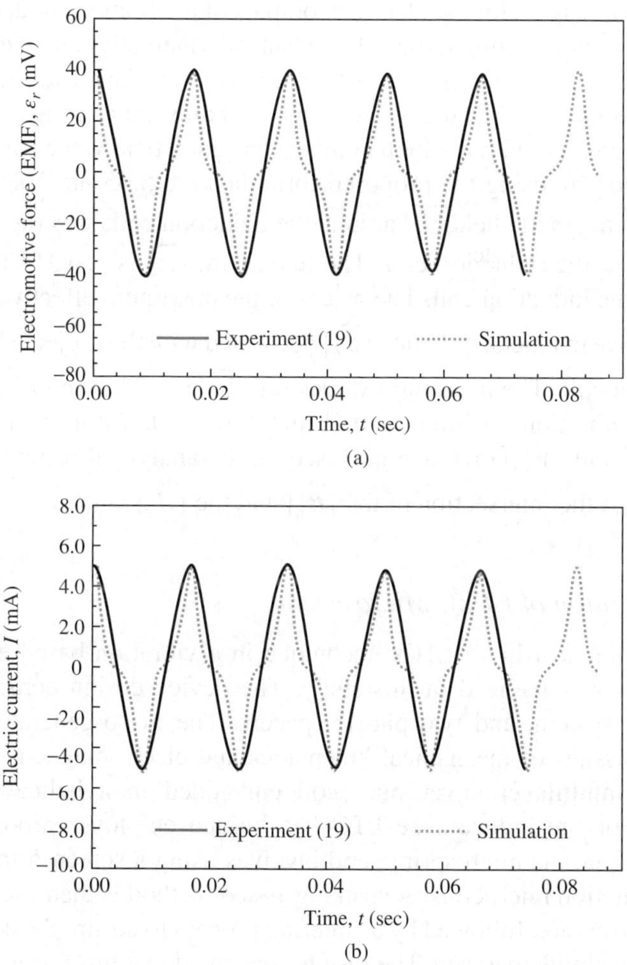

Figure 3.36 Comparisons between analytical and experimental EMF and electric current against time: (a) electromotive force and (b) electric current. $\omega_n = 120\pi$ rad s^{-1}, $Y = 1$ mm, and $R = 8\Omega$ [18]

zero EMF and I. The reason for the curvatures possibly resulted from the simplifying model and the calculating error in the non-linear effective magnetic field and effective magnetic field rate. This indicates that the effective magnetic field and magnetic field rate are more sensitive to a change in EMF and I output than other governing parameters such as vibration position and velocity in this case.

3.7 Conclusions

3.7.1 Analysis of Microgenerators and Vibration Mode and Simulation of the Magnetic Field

An analytical expression for the electrical outputs of a vibration-induced microgenerator associated with coupling issues of mechanical kinematics and electromagnetism was established based on Castigliano's theorem, Hooke's law, Biot-Savart's law, and Lenz's law. The analytical results showed that the equivalent spring stiffness k_e of the microspring decreased with the loop number. The analytical values of k_e for 12 loops and a single loop by using the proposed formula were 5.05 and 108 N m^{-1}, respectively. A higher magnetic field $|\vec{B}|$ around the induction coils was found, whereas the PM was close to the induction coil. The maximum $|\vec{B}|$ is *c.* 0.314 T when the PM was closest to the induction coil. The values of the maximum effective magnetic field $|\vec{B}_e|$ and effective magnetic field rate $|\dot{\vec{B}}_e|$ decreased with the projected radial distance of the induction coil. The maximum values for $|\vec{B}_e| = 0.21$ T and $|\dot{\vec{B}}_e| = 227.8$ T s^{-1} existed at the vibration position $z = 0.96$ mm and $z = 0.72$ mm, respectively, at the innermost turn coil of $|\vec{p}| = 0.8$ mm. Based on the analytical results, the maximum EMF occurred at the intersection of the $|\vec{B}_e|$ and the $|\dot{\vec{B}}_e|$ curves.

3.7.2 Fabrication of LTCC Microsensor

A new development using a LTCC technique in a vibration-based electromagnetic meso-generator was realized in this study. The device design consists of a LTCC inducer, a PM system, and two plastic spacers. The meso-generator is associated with coupling issues of mechanical kinematics and electromagnetism. The inducer is made of a multilayer silver microcoil embedded in a helical ceramic-based microspring. For the inducer, the LTCC technique employs carbon-dioxide laser cutting to fabricate the microspring, and involves using a screen-printing method to pattern the induction microcoils. A stacking-based method is then used to construct a six-layer coil structure, followed by a sintering process to co-fire the ceramic material and silver metal simultaneously. The results confirmed that this fabrication procedure of the inducer using LTCC is valid. The fired microspring has a beam width of 100 μm, a beam interspace of 100 μm, an innermost loop radius of 1.7 mm, and 5.5

unequal radius loops. In addition, the fired microcoil has a cross-sectional area of 90 μm (line width) × 10 μm (line thickness). The entire volume of the meso-generator is 10 mm × 10 mm × 7.18 mm (*c.* 718 mm^3). The measurements were performed at a natural frequency of 69 Hz with maximum amplitude of 0.03 mm. The safe design of the double supporting beam was confirmed: the supporting beam has a maximum bending stress of 32 MPa and a lifetime of 1.95×10^8 cycles when the vibration amplitude is at its maximum.

3.7.3 Measurement and Analysis Results

The developed generator has a 44.5 mV maximum voltage output, an 83.1 mA maximum current output and a 1.58 mW maximum power output. The PD is calculated at 2.17 μW mm^{-3}, with a normalized PD of 6.46 μW mm^{-3} g^{-2}. Finally, the experimental results indicated that the EMF of the generator is more sensitive to the number of magnets than to the number of coil layers. The induced current and power output are proportional to both the number of coil layers and the number of magnets. Increasing the number of magnets and multilayer coils are both possible approaches toward improving the generator performance.

An analytical maximum EMF ε_t of 40.1 mV and electrical current of 5.0 mA with an electric resistance of 8.0 Ω, an excitation amplitude of 1 mm, and a frequency of 60 Hz were obtained. The differences between the analytical and experimental values of the EMF and electric current are all smaller than 2%. The analytical power output of 99 mW is similar to the experimental value of 100 mW. These results have validated the proposed analytical models and formulae.

3.8 Summary

This chapter introduced the development of a vibration-based electromagnetic microgenerator. The manufacturing technology of LTCC applied to the construction of multilayer silver microinduction coils and spiral ceramic microspring in a microgenerator has also been introduced. Compared with silicon-based generator manufacture processes, the LTCC has advantages such as low cost and short manufacturing time. Further, the combined ceramic and silver is not easily broken or fractured after sintering. The microgenerator developed by this study has the feature of high current output due to the low resistance of the multilayer structure and silver conductor.

Future studies can improve microgenerator performance as follows.

1. **Designing a microsystem circuit to increase the output electricity:** The voltage of the direct current electricity is quite low for normal application. If multiple generators are series-connected with a special microsystem circuit design, the total electricity voltage can be raised to the level of normal application.

2. **Selecting low-magnetic-loss materials:** Using a low-magnetic-loss material such as permalloy as the core element can reduce the magnetic reluctance effect on the magnetic field intensity.
3. **Improving the stacking precision:** Improving the stacking precision can enhance the manufacturing performance and increase the number of layers of the induction coils within the same space. For example, if the shifting amount is smaller than 15 μm, a more than six-layer structure can be achieved. When the coils number increases, a larger electricity output can be obtained.
4. **Measuring the spring constant, resonance frequency, and damping coefficient:** The resonance frequency, the spring constant, and damping coefficient used in the analyses were all predicted values, which were verified indirectly by the measurement of the oscillation amplitude. They have reference values for the design of a microsystem. Nevertheless, when they are obtained by measurement, the precision of the analytical results increases.
5. **Measuring the bending stress and fatigue life:** The bending stress, fatigue life, and safety factor are all analytical values that can be verified indirectly by observing the generator after oscillation. However, if these values are applied to different environment conditions in the future, the actual parameters should be measured and verified. Such values can be adjusted to obtain the maximum amplitude and, accordingly, the maximum power output.

References

1. Glynne-Jones, P. and White, N.M. (2001) Self-powered systems: a review of energy source. *Sensor Review*, **21** (2), 91–97.
2. Glynne-Jones, P., Beeby, S.P. and White, N.M. (2001) Towards a piezoelectric vibration-powered microgenerator. *IEE Proceedings: Science, Measurement and Technology*, **148** (2), 68–72.
3. Beeby, S.P., Tudor, M.J. and White, N.M. (2006) Energy harvesting vibration sources for microsystem applications. *Measurement Science and Technology*, **17**, R175–R195.
4. Arnold, D.P. (2007) Review of microscale magnetic power generation. *IEEE Transactions on Magnetics*, **43** (11), 3940–3951.
5. El-Hami, M., Glynne-Jones, P., White, N.M. *et al.* (2001) Design and fabrication of a new vibration-based electromechanical power generator. *Sensors and Actuators A: Physical*, **92** (1–3), 335–342.
6. Ching, N.N.H., Wong, H.Y., Li, W.J. *et al.* (2002) A laser-micromachined multi-modal resonating power transducer for wireless sensing system. *Sensors and Actuators A: Physical*, **97**, 685–690.
7. Shearwood, C. and Yates, R.B. (1997) Development of an electromagnetic microgenerator. *Electronics Letters*, **33** (22), 1883–1884.
8. Williams, C.B. and Yates, R.B. (1996) Analysis of a microelectric generator for microsystems. *Sensors and Actuators A: Physical*, **52** (1–3), 8–11.
9. Williams, C.B., Shearwood, C., Harradine, M.A. *et al.* (2001) Development of an electromagnetic microgenerator. *IEE Proceedings: Circuits Devices and Systems*, **148** (6), 337–342.
10. Beeby, S.P., Torah, R.N., Tudor, M.J. *et al.* (2007) A micro electromagnetic generator for vibration energy harvesting. *Journal of Micromechanics and Microengineering*, **17** (7), 1257–1265.
11. Buren, T.V. and Troster, G. (2007) Design and optimization of a liner vibration-driven electromagnetic micro-power generator. *Sensors and Actuators A: Physical*, **135** (2), 765–775.

12. Amirtharajah, R. and Chandrakasan, P. (1998) Self-powered signal processing using vibration-base power generation. *IEEE Journal of Solid-State Circuits*, **33** (5), 687–695.
13. Li, W.J., Ho, C.H. and Chan, M.H. (2000) Infrared signal transmission by a laser-micromachined vibration-induced power generator. Proceedings of the 43rd IEEE Midwest Symposium on Circuits and Systems, vol. **3**, pp. 236–239.
14. Pan, C.T., Hwang, Y.M. and Hu, H.L. (2006) Fabrication and analysis of a magnetic self-powered microgenerator. *Journal of Magnetism and Magnetic Materials*, **304** (1), 394–396.
15. Li, W.J., Wen, Z., Wong, P.K. *et al.* (2000) A micromachined vibration-induced power generator for low power sensors of robotic systems. 8th International Symposium on Robotics with Applications, Hawaii, 16–21 June.
16. Williarms, C.B., Pavic, A., Crouch, R.S. and Wood, R.C. (1998) Feasibility study of vibration-electric generator for bridge vibration sensors. Proceedings of the 16th International Modal Analysis Conference, IMAC, vol. **3243**, pp. 1111–1117.
17. Lu, W.L., Hwang, Y.M., Pan, C.T. and Shen, S.C. (2011) Analyses of electromagnetic vibration-based generators fabricated with LTCC multilayer and silver spring-inducer. *Microelectronics Reliability*, **51** (3), 610–620.
18. Lu, W.-L. (2010) Design and Fabrication on Vibration-Induced Electromagnetic Micro-Generators Using LTCC Technology. PhD thesis, Department of Mechanical Electro-Mechanical Engineering National Sun Yat-sen University.
19. Cheng, D.K. (1997) *Fundamentals of Engineering Electromagnetics*, Addison-Wesley, Reading, MA.
20. Shigley, J.E. and Mischke, C.R. (1989) *Mechanical Engineering Design*, 4th edn, McGraw-Hill, Book Company.

4

Design and Fabrication of Rotary Electromagnetic Microgenerator

4.1 Introduction

Developing microelectromechanical systems (MEMS) offers many advantages. For example, microsystems can be downscaled and combined with an integrated circuit (IC). The MEMS technique has been successfully applied to microapparatus but problems still exist, such as the lack of an independent and compact power supply system. The development of several microsystems, such as micromotors, microsensors, micropumps, and microelectric apparatus, has grown substantially in the field of MEMS applications, requiring independent microscale power generation units in the range of milliwatts (mW) to microwatts (μW). There is currently considerable interest in mechanisms that can harvest energy from the environment for use in portable and non-portable systems. Self-powered devices supply their own energy and remove the need for periodic recharging or batteries. Electromagnetic microdevices that convert mechanical energy into electrical energy through relative motion are favorable candidates for power generation. A generator with a rotary-induced energy source is an attractive example of human motion being used to drive the system. A high-power rotary-type electromagnetic microgenerator consists mainly of a multilayer planar microcoil and a multipolar hard magnet, which may interact

Design and Fabrication of Self-Powered Micro-Harvesters: Rotating and Vibrating Micro-Power Systems, First Edition.
C. T. Pan, Y. M. Hwang, Liwei Lin and Ying-Chung Chen.

with each other to induce a voltage in the coil according to Faraday's law. This developed microgenerator currently requires an input power to rotate the rotor and generate power. In the future, microgenerators will be integrated within bicycles to generate power for light-emitting diode (LED) systems that are used in lighting or warning signs, and within cars as power sources in tires for tire pressure monitoring systems. This chapter introduces the design, manufacture, and analysis of two cases of in-plane rotary electromagnetic microgenerators.

Case 1 consist of an in-plane rotary electromagnetic microgenerator, which consists mainly of a multilayer planar copper (Cu) microcoil and a multipolar hard magnet made of NdFeB, the entire volume of which is approximately $5 \times 5 \times 2\,\text{mm}^3$. This study focused on the design and manufacturing processes required to obtain a high-power generation output, and an analytical model was developed to predict the power output for designing different types of microgenerators. The geometric pattern design of the Cu planar microcoil was manufactured using the filament winding method. Both the line width and spacing of the microcoil were 30 μm. The multipolar hard NdFeB magnet was molded and sintered, and a specially designed piece of equipment (a yoke) was used to magnetize the NdFeB magnet to produce an external magnetic field. After magnetization, an anisotropic residual induction B_r of 1.44 T is produced. The theoretical model of this power microgenerator was evaluated and compared with the experimental results, and it was found that the analytical simulation agreed favorably with the experimental results. The induced EMF was 111.2 mV, and a maximum power output of 0.412 mW at a frequency of 149.3 Hz was obtained.

In Case 2 we describe the design, fabrication, testing, and application of an in-plane rotary electromagnetic microgenerator to obtain a high-power output. The microgenerator comprises multilayer planar low-temperature co-fired ceramic (LTCC) Ag microcoil and a multipole hard magnet of NdFeB. Finite-element simulations were performed to observe electromagnetic results. This study also established analytical solutions for the microgenerator to predict the induced voltage. Three configurations of planar microcoils were investigated: sector-shaped, circular, and square microcoils. A prototype of the microgenerator is as small as $9 \times 9 \times 1\,\text{mm}^3$ in volume. The experimental results showed that the microgenerator with a sector-shaped microcoil has the highest power output of 1.89 mW, and an effective value of the induced voltage of 205.7 mV at 13 325 rpm is achieved.

The proliferation of microelectronic and MEMS devices in recent years has resulted in increasing interest in alternative power sources that might eliminate the need for chemical batteries. The motivation for developing such sources depends on the application field and power requirements. Energy-scavenging power generators based on MEMS technology have been researched extensively over the past 10 years. Computer, communication, and consumer (3C) electronic devices have led to a trend of small and portable electronic devices and miniature systems. MEMS processes have been successfully applied to microapparatus but problems such as a lack of independent and compact power supply systems have prevailed. Determining how to supply power efficiently for such apparatus is an important engineering challenge. Traditionally, power supply systems involve battery-supplied energy. Batteries have been widely used as a system of power; however, they entail many problems such as a short lifetime, periodic recharging, and pollution of the environment. Batteries, which convert chemical energy into electrical energy, consume much of Earth's natural resources and are never manufactured in environmentally friendly electronic containers. The energy storage of the battery is limited by its volume, environment, and operating temperature. A battery has a limited lifetime and must be renewed or recharged. Although many electronic devices can now be manufactured to be smaller, batteries cannot keep pace with today's technology as smaller power sources. Many studies have addressed the problems created by batteries [1–5], but many of them have failed. Developing an inexhaustible replacement for batteries is therefore crucial. As this study demonstrates, a self-powered system is a favorable solution. Self-powered systems extract energy from the surroundings [6] and are driven by energy conversion without causing serious pollution. Self-powered systems that have compact structures and eco-friendly features are defined as harnessing ambient energy within the environment of the system. The energy-harvesting devices convert kinetic energy into electrical energy. Self-powered systems have been microminiaturized using MEMS technology, and are more favorably suited to MEMS than most batteries.

4.1.1 Piezoelectric, Thermoelectric, and Electrostatic Generators

A piezoelectric generator consists of a piezoelectric laminated cantilever and a mass, and harvests vibration energy through the resonant inertial oscillation of the cantilever. The power of a piezoelectric generator is generated by the inherent piezoelectric effect. The application of stress to a piezoelectric material generates a corresponding electric charge [7]. Fang *et al.* designed an MEMS-based piezoelectric power generator to obtain a power output of 1.15 μW at 609 Hz [8]. In the following year, the same research team improved the power generator to obtain a power output of 2.16 μW [9]. The mechanical energy of a piezoelectric generator is converted into electrical energy. A piezoelectric generator must operate at the resonance frequency to harvest higher power. The friction and damping of generators reduce its output. These obstacles must be addressed.

A thermoelectric generator converts thermal energy into electrical energy [10, 11]; its conversion efficiency is however lower than that of a rotary electromagnetic generator, preventing its widespread use. An electrostatic generator converts ambient mechanical energy into electrical energy [12–14]. The obstacles to operating the electrostatic generator at low frequencies have been surmounted, but the frequency of induction remains difficult to measure.

4.1.2 Vibrational Electromagnetic Generators

Microelectromagnetic generators demonstrate the advantages of power output and IC package processing; microelectromagnetic generators therefore save energy and improve generator efficiency [15]. An electromagnetic generator converts mechanical energy into electrical energy. According to Faraday's law, a permanent magnet moving relative to a coil creates an electromotive force (EMF). Electromagnetic transduction mechanisms include vibrational and rotational power generators [16]. When a permanent magnet installed on a mass of a vibrational electromagnetic generator moves relative to a coil, it induces an EMF [17, 18]. Kulkarni *et al.* designed a microscale vibration-based electromagnetic generator to obtain a power output of 85 μW [19]. The power of the vibrational generator is, however, limited by natural resonance frequencies and the operation of the vibrational source [20–22], and is easily influenced by air resistance caused by motion intended to lower the electric power [23].

4.1.3 Rotary Electromagnetic Generators

The recycling of mechanical energy has been the focus of many previous studies. Rotary electromagnetic generators show potential for the development of self-sufficient energy sources. Machines using planar rotary permanent magnets are developed for various applications because they eliminate field excitation losses [24]. The frequencies of rotary electromagnetic generators do not have specific restrictions, making them environmentally friendly. Furthermore, as long as there is movement between the coil and the magnet, energy can be continuously created, thereby ensuring performance in machines of all sizes. Recently, hub dynamos have become popular for powering bicycle lights [25–27]. Rotary electromagnetic generators are therefore suitable for use in bicycle dynamos. The principle of a rotary electromagnetic generator, which can be easily installed because of its rotary motion [28–31], is similar to that of a vibrational electromagnetic generator and is not limited to use at the resonance frequency. The rotary electromagnetic generator has been revised to a millimeter-scale axial-flux electromagnetic generator, such as the 2-mm-thick generator presented by Pan and Wu [28]. This small volume reduces air resistance.

Liao *et al.* examined axial-flux electromagnetic microgenerator technologies and reported constructing a prototypical axial-flux electromagnetic microgenerator with a diameter of 10 mm to deliver a 7.23 mW power output at a rotational speed of 10 krpm [30]. Arnold *et al.* designed an optimal axial-flux microturbine electromagnetic microgenerator at a rotational speed of 305 krpm by extracting air flow energy to drive 8 W of output [32, 33]. Raisigel *et al.* presented the first integrated planar microturbo-generator prototype with a permanent magnet to generate a power output of 14.6 mW at a rotational speed of 58 krpm [34, 35]. Das *et al.* fabricated a multipole permanent magnet rotor with a power output of 2.6 W at a rotational speed of 120 krpm [36].

Arnold *et al.* designed a rotary-type microelectromagnetic generator to obtain a voltage of 0.9 V and a power output of 3.3 W at 225 krpm in three phases using eight poles and a large permanent magnet [37]. Subsequently, Arnold *et al.* obtained a power output of 12.7 W by improving the structural dimension and coil shape. However, because of the internal resistance of the coil, the actual power output was 8 W.

Rotary motion is combined with common driving sources, which are applied to turbine devices [38]. When a turbine is rolled by an external force (such as wind or water), the generator on the rotary axis produces electric power. The higher the rotary speed, the larger the power output which can be achieved [39]. Voltage can be induced in a closed circuit in the changing magnetic field. The power generation efficiency and the volume size of rotary electromagnetic generators have been studied in depth. Herrault *et al.* fabricated a microgenerator with a diameter of 2 mm and a thickness of 0.5 mm that powered a microchip [40]. Zwyssig and Kolar designed a generator with an output power of 100 W with a volume of 3 cm^3 at 500 krpm [41]. In addition, as long as the magnet and the generator have an appropriate structure, rotary electromagnetic generators offer improved resistance to the fall in capacity generation at high temperatures [42]. The induced EMF of the rotary electromagnetic generator can continue to induce as long as the change of the magnetic flux flows through the circuit. Furthermore, the EMF is not restricted by rotary speed. For example, Krähenbühl *et al.* fabricated a mesoscale electric power generation system from a pressurized gas flow that had a rotary speed of up to 60 million rotations per minute [43]. Corderol *et al.* designed an axial-flux permanent-magnet generator with a rotary speed of 4 krpm [44]. Rotational electromagnetic generators are highly adaptable. The structures of rotary electromagnetic generators vary according to application type. Boland *et al.* used a flat-shaped conductor for the induced current output, finding that the induced current was difficult to control and caused low power [45].

Rotary electromagnetic generators usually use winding coils that are easy to fabricate and achieve high power output [46]. However, it is difficult to control the flatness of winding coils and make them thinner. Many processes, such as lithography technology, have been investigated to improve the problem of reduced generator volume [47]. Lithography technology is complex however, and requires expensive masks and an exposure source that cannot be easily attained. Reducing the generator's volume leads to a low power output, making it difficult to power even small machinery [48]. Most studies have focused on maximizing power generation, although the optimal solution is to increase the rotary speed. To achieve the required amount of power, the rotary speed of the rotary electromagnetic generator must be distributed in the range of tens of thousands to hundreds of thousands of rotations per minute [49]. However, generators requiring a high rotary speed are difficult to apply in many environments.

4.1.4 Generator Processes

Conventional microcoil fabrication process include LIGA (a German acronym for *Lithographie, Galvanoformung, Abformung*), LIGA-like, filament winding, LTCC (low-temperature co-fired ceramic), and PCB (printed circuit board) processes. LIGA and LIGA-like processes consist of a thick photoresistant exposure, thin film, and a high-aspect ratio electroplating process. The LIGA process also requires expensive masks and X-ray synchrotron facilities, which are unavailable and inaccessible in many companies and research centers [50]. There are some disadvantages to the filament-winding method: the multifold-winded coil is difficult to use to control its flatness and to lower its thickness [51]. The benefit of using LTCC for manufacturing IC devices was reported by Wi *et al.* [52] and Ponchak [53]. In addition, some LTCC technology applications for microdevices have been explored [54–56]. However, in the field of microgenerator applications, LTCC processes are a new direction.

4.1.5 Lithographie Galvanoformung Abformung Process

Microfabrication with deep X-ray lithography (DXRL), also known as LIGA, is a sequence of X-ray lithography, electroforming, and injection-molding steps [57–59] for fabricating microstructures with high aspect ratios. The LIGA technique surpasses the possibilities of conventional precision machining, especially in mass production and precision of replication [60, 61]. Possible fields of application for the LIGA process covers different technological disciplines such as integrated optics, micromechanics, sensor and actuator technology, chemistry, medicine, and bioengineering. Industrial applications and concepts for microstructure production using the LIGA process are rapidly emerging [62]. LIGA techniques have the potential of fabricating highly accurate electrical and optical fiber connector alignments.

This study used a Taiwan light source (TLS) as the exposure light, with an operating electron beam energy of 1.5 GeV and an initial current of approximately 200 mA. The micromachining beamline of the TLS is located at 18 B, where the X-ray is extracted at 6° from the first magnet of the triple bending magnets immediately after the long straight section of the 1.8 T, 25-pole wigglers [63]. The synchrotron light channeled from the storage ring passed through a Be window (125 μm thick) and directly into an X-ray scanner. The polymethylmethacrylate (PMMA) photoresist and mask were then placed inside the X-ray scanner in preparation for exposure. A conformal mask was required for use in the TPHARM (thick photoresist with high aspect ratio microstructure) experiments, and the procedure can be broken down into the following steps.

1. Sputter Cu as a seed layer onto the PMMA substrate.
2. Coat a negative-tone photoresist (JSR THB-37) on the Cu at 500 rpm for 60 seconds.
3. Expose the photoresist to a UV light source at a wavelength of 435 nm for 110 s, and then develop the sample for 120–140 s at room temperature.
4. Electroplate Au as the absorber.
5. Remove JSR using NaOH (6 wt%), $Na_2S_2O{\cdot}5H_2O$ (6 wt%), and NP-A (2 vol%).
6. Strip off the Cu seed layer using $FeCl_3$.

The incident X-ray dose of 800 mA min cm^{-1} passed through the Be window and directly exposed the 3-mm-thick PMMA. The PMMA exposed to the X-ray was then placed into a beaker filled with GG developer in an isothermal bath at 28 and 36 °C, respectively, for development. Developing depth measurement was performed by interrupting the development and using a 50-μm-diameter probe associated with a mechanical altitude measurement platform. The PMMA photoresist was first removed from the developing solution and washed with deionized (DI) water to remove any residual glue before the measurement was conducted on the measurement platform. Four corners and a central point were measured and averaged for each specimen. The feasibility of using a graphite membrane supported gold absorbers in the X-ray micromachining process.

4.1.6 Winding Processes

The filament-winding method, which involves manufacturing planar microcoils for fast prototyping, is unique in MEMS field applications. Filament-winding equipment requires precise $x-y$ positioning tables and z-axis height-controlling systems. An 8-pole, 12-turn planar microcoil with a single filament of Cu wire was successfully manufactured. Using the filament-winding method, a precise geometric pattern of microcoil was impregnated on the 50-μm-thick PET (polyethylene terephthalate) substrate prepared for use in this study. This polymer can be transparent (when in its amorphous state) or translucent (in its semi-crystalline state). By demonstrating excellent barrier (CO_2 permeation) and mechanical properties, PET is becoming the polymer of choice for MEMS applications, even replacing traditional materials such as metal or glass. Using the flexible PET substrate enables folding and stacking, and four layers of microcoil are stacked.

4.1.7 LTCC

The conventional microcoil fabrication process includes LIGA, LIGA-like, and filament-winding processes. LIGA and LIGA-like processes consist of thick photoresist exposures, thin films, and high-aspect-ratio electroplating processes. In addition, LIGA processes require expensive masks and synchrotron X-ray facilities, which are unavailable and inaccessible to many companies and research centers [64]. The filament-winding method has some disadvantages: the multifold-winded coil is difficult to control, particularly regarding its flatness and thickness [65]. The benefits of using LTCC for manufacturing IC devices have been reported by many authors [52, 53, 66, 67]. Furthermore, some LTCC technology applications for microdevices have been explored [68–70]. Many studies have examined LTCC [71–73], investigating its feasibility in applications such as oscillators, sensors, and chip integration [59, 60]. Furthermore, LTCC technology enables stacking multilayer microstructures with green sheets, which are then laminated together and fired, all in the same process. Every green sheet layer can be inspected for inaccuracy before firing. This saves fabrication time and cost, and facilitates the fabrication of planar multilayer microcoils [61]. However, in the field of microgenerator applications, LTCC processes are a new attempt.

This chapter presents a new method for fabricating an in-plane rotary electromagnetic microgenerator, by focusing on the fabrication of an LTCC microcoil and testing this powering system. LTCC is a ceramic-based system demonstrating the advantages of high-temperature co-fired ceramics (HTCCs). The firing temperature of LTCC is below 800–900 °C. The process has the advantage of being able to use low-resistivity silver (Ag) as a conductive material. LTCC slurry is a mixture of recrystallized glass and ceramic powder in binders and organic solvents, whereas HTCC at firing temperatures of 1500–1600 °C is an aluminum oxide substrate, which allows for the co-firing of conductors only, such as tungsten and molybdenum. LTCC technology enables the stacking of multilayer microstructures with green sheets, which are laminated together and fired all in one process. Every layer of green sheet can be inspected in case of inaccuracy before firing, thus saving fabrication time and cost.

4.1.8 *Printed Circuit Board Processes*

Printed circuit boards for computer main boards are a well-established industry; this PCB-based harvester therefore has high potential for integration in IC-related industries. Moreover, PCBs are made of plastic and copper, materials that are relatively inexpensive. This is a considerable advantage for industry and mass production.

Selectively removing the copper by performing etching processes requires etching the unwanted copper while protecting the wanted copper from the etchant.

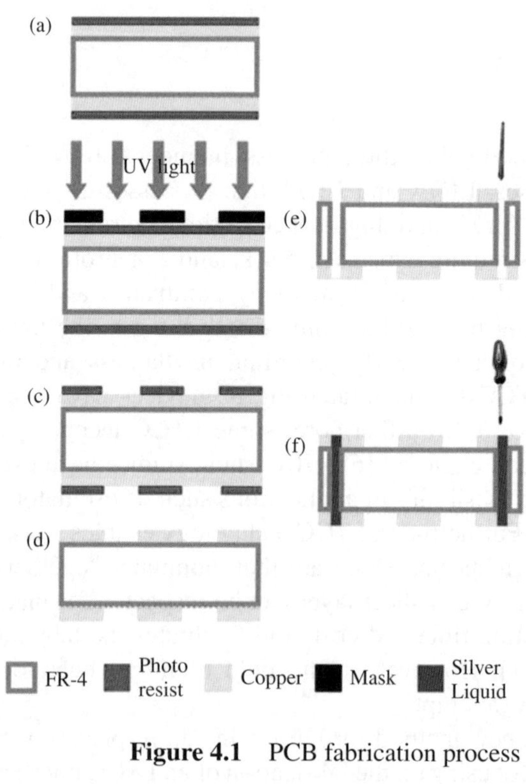

Figure 4.1 PCB fabrication process

This protection is provided by a polymer coating (called a 'photoresist') deposited onto the surface of the copper cladding, as shown in Figure 4.1a. The photoresist is patterned in the shape of the desired printed circuit through a process called 'photolithography'. The patterned resist protects selected areas of the copper from the etchant and exposes the copper to be etched. The two steps of photolithography are patterning and developing the photoresist. Patterning is accomplished by exposing the resist to light (typically UV), and developing it is accomplished by washing it in a chemical bath. The two types of photoresist are positive resists and negative resists. When a positive resist is exposed to UV light, the polymer breaks down and can be removed from the copper. Conversely, a negative resist that is shielded from UV light is removed. A mask is used to expose the desired part of the photoresist. A mask is a specialized black and white photographic film or glass photoplate on which a picture of the traces and pads is printed using a laser photoplotter. The mask is placed on top of the photoresist, and the assembly is exposed to UV light. The dark areas block UV light, and the white (transparent) areas allow the UV light to reach the photoresist, which imprints the circuit image into the photoresist, as shown in Figure 4.1b. A separate mask is used for each layer of a circuit board. The software OrCAD PCB Editor generates the data that the photoplotter uses to manufacture these masks. Another way to expose the photoresist is by using a programmable laser to 'draw' the pattern directly onto the photoresist. This is a newer technique, called laser direct imaging (LDI). A benefit of the LDI process is that it uses the same data as photoplotters, but without the requirement of masks. After the photoresist has been exposed (either with the mask and UV or with the laser), it is washed in a chemical called the developer. Regarding the positive resist, the resist breaks down during exposure and is removed by the developer. In the case of the negative resist, the UV light cures the resist and only the unexposed resist is removed by the developer. Common developers are sodium hydroxide (NaOH) for positive resists and sodium carbonate (Na_2CO_3) for negative resists. Once the resist has been exposed and developed, a circuit image made of the photoresist is left on the copper, as shown in Figure 4.1c. The board is then etched in a corrosive solution, such as alkaline ammonia or cupric chloride. The etching solution does not significantly affect the photoresist but attacks the bare copper and removes it from the substrate, thus leaving behind the resist-coated copper, as shown in Figure 4.1d. Some processes involve the use of a plated tin alloy as the etch resist. Thin alloy plating is more resistant to etchants and is used to prepare copper surfaces for soldering processes. In this case, photolithography processes are used to plate the circuit pattern selectively onto the copper surfaces prior to etching. When polymer etch resists are used, the photoresist is cleaned from the copper with a resist stripper, leaving behind the copper traces. When metal etch resists are used, the plating is typically left in place. If it is a multilayer PCB, it also requires a via to conduct the layers. In this case, plated through hole or other processes are applied to the board. A computer numerical control machine can be used to drill the via. After drilling, liquid silver is injected into the via, as shown in Figure 4.1e, f.

4.1.9 Finite-Element Simulation and Analytical Solutions

Based on previous studies, the micro power system is used for small-volume and large-energy density to match the economy of the environmental protection section [74]. Under microscale conditions, the power of the generator is greatly affected by a small change in parameters [75]. Estimation and optimization of power are even more important [76]. Finite-element simulation is an extensively used analysis tool used in the research fields of electrical engineering, chemical engineering, electromagnetics, acoustics, mechanical engineering, fuel cell, electronic optics, MEMS, heat transfer, and wave propagation [65, 77]. We obtained the initial result in a planar rotary-type microelectromagnetic generator [78]. This study focused on analyzing currents generated by electromagnetic induction by performing a quasi-static simulation. An electromagnetic module was used in the finite-element simulations, and quasi-statics was used in simulating electromagnetic induction. Many factors, such as electromagnetism, kinematics, and geometric parameters, may influence power generation. It was is necessary to couple the mutual effect on these factors and, more important, establish analytical solutions for a microgenerator [79–81] to predict the induced voltage.

4.2 Case 1: Winding Generator

This section describes the design, manufacture, and analysis of an in-plane rotary electromagnetic microgenerator. An analytical model was developed in this study to predict the power output, which aids in understanding different designs of microgenerators. This direct physical modeling approach is capable of revealing explicit guidelines for quality control and technology extension. The microgenerator consists of a Cu planar microcoil and a NdFeB magnet. The development process includes the following steps.

First, different patterns of Cu planar microcoils are manufactured using the filament-winding method, a new process that involves winding resin-impregnated Cu enameled wire, 30 μm in diameter, onto a PET substrate in precise geometric patterns. This is accomplished by moving the PET substrate while a wire-delivery head continuously delivers and precisely positions the Cu-enameled wire onto the substrate surface, producing planar microcoils in an extremely short time compared to the traditional MEMS lithography process. Second, the multipolar NdFeB magnets that provide the external field are composed of raw powder materials and are manufactured by molding, sintering, heat treatment, and anisotropic magnetization processes. A specially designed piece of equipment (a yoke) is used to magnetize this sintered NdFeB magnet. Third, the microcoil and multipolar magnet are assembled. The entire volume is approximately $5 \times 5 \times 2\,\text{mm}^3$. While the microcoil is rotated relative to the multipolar magnet, an electric current is induced, and a maximum power output of 0.412 mW at 149.3 Hz is achieved. Finally, the analytical evaluation and experimental results are compared.

4.2.1 Design

Figure 4.2a is an illustration of the concept of an in-plane rotary electromagnetic microgenerator consisting of a permanent NdFeB magnet and a planar microcoil. The rotor is a permanent magnet with an outer radius of 2.5 mm and an inner radius of 1 mm, whereas the dimensions of the planar microcoil functioning as the stator are the same as those of the rotor. Because the degree of miniaturization of the magnet is limited by current yoke manufacture technology, the total volume of the magnet should be considered carefully in this type of design. For the stator, a sector is defined as one pole, as schematically shown in Figure 4.2b. A sector-shaped area accommodates one pole of the microcoil. To manufacture a microcoil with as many turns as possible in this area, the pattern of the coil is designed as a sector-shaped loop in each pole, as shown in Figure 4.3a. A detailed characterization is schematically shown in Figure 4.3b. The via pad is used to connect the interlayer of the coil, as shown in Figure 4.3c. The rotor and stator both have an outer radius of 2.5 mm and are approximately 1 mm in total thickness, thereby enabling the manufacture of the desired compact integrated system. The numbers of poles P for both the stator and rotor were set at $P = 8$ in this study.

The planar microcoils, 30 μm in line width, are manufactured using the filament-winding method. Because the filament-winding method is a new approach for making planar microcoils for fast prototyping, it is unique in the field of MEMS applications. Filament-winding equipment requires precise x–y positioning tables and z-axis height-controlling systems. An 8-pole, 12-turn planar microcoil with a single filament of Cu wire was successfully manufactured, shown in Figure 4.4a. By using the filament-winding method, a precise geometric pattern of microcoil was impregnated on the 50-μm-thick PET substrate prepared for use in this study. This polymer can be transparent (when in its amorphous state) or translucent (in its semi-crystalline state). With its excellent barrier (CO_2 permeation) and mechanical properties, PET is becoming the polymer of choice for MEMS applications, even

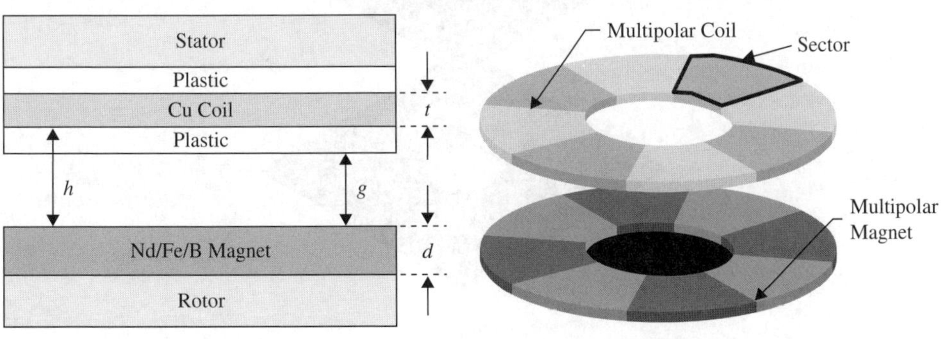

Figure 4.2 Schematic diagram of the electromagnetic microgenerator: (a) cross-section view and (b) a 3D schematic view of the eight-pole rotors and stators [28]

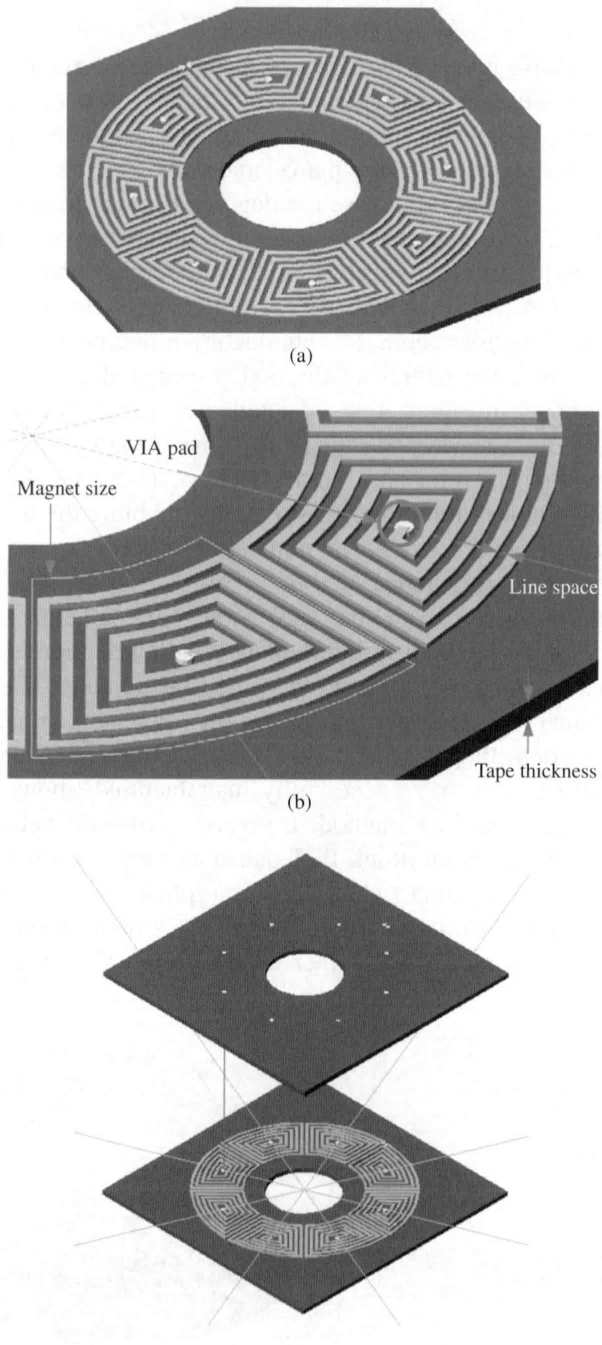

Figure 4.3 (a) The geometric pattern of the coil is designed as a sector-shape loop in each pole; (b) detailed characterization; and (c) the via pad is used to connect to the next layer of the coil [28]

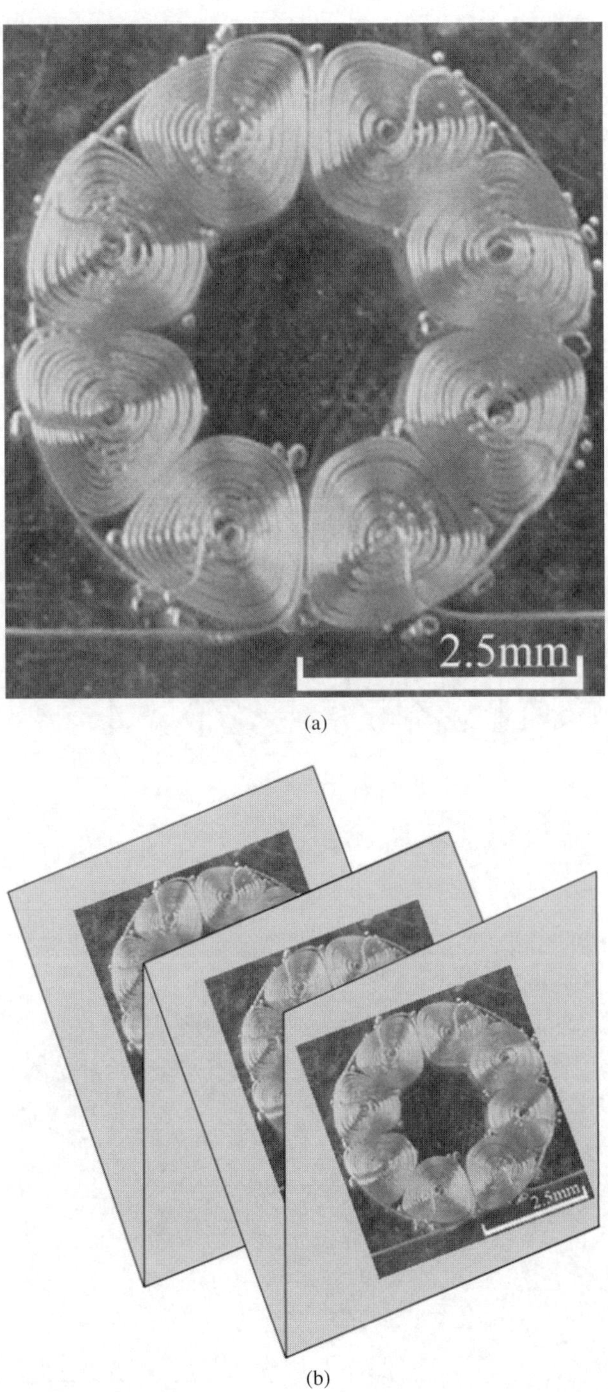

(a)

(b)

Figure 4.4 Planar microcoil of (a) an eight-pole and (b) a four-layer structure, manufactured using the filament winding method [28]

replacing traditional materials such as metal or glass. Using the flexible PET substrate enables folding and stacking, and four layers of microcoil are stacked as shown in Figure 4.4b. The properties of the Cu planar microcoil are listed in Table 4.1.

Sintering magnet technology is used to manufacture eight-pole magnets, and the custom-designed yoke is used to magnetize the NdFeB magnet. Figure 4.5a

Table 4.1 List of Cu planar microcoil properties

Conductivity	$5.8 \times 10^7\ \Omega\,m^{-1}$
Resistance	$30\ \Omega$
Relative permeability	1
Resistivity	$1.71 \times 10^{-8}\ \Omega m$
Line width	30 μm

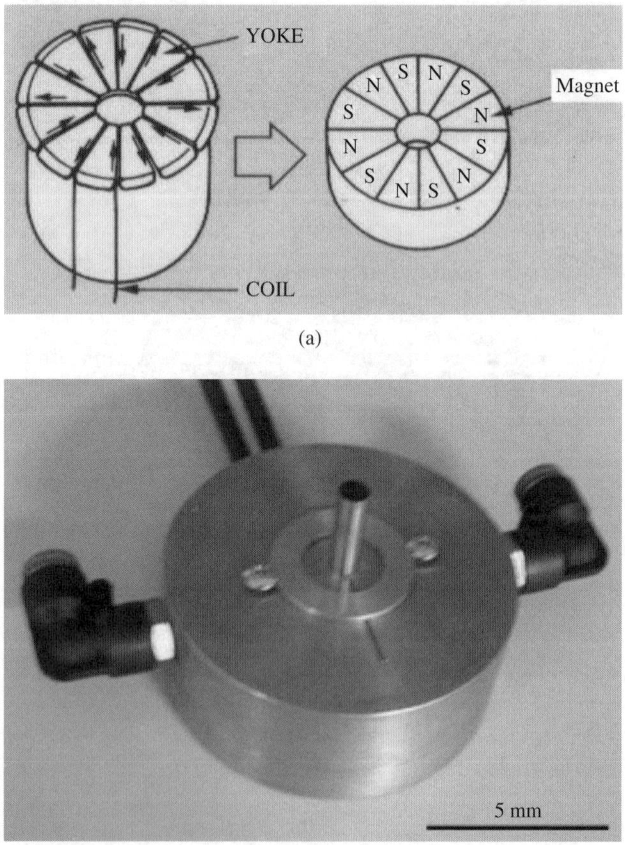

(a)

(b)

Figure 4.5 Illustration of (a) the magnetization concept and (b) the specially designed yoke used to create the multipolar magnet [28]

is a schematic representation of multipolar magnetization equipment and concepts, whereas Figure 4.5b shows the prototype. This compact-sized yoke is custom-designed for an eight-pole magnet 5 mm in diameter. The magnetized magnet of NdFeB is shown in Figure 4.6a. The eight-pole magnet is shown clearly with the help of magnetic fluid (Figure 4.6b). The multipolar magnet is composed of S-pole and N-pole magnetic segments; the segment-shaped magnets are easier to manufacture than contiguous annular magnets, and the yield rate of these magnetic segments is higher than that of contiguous annular magnets. Furthermore, the equipment required to create a uniform field to magnetize these magnets anisotropically is more expensive than that required for the yoke.

Each sector shows a dark color, meaning that this magnet produces a dense magnetic flux that clusters magnetic fluidic particles. The B–H curve (Figure 4.7) is measured

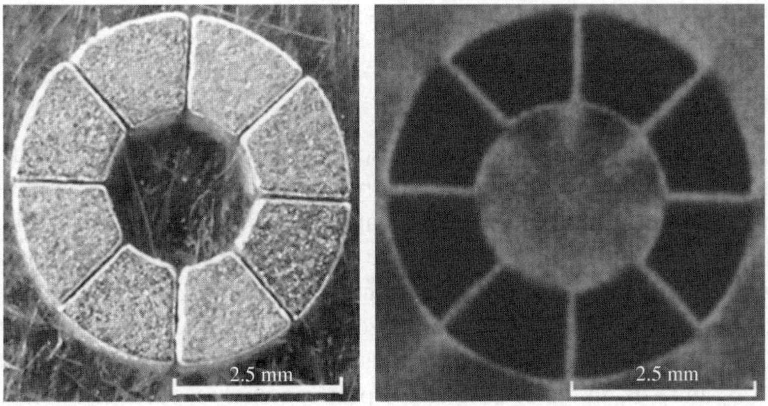

Figure 4.6 The sintered NdFeB permanent magnet of (a) a ring shape (outer radius × inner radius × thickness is $2.5 \times 1 \times 0.7\ mm^3$) and (b) an eight-pole magnetic image [28]

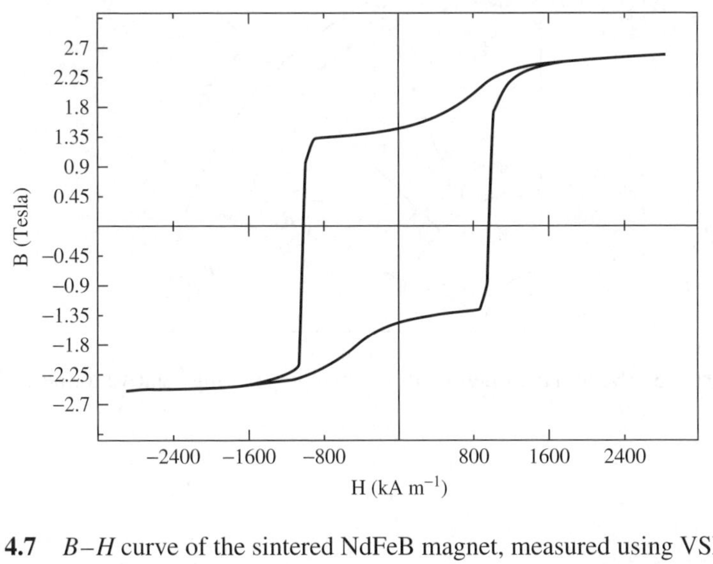

Figure 4.7 B–H curve of the sintered NdFeB magnet, measured using VSM [28]

Table 4.2 List of sintered NdFeB magnet properties

Residual induction	1.44 T
Relative permeability	1
Coercive force	$975\,\mathrm{kA\,m^{-1}}$
Maximum energy product	$367\,\mathrm{kJ\,m^{-3}}$

using a vibrating sample magnetometer (VSM) system, and shows that the sintered NdFeB is a favorably hard magnet with a residual induction B_r of 1.44 T and a coercive force H_c of $975\,\mathrm{kA\,m^{-1}}$. The properties of the sintered NdFeB magnet are listed in Table 4.2.

4.2.2 *Analytical Formulation*

A direct physical modeling approach is described in this section, and the design and manufacture are based on the analytical formulations of electromagnetic induction. Regarding the development of the analytical model, one segment of the closed-loop microcoil is schematically shown in Figure 4.8, illustrating that a particular turn of the microcoil, 1234, rotates at a constant angular velocity ω ($\mathrm{rad\,s^{-1}}$) in a constant magnetic field B_r (Tesla). This is an idealized model that accounts for the true spatial B-field distribution in the air gap because of flux leakage; it assumes a perfectly constant $+B_r$ and $-B_r$ in the N-pole and S-pole regions, respectively. When the turn of the microcoil 1234 crosses the region surrounded by the dotted line, as shown in

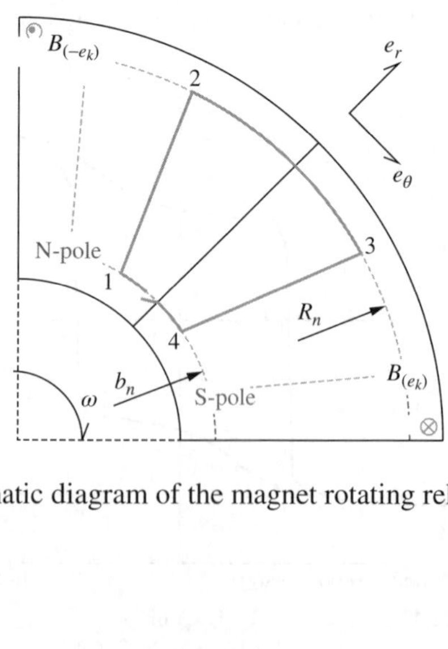

Figure 4.8 Schematic diagram of the magnet rotating relative to the coil [28]

Figure 4.8, the coil 1234 induces a current. According to Fleming's right-hand rule, the current induced through Segment 34 moves from Point 4 to Point 3, and the current induced through Segment 12 moves from Point 2 to Point 1. By contrast, because Segments 23 and 41 are parallel to the tangential velocity of the rotor, these segments do not induce any current. When the microcoil 1234 rotates relative to the multipolar magnet, as schematically shown in Figure 4.8, the induced voltages of Segments 34 and 12 are respectively:

$$V_{43} = \int_3^4 \left[(\vec{e_\theta} r\omega) \times \vec{e_k} B_\text{r}\right](\vec{e_r} dr) = \omega B_\text{r} \int_{R_n}^{b_n} r\,dr = -\frac{1}{2}\omega B_\text{r}\left(R_n{}^2 - b_n{}^2\right) \tag{4.1}$$

$$V_{21} = \int_1^2 \left[(\vec{e_\theta} r\omega) \times (-\vec{e_k} B_\text{r})\right](\vec{e_r} dr) = -\omega B_\text{r} \int_{b_n}^{R_n} r dr = -\frac{1}{2}\omega B_\text{r}\left(R_n{}^2 - b_n{}^2\right) \tag{4.2}$$

where e_θ, e_k, and e_r are the unit vectors in the directions of θ, the z axis, and r, respectively and R_n and b_n are the outer and inner radii of the nth microcoil, respectively.

Based on Equations (4.1) and (4.2), the total induced voltage of coil 1234 in the magnetic field can be derived using

$$V_{43} + V_{21} = -\omega B_\text{r}\left(R_\text{n}{}^2 - b_\text{n}{}^2\right) \tag{4.3}$$

Figure 4.9 depicts a coil of outer radius R_n and inner radius b_n, such as microcoil 1234. R_f (μm) and b_f (μm) are the outer and inner radii, respectively, of the fth turn of the microcoil. The induced voltage of the nth turn is denoted V_n. In Figure 4.9, both the line width and spacing of the microcoil are assumed to be λ (μm). When a microcoil with a single layer, a single pole, and N turns rotates in the magnetic field, the induced

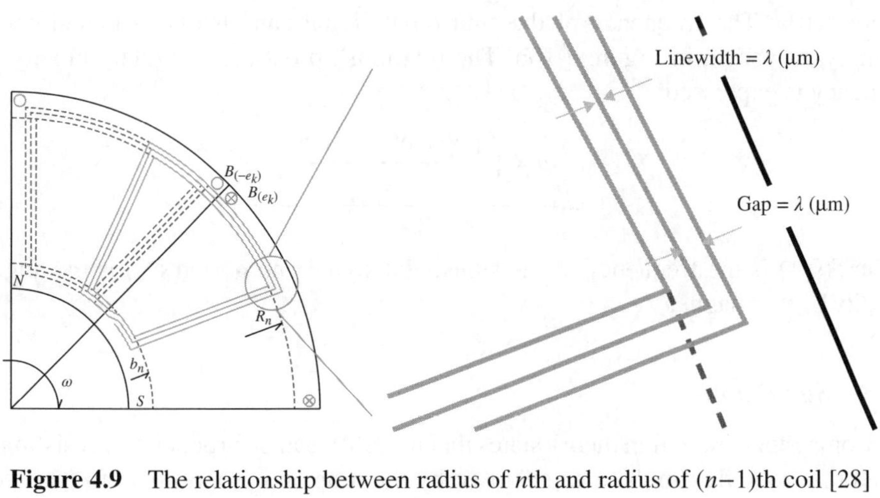

Figure 4.9 The relationship between radius of nth and radius of $(n-1)$th coil [28]

EMF voltage is denoted $|V_t|$. Likewise, for a microcoil with L layers, P poles, and N turns, the induced voltage is expressed as $|V_t| \times L \times P$. The relationship of the radius between the nth and $(n-1)$th coil is expressed

$$\begin{aligned} R_{(n-1)} &= R_n - (1 \times 1 \times \lambda + 1 \times 1 \times \lambda) \\ b_{(n-1)} &= b_n + (1 \times 1 \times \lambda + 1 \times 1 \times \lambda) \end{aligned} \tag{4.4}$$

R_n and b_n can therefore be expressed in terms of R_f, b_f, and m as follows:

$$\begin{aligned} R_n &= R_f - 2m\lambda \\ b_n &= b_f + 2m\lambda \end{aligned} \tag{4.5}$$

where $n = 1, 2, \ldots, N$; $m = 0, 1, 2, \ldots, (N-1)$; and $m = n-1$. Furthermore, V_n is calculated from Equations (4.3) and (4.5):

$$V_n = -\omega B_{\mathrm{r}} \left\{ \left[(R_f - 2m\lambda) \times 10^{-6} \right]^2 - \left[(b_f + 2m\lambda) \times 10^{-6} \right]^2 \right\} \tag{4.6}$$

V_t of a microcoil with a single layer, a single pole, and N turns is calculated by $V_1 + V_2 + \ldots + V_N$. Hence, the induced voltage of the microcoil in the case of L layers, P poles, and N turns is defined:

$$V_t \times L \times P = \sum_{m=0}^{N-1} \left(-LP\omega B_{\mathrm{r}} \left\{ \left[(R_f - 2m\lambda) \times 10^{-6} \right]^2 - \left[(b_f + 2m\lambda) \times 10^{-6} \right]^2 \right\} \right) \tag{4.7}$$

Equation (4.7) is used to model electromagnetic induction. Based on Equation (4.7), the relationship between the frequency of the induced voltage and the angular velocity of the magnet can be investigated. In the case of P-pole magnets, when a powering cycle is completed $P/2$ sinusoidal signals are obtained, as schematically shown in Figure 4.10a. The frequency of the sinusoidal signal can also be obtained experimentally, as shown in Figure 4.10b. The relationship between angular velocity and frequency is expressed:

$$f = \frac{\omega \times \left(\dfrac{\text{poles of magnet}}{2} \right)}{2\pi} \tag{4.8}$$

where f (s^{-1}) is the frequency of the sinusoidal signal and ω ($\mathrm{rad\,s}^{-1}$) is the angular velocity of the magnet.

4.2.3 *Simulation*

Electromagnetic induction theory states that the EMF can be produced by resisting the varied magnetic flux in the area enclosed by the loop and proportional to the rate of the varied magnetism. In this study, the induced current of the rotary electromagnetic microgenerator is simulated by combining FEMLAB® with MATLAB® software.

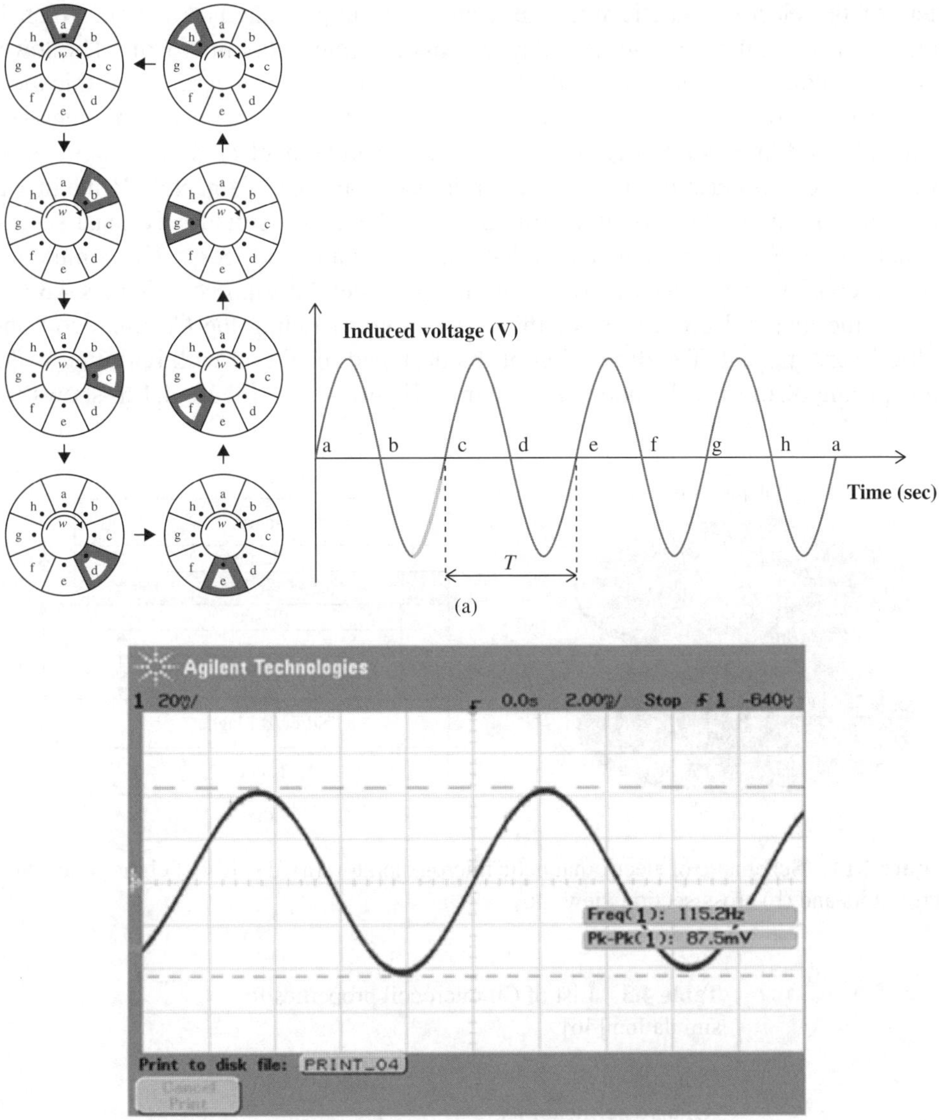

Figure 4.10 (a) Schematic diagrams of the in-plane rotary microgenerator for the coil rotating relative to the eight-pole magnets and (b) the sinusoidal waveform of the induced voltage, obtained using an oscilloscope [28]

FEMLAB has many advantages, most notably that it can be used to process multiphysics models including electromagnetic problems. MATLAB® can be used for further processing. Based on Ohm's law, the EMF can then be calculated. The schematically designed microgenerator, which includes mainly the multipolar micro-coil and multipolar magnet, is shown in Figure 4.11a. Figure 4.11b is a cross-sectional view of the electromagnetic microgenerator. The upper part is a stator, which is composed of a Cu microcoil and plastic films to protect this microcoil. The lower part is a rotor, which is composed of an eight-pole NdFeB magnet. The thickness of the Cu microcoil is t and the thickness of the NdFeB magnet is d. The thickness of the plastic film is so thin compared to t that it can be neglected in the simulation process. The gap between the rotor Cu microcoil and the stator NdFeB magnet is denoted g. In the FEMLAB simulation process, the analysis mode of 'quasi-statics, in-plane currents' is chosen to conduct this simulation process. The simulation process consists of a preprocessor, solution, and general postprocess. Tables 4.3 and 4.4 list the material properties for this simulation, including the Cu microcoil and NdFeB hard magnet. The dimension of the designed microcoil with both line width and spacing 30 μm is schematically shown in Figure 4.12a. Figure 4.12b shows the

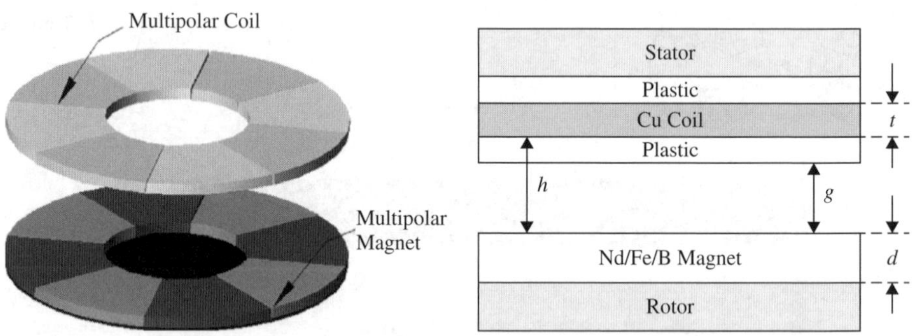

(a) (b)

Figure 4.11 Schematic of electromagnetic microgenerator: (a) 3D view of eight-polar rotors and stators and (b) cross-section view [46]

Table 4.3 List of Cu microcoil properties for simulation [46]

Conductivity	$5.8 \times 10^7\ \omega\,m^{-1}$
Relative permeability	1
Line width	30 μm

Table 4.4 List of NdFeB magnet properties for simulation [46]

Residual induction	1.44 T
Coercive force	$975\ kA\,m^{-1}$
Maximum energy product	$367\ kJ\,m^{-3}$
Ring (outer × inner × thickness)	$2.5 \times 1 \times 0.7\ mm^3$

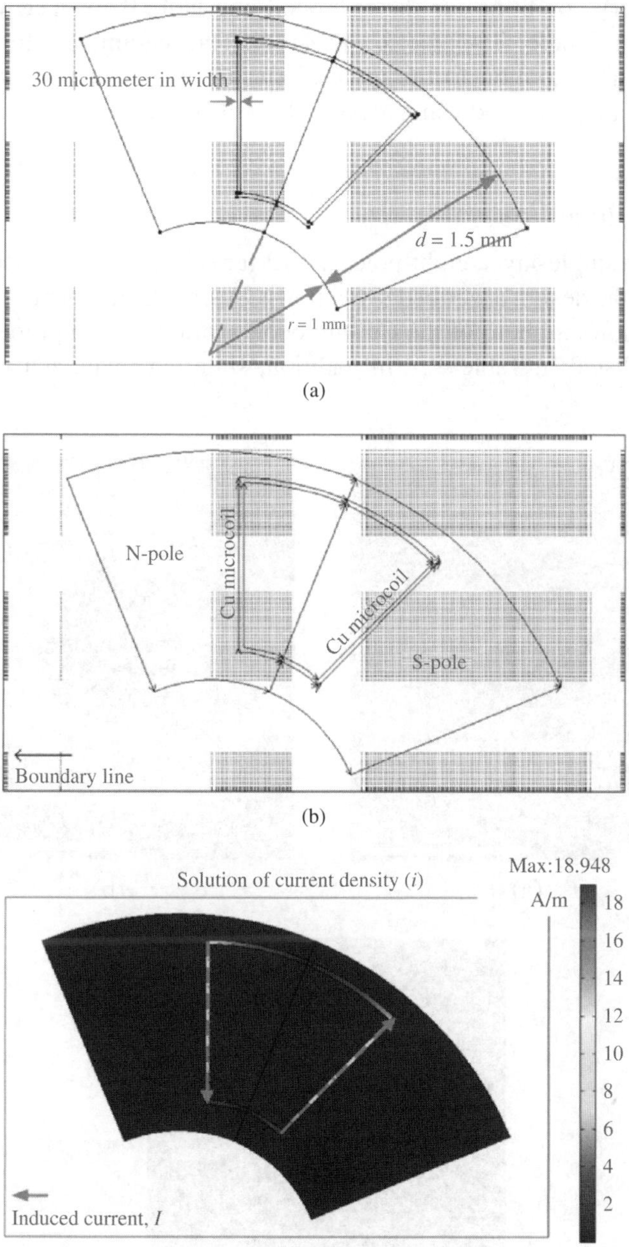

Figure 4.12 Simulation models consist of (a) the dimension of one turn of microcoil with 30 μm line width and spacing; (b) the boundary condition of N- and S-pole; and (c) the induced current distribution [46]

defined boundary condition for the microcoil and NdFeB magnet, which includes a N-pole and a S-pole. Tables 4.3 and 4.4 list the parameters for the boundary condition, and the solution to the induced current is shown in Figure 4.12c. The simulated data are processed using MATLAB® software.

4.2.4 Fabrication Process

The pattern of a single-layer, eight-pole, and 12-turn microcoil was successfully fabricated based on the design described in Section 4.2 using the filament-winding method. Figure 4.13a shows a photo of the Cu microcoil prototype. The properties of the Cu microcoil are listed in Table 4.3. In addition, sintering manufacturing technology

Figure 4.13 The microgenerator consists of (a) the planar microcoil; (b) the magnet; and (c) eight-polar magnetic image obtained using magnetic fluid [46]

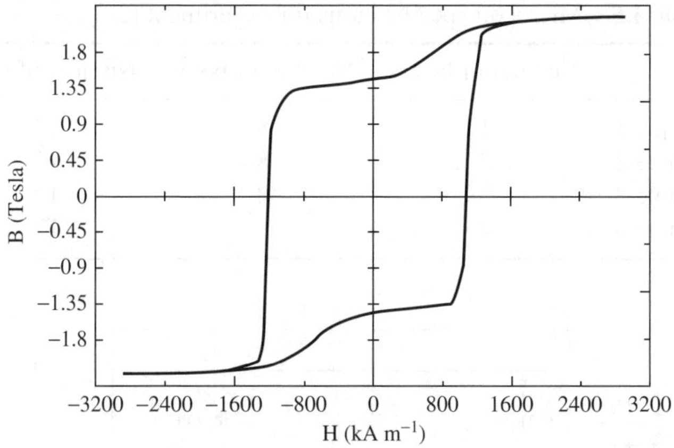

Figure 4.14 Hysteresis loop of sintered NdFeB magnet [46]

was applied to fabricate the eight-pole NdFeB hard magnet. The process of sintered technology consists mainly of an anisotropic magnetic field orientation, sintering, and product magnetizing. The magnetized products are shown in Figure 4.13b. The eight-pole magnetic image is displayed clearly with the help of magnetic fluid, as shown in Figure 4.13c. Table 4.4 lists the measured properties of the sintered NdFeB hard magnet after VSM. The sintered NdFeB exhibits a favorably hard magnet with a B_r of 1.44 T and an H_c of 975 kA m^{-1}, as shown in Figure 4.14.

4.2.5 Results and Discussion (1)

The planar microcoils with line width and spacing of 30 μm were used as a stator to generate power in this experiment. Table 4.5 lists the specifications of the different geometric designs of microcoils for microgenerators. The assembly of the microgenerator is schematically illustrated in Figure 4.15. First, the magnet is fixed onto a spindle and the microcoil is attached to a platform. The gap between the microcoil and magnet is set at 1 mm. Second, the spindle is driven to rotate the multipolar magnet relative to the microcoil, thereby inducing an electrical current. A sinusoidal signal of voltage is induced in the planar microcoil and can be observed using an oscilloscope, as shown in Figure 4.10b.

To explain the effect of the number of microcoil layers on power generation, four samples with different numbers of layers were tested, as listed in Table 4.5. Figure 4.16 shows the induced voltage as a function of time, measured at the same angular velocity. Sample 4 with four layers has the largest voltage output; as shown in Figure 4.16, the more layers the microgenerator microcoil has, the larger the induced voltage.

The relationship between induced voltage and angular velocity has been investigated and extensive experiments investigating the effects on power generation have been performed. The experimental voltage generation results for different angular

Table 4.5 Microcoil specifications for experiment [28]

	Number of layers	Number of poles	Number of turns
Sample 1	1	8	12
Sample 2	2	8	12
Sample 3	3	8	12
Sample 4	4	8	12

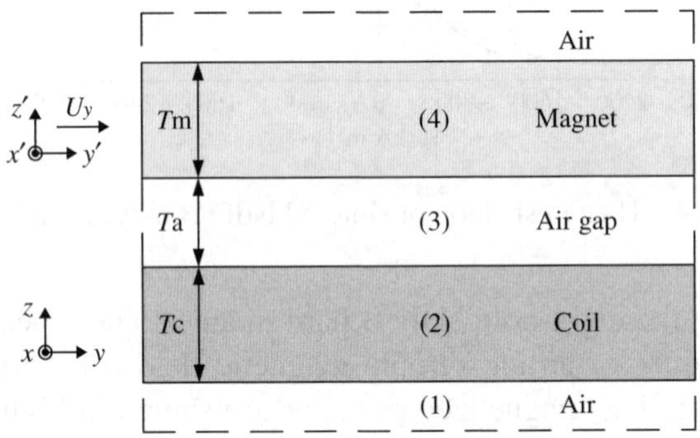

Figure 4.15 The geometric diagram of the generator

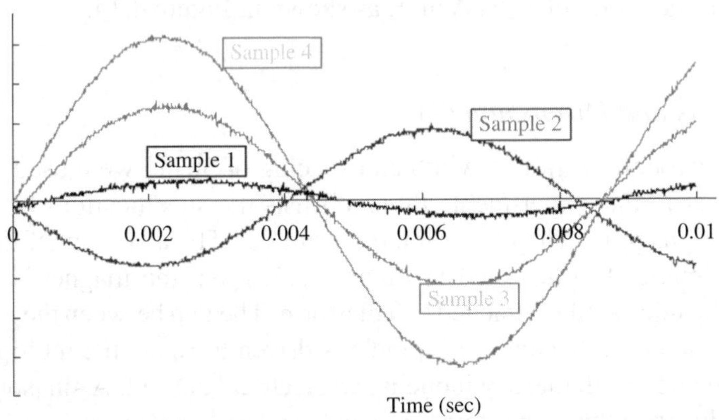

Figure 4.16 Plot of induced voltage for microcoils with different numbers of layers, measured at the same frequency [28]

velocities are shown in Figure 4.17, indicating that the induced voltage is proportional to the angular velocity (i.e. induced voltage increases with angular velocity). Figure 4.17 also explains the effect of the number of microcoil layers on the induced

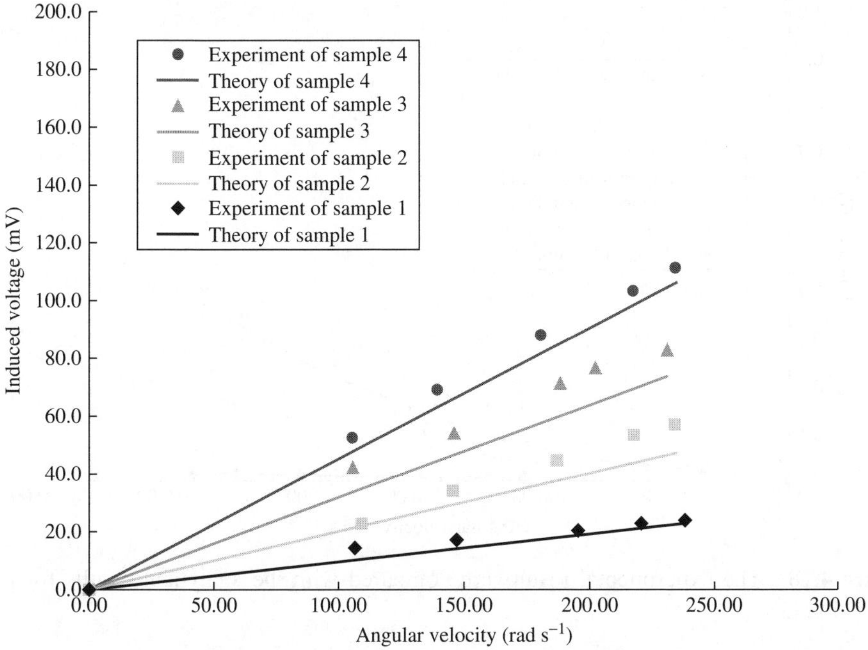

Figure 4.17 The experimental results are compared with the analytical results for induced voltage generation [28]

voltage, revealing that the more layers on the microcoil, the greater the induced voltage. The parameters listed in Tables 4.1 and 4.2 were input into Equation (4.7), enabling comparison of the analytical evaluation with the experimental results. The predicted and experimental results were found to be in good agreement. Power generation is expressed as

$$\text{Power} = \frac{V^2}{R} \tag{4.9}$$

As expected, the power is proportional to the square of the induced voltage. To verify Equation (4.9), the experimental power generation results were compared with the analytical results as shown in Figure 4.18, revealing favorable agreement. The circuit schematically shown in Figure 4.19 was used to induce current. The load resistance R is in parallel with the conductor internal resistance (Cu coil), and I is the induced current. When I passed through R, the voltage V of R was measured by the output terminals of the device. The volume of this eight-pole magnet, four-layer Cu planar microcoil generator was approximately $5 \times 5 \times 2\,\text{mm}^3$. A maximum power output of 0.412 mW at 234.5 rad s^{-1} or 149.3 Hz, calculated using Equation (4.8), was achieved.

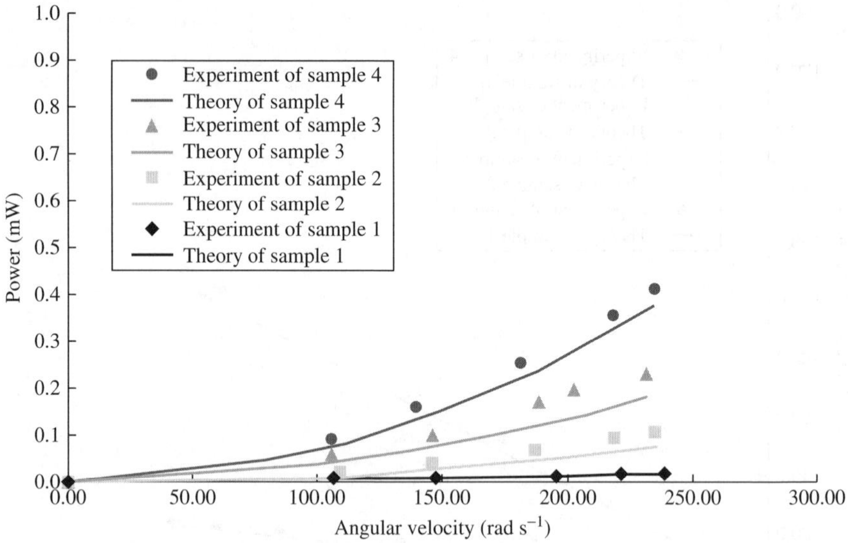

Figure 4.18 The experimental results are compared with the analytical results for power generation [28]

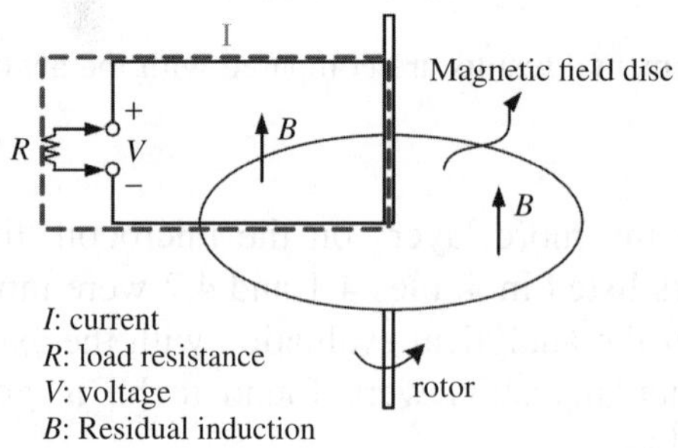

Figure 4.19 The schematically shown circuit used to induce current [28]

4.2.6 Results and Discussion (2)

In this study, five types of microcoils were designed and fabricated for a power generation experiment. For example, Sample IV includes four layers of microcoils. Each layer has eight-pole microcoils, and each pole consists of 12 turns of Cu coils. The specifications of the microcoil and magnet for simulations and experiments are listed in Table 4.6. The experimental set-up for power generation is illustrated in Figure 4.20. First, the magnet is fixed on a rotary mechanism and the microcoil is attached to a static platform. The gap between the microcoils and the magnet is set at 1 mm. Second, the rotary mechanism is driven by a power supply to cause the multipolar magnet to rotate at different angular velocities or frequencies. When the eight-pole magnet is rotated relative to the Cu microcoil, the planar microcoil induces voltage and current.

Table 4.6 Planar Cu microcoil specification for experiment [46]

	Specification		
	Layer	Pole	Turn
Sample I	1	8	12
Sample II	2	8	12
Sample III	3	8	12
Sample IV	4	8	12

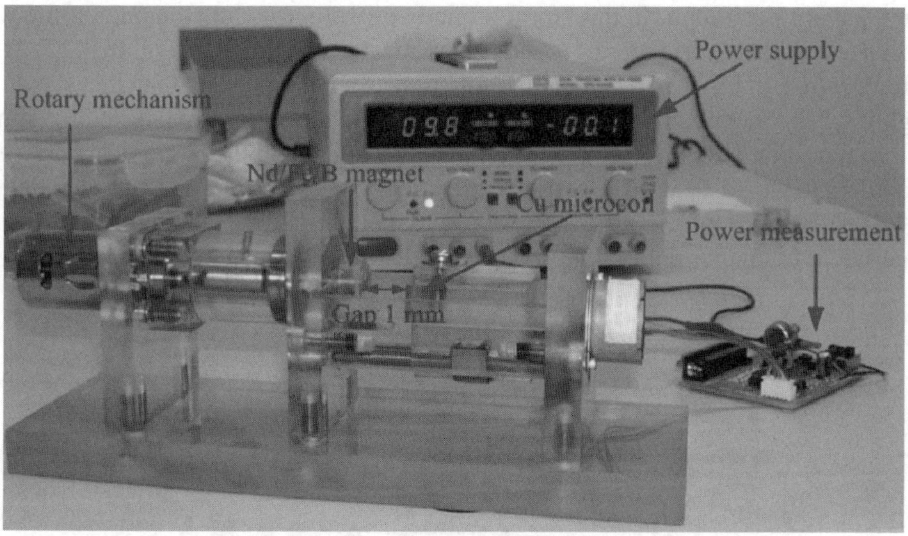

(a)

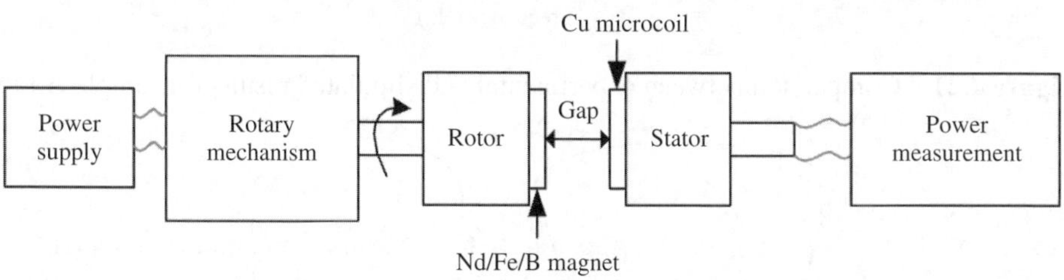

(b)

Figure 4.20 (a) Photograph of measurement set-up and (b) schematic illustration of measurement set-up [46]

In this study, the experimental data of Sample with 1 layer, 1 pole, and 12 turns were acquired and compared with the simulated results, as shown in Figure 4.21. As expected, the faster the angular velocity of the magnet, the greater the induced output voltage. The figure shows that the induced voltage is proportional to the angular velocity, but the deviation between the experimental and the simulated results is shown in Figure 4.21. This error may be a result of the following two points. First, the pattern of the fabricated microcoil is slightly different from that of the simulation model. The pattern of the one-pole microcoil for the simulation model is shown in Figure 4.22a, and the pattern for the fabricated microcoil is shown in Figure 4.22b. Based on Ampere's right-hand rule, when the moving tangent direction of a magnet is perpendicular to that of the microcoil, maximum electric current is induced. Figure 4.22a shows an ideal microcoil design; however, fabricating such an angled and acute pattern is difficult. In this study, a microcoil with a smooth, curved profile was fabricated as shown in Figure 4.6 and schematically illustrated in Figure 4.22b. The pattern shown in Figure 4.22a can therefore induce a larger current than that shown in Figure 4.22b. Second, the gap between the rotor and the stator is neglected for simulation. Based on field and wave electromagnetics, magnetic flux significantly decays according to gap distance. The residual induction of the NdFeB magnet for simulation is 1.44 T. Because of the gap, the residual induction of the NdFeB magnet for the experiment is smaller than 1.44 T; this also reduces the induction efficiency.

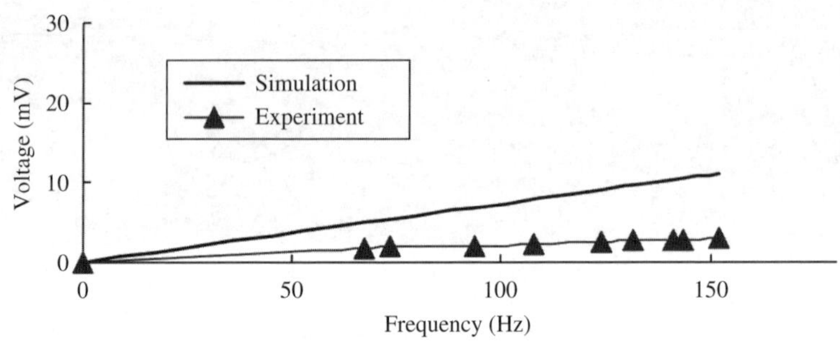

Figure 4.21 Comparison between experimental and simulated results for sample A [46]

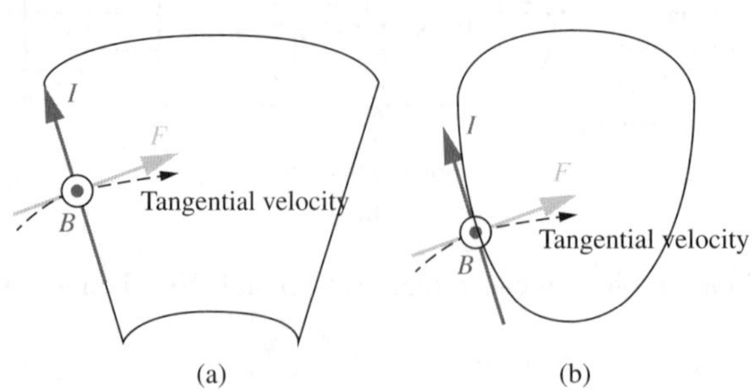

Figure 4.22 The influence of pattern of microcoil (a) for simulation and (b) for experiment on induced voltage of microgenerator [46]. (*B* is magnetic flux, *I* is induced current, and *F* is tangent direction of angular velocity.)

In addition, intensive experiments of multipolar and multilayer Cu microcoils were conducted. The fabricated multilayer Cu microcoil is shown in Figure 4.23. The induced voltages for Samples I–IV were compared, as shown in Figure 4.24, indicating that when the rotary velocity of the magnet increases, the induced voltage can be enhanced for each sample. Furthermore, the induced voltages at the same frequency of 150 Hz for different samples are shown in Figure 4.25. Sinusoidal waveforms regarding frequency can be observed. Figure 4.25 also shows that the greater the number of layers of a microcoil, the larger the output voltage. The figure shows that the increased rate of output voltage is proportional to the number of layers. For example, the output voltage of Sample IV is nearly four times larger that of Sample I. The power output of Sample IV is nearly 16 times larger than that of Sample I. Regarding Sample IV, the experimental result shows that the maximum induced voltage of 111.2 mV and maximum power output of 386.42 μW can be achieved.

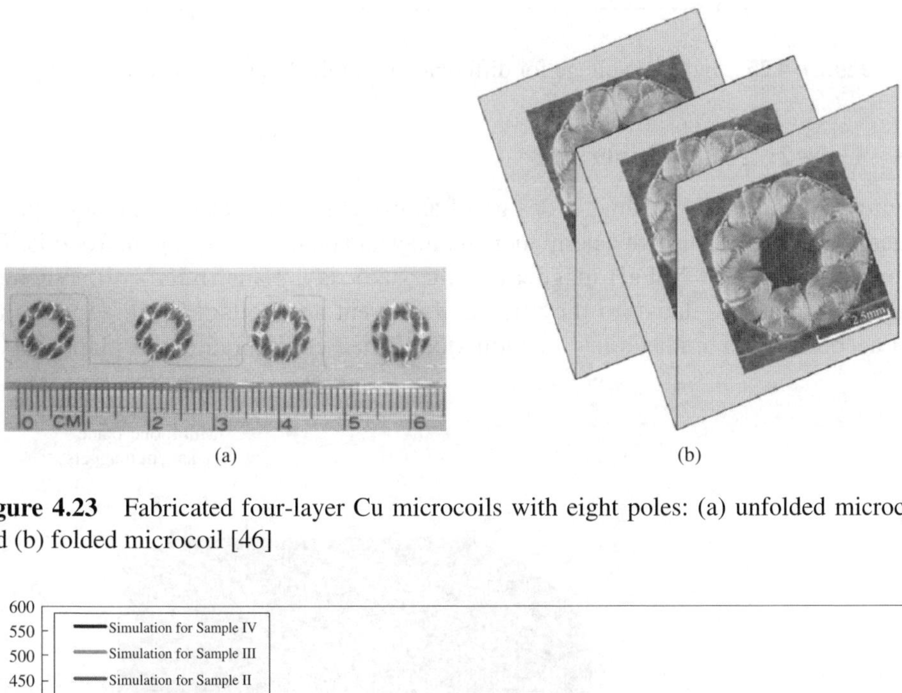

Figure 4.23 Fabricated four-layer Cu microcoils with eight poles: (a) unfolded microcoil and (b) folded microcoil [46]

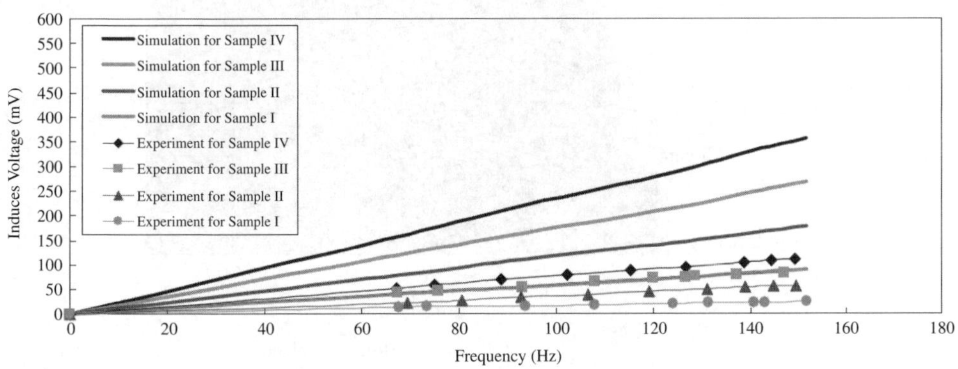

Figure 4.24 Comparison between experimental and simulated results for different samples [46]

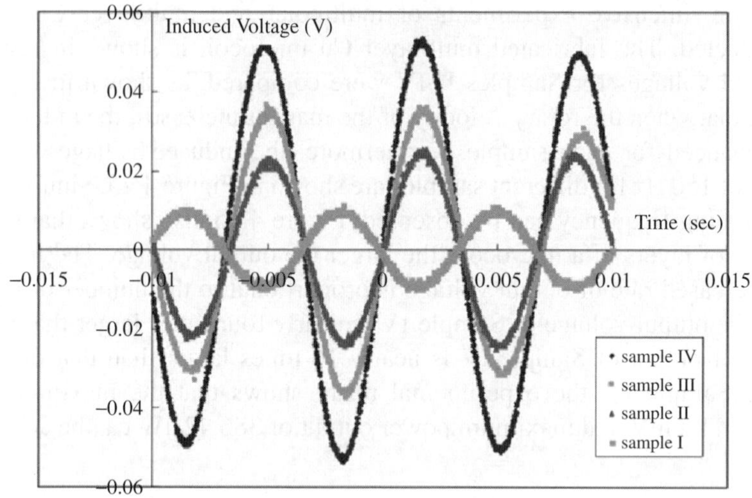

Figure 4.25 Induced voltage for different coils at the frequency of 150 Hz [46]

4.3 Case 2: LTCC Generator

Figure 4.26 shows the schematic design of an in-plane rotary electromagnetic power generator, which consists of a permanent magnet and planar LTCC Ag microcoils. The rotor is a permanent NdFeB magnet and the stator is a planar microcoil. When the magnet is rotating relative to the microcoil, current and voltage are induced in the coil according to Faraday's law of induction. Three configurations of planar LTCC

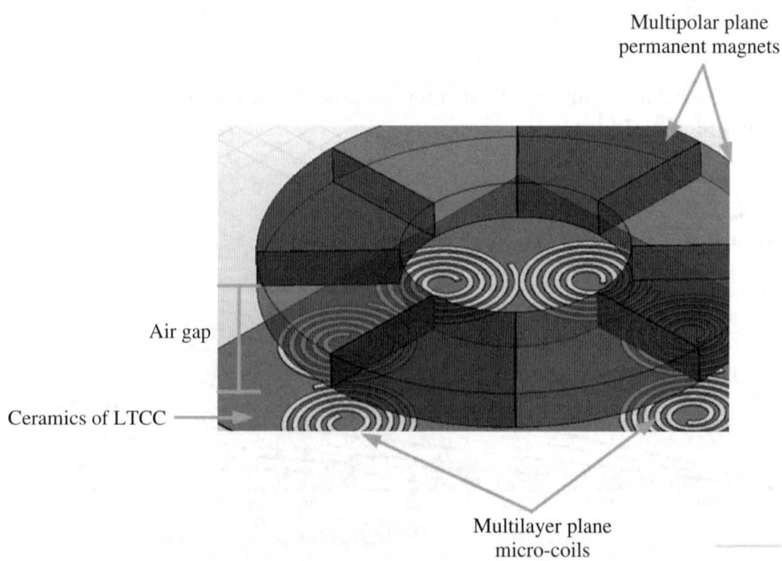

Figure 4.26 Schematic electromagnetic microgenerator

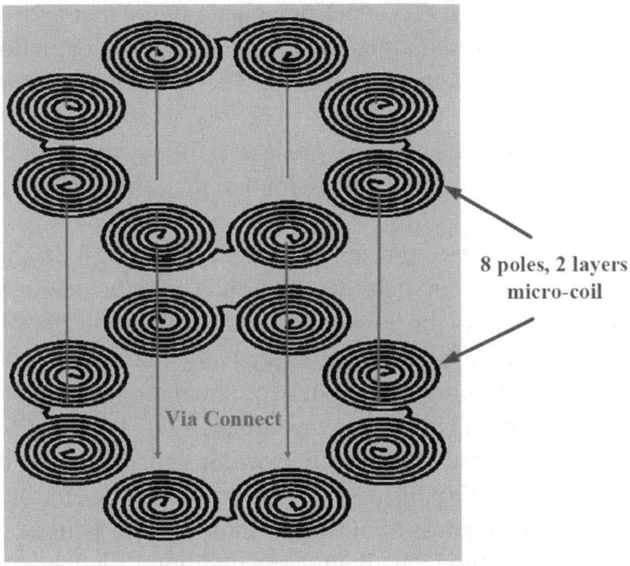

Figure 4.27 Schematic illustration of the winding microcoils

Ag microcoils were investigated in this study. Both the printing line width and interspace of these microcoils are 100 μm. A greater number of coils can be packed into the substrate when they are designed and sintered more tightly. Figure 4.27 shows the schematic design of winding microcoils. When designing the coils, the direction between each pole should be considered. Because the magnetite changes poles from N to S, the current direction of the N-pole coils is the reverse of that of the S-pole coils. The wires are printed on the substrate by performing LTCC fabrication, and are not intersected. The current is conducted to the next layer and to the next pole by the vias. The multipolar hard NdFeB magnet with an outer diameter of 9 mm and a thickness of 700 μm is molded and sintered. A prototype of the microgenerator has a volume of less than 1 cm^3.

4.3.1 Simulation

The microcoils are driven to rotate in a uniform magnetic field with an angular velocity by an external force in the rotary-type generator system. When the microcoils are rotating, the induced voltage is output in sine waves. According to Faraday's law, the induced EMF in any closed circuit is equal to the time rate of change of the magnetic flux through the circuit. The magnet (rotor) rotating relative to the microcoil (stator) can induce a change in the magnetic flux in the microcoil. When the electric charge is accumulated, the EMF can be produced. Two- and three-dimensional FEA simulations are used to determine the voltage and power that can be generated from different designs.

The induced current of the rotary electromagnetic microgenerator is simulated by conducting finite-element simulations. The finite-element simulation environment facilitates all steps in the modeling process. There are many characteristics and advantages of finite-element simulations. The major advantage, for example, is that such simulations can be used to process multiphysics modeling, including electromagnetic problems. Finite-element simulations can be divided into four parts as follows.

First, sketch the outer shape of the simulated object. Figure 4.28 shows three types of microcoil shape and magnetic field distribution. The right side of the figure is the N pole, and the left side shows the S pole. Positive and negative signs in the middle of the graph indicate the N pole and S pole, respectively. Second, establish the material properties and actual environmental physics for simulation. The material properties of the microcoil and the hard NdFeB magnet are then set. Table 4.7 lists the material properties for this simulation. Both the line width and interspace of these microcoils are 100 μm. The hard NdFeB magnet with an outer diameter of 9 mm, an inner diameter of 4 mm, and a thickness of 700 μm is thus defined. Third, define the boundary condition of the simulation. The boundary condition is defined for the microcoil and the hard magnet Nd/Fe/B. Finally, the induced current is resolved as shown in Figure 4.28. The induced voltages of differently shaped coils (square-shaped, circular and sector-shaped) are simulated separately.

4.3.2 Analytical Theorem of Microgenerator Electromagnetism

This study used the electromagnetic, kinematic, and geometric parameters of the electromagnetic microgenerator to analyze power. This study also determined an analytical model and solutions, and validated the proposed analytical solutions according to the experimental results. Figure 4.29 shows a block diagram of the analytical process in this study. The operating principle of a rotary electromagnetic generator involves a permanent magnet moving relative to a microcoil, thus inducing an EMF. Although the structure of this type of generator is similar to previous designs described in relevant literature, the analytical theorem is based on electromagnetism [71]. This study used the theorem of electromagnetics, Ampere's law, Faraday's law, and the Lorenz gauge condition to derive the formula for the electric field and magnetic field, which is expressed as follows:

$$\nabla^2 \widetilde{A} + \nabla \times \mu \widetilde{M} = \mu\sigma \left(\frac{\partial \widetilde{A}}{\partial t} - \widetilde{v} \times \nabla \times \widetilde{A} \right) \tag{4.10}$$

where $\widetilde{A}$ is the vector magnetic potential, $\widetilde{M}$ is the magnetization, μ is the permeability, σ is the conductivity, t is time, and $\widetilde{v}$ is velocity. It is compressed in the direction of the rotary axis (z-axis) to minimize the size. Because of its simple rotary axial

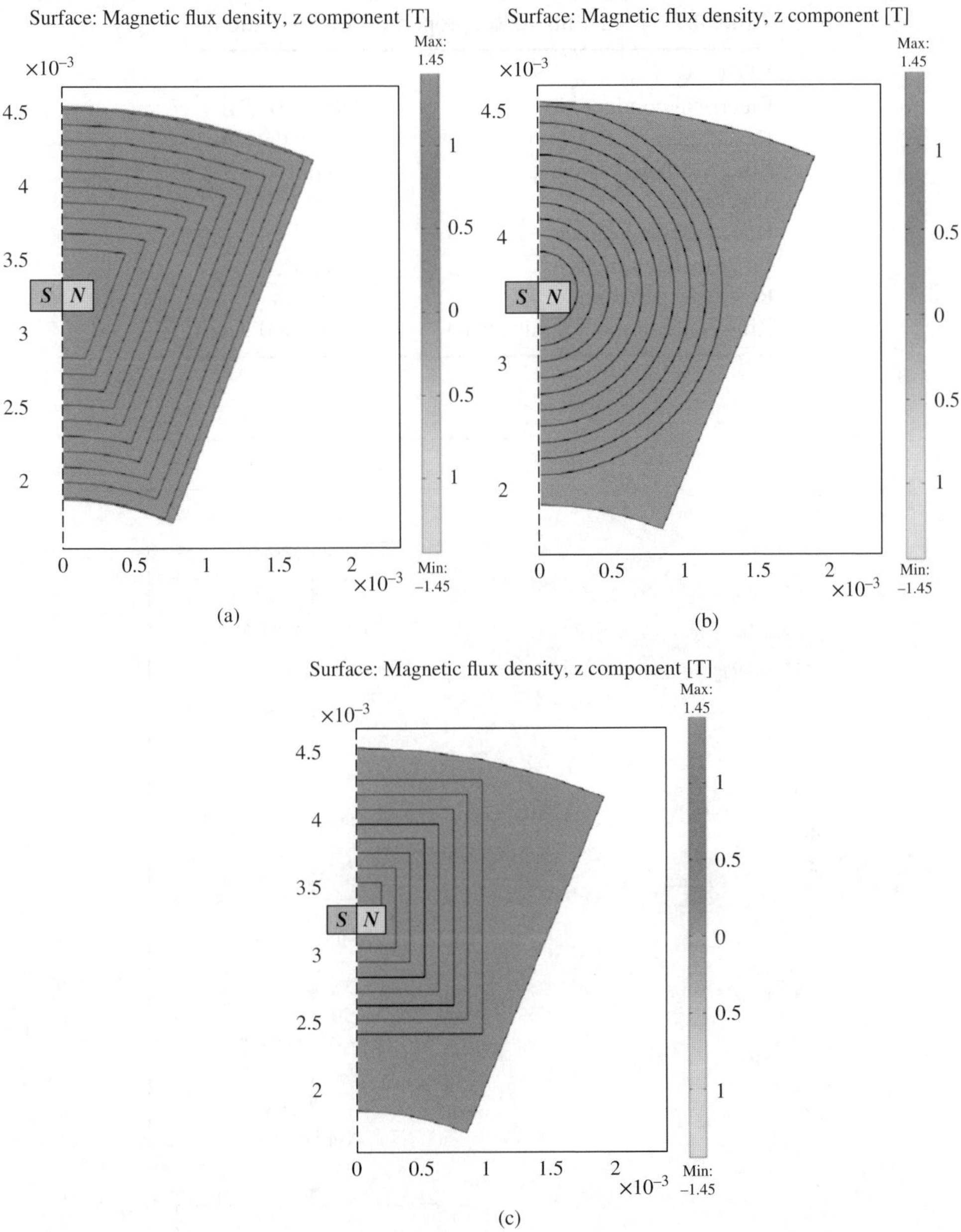

Figure 4.28 (a) Sector-shaped, (b) circle-shaped, and (c) square-shaped microcoil shape and distribution of magnetic field

Table 4.7 Planar microcoil properties for simulation

LTCC Ag	
Electrical conductivity	2.94×10^4 S m^{-1}
Resistivity	3.4×10^{-5} Ωm
Line width and spacing	100 μm
Thickness	30 μm
Hard NdFeB magnet	
Electrical conductivity	97.1 S m^{-1}
Residual induction	1.4 T
Ring (outer × inner × thickness)	$4.5 \times 2 \times 0.7$ mm^3

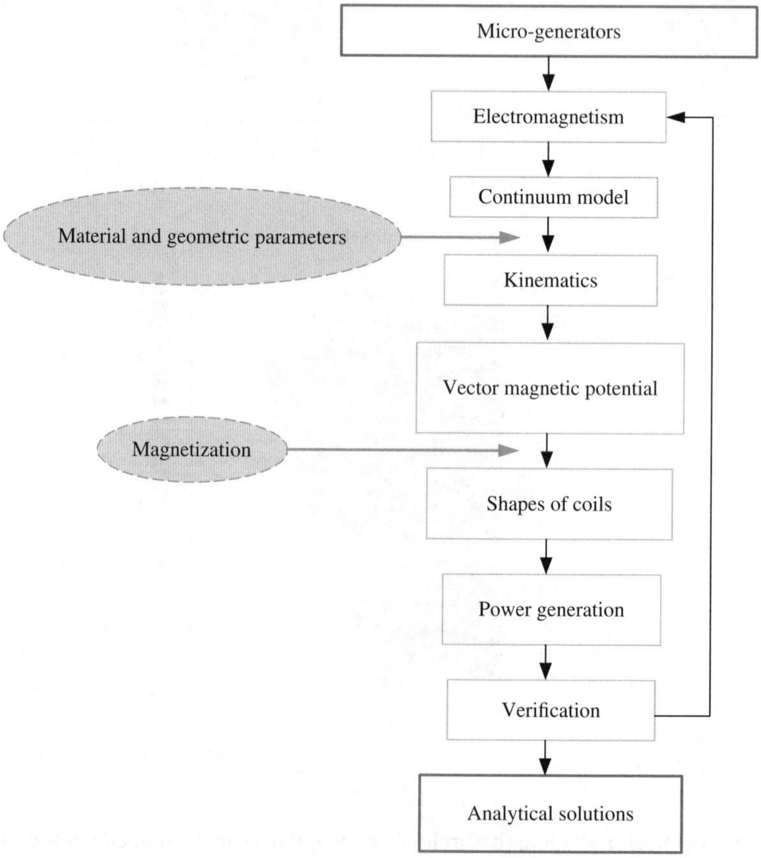

Figure 4.29 The block diagram of the analytical process

structures, this microgenerator structure is deemed as a set of planar continuum layers (Figure 4.30). Cartesian geometry was therefore used in this study to simplify the analysis. The coordinate system of the rotor (the magnets) is defined as (x', y', z') and the stator (the coils) as (x, y, z). The relationship between these is

$$y' = y - U_y t \tag{4.11}$$

where U_y is the velocity of the rotary layer in the y-direction. Equation (4.10) can be simplified as

$$\nabla^2 A_x + \mu \frac{\partial M_z}{\partial y} = \mu\sigma \left\{ \frac{\partial A_x}{\partial t} + U_y \frac{\partial A_x}{\partial y} \right\} \tag{4.12}$$

The magnetization vector moves with velocity U_y and has the vector magnetic potential of the x-direction. Magnetization can be derived by performing Fourier analysis on the (x', y', z') system, which can be shown to be

$$M_z(y') = \mathrm{Re}\left(\sum_{n=-\infty}^{\infty} \widetilde{M}_{zn} e^{-jnky'} \right) \tag{4.13}$$

where Re is the real part of the Fourier transform and k is the wave number. Substitute Equation (4.11) into Equation (4.13) to convert to an (x, y, z) system. The velocity U_y is transformed to $U_y = W_m R$ where W_m is the angular velocity of the rotor and R is radius. $\omega_m = PW_m$ is then defined and substituted into Equation (4.13), where P is the number of poles. Equation (4.13) can be rewritten as

$$M_z(y, t) = \mathrm{Re}\left(\sum_{n=-\infty}^{\infty} \widetilde{M}_{zn} e^{jn(\omega_m t - ky)} \right) \tag{4.14}$$

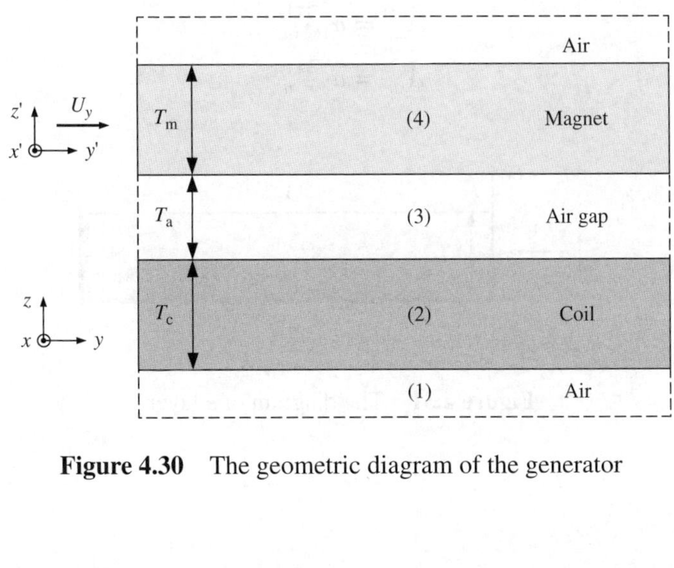

Figure 4.30 The geometric diagram of the generator

The partial derivative of the vector magnetic potential in the z-direction is the magnetization. The vector magnetic potential has the same form as the magnetization:

$$A_x(y,z,t) = \mathrm{Re}\left(\sum_{n=-\infty}^{\infty} \widetilde{A}_{xn}(z)\, e^{jn(\omega_m t - ky)}\right) \tag{4.15}$$

4.3.3 Simplification

Equation (4.12) can be simplified and rewritten as

$$\nabla^2 A_x = \mu\sigma \left\{ \frac{\partial A_x}{\partial t} + U_y \frac{\partial A_x}{\partial y} \right\} \tag{4.16}$$

The ordinary differential equation can then be used to obtain the general solution of the vector magnetic potential:

$$\widetilde{A}_{xn}(z) = A_{xn}^{(b)} \frac{\sinh(\gamma z)}{\sinh(\gamma T)} + A_{xn}^{(a)} \frac{\sinh[\gamma(T-z)]}{\sinh(\gamma T)} \tag{4.17}$$

where T is a layer thickness, a and b are the boundary conditions as shown in Figure 4.31, and $\gamma = \sqrt{n^2k^2 + jn\mu\sigma(\omega_m - kv)}$.

The magnetic field intensity $\widetilde{H}_{yn}$ can be obtained according to the relationship between magnetic field intensity and the vector magnetic potential:

$$\widetilde{H}_{yn} = \frac{1}{\mu}\left(\frac{d\widetilde{A}_{xn}(z)}{dz}\right) \tag{4.18}$$

The relationships between the amplitudes of the vector potential $\widetilde{A}_{xn}^{(1)}$ and the magnetic field intensity $\widetilde{H}_{yn}$ can be obtained using the following calculations:

$$\widetilde{A}_{xn}^{(2)} = \alpha_1 \widetilde{A}_{xn}^{(1)} \tag{4.19}$$

$$\widetilde{A}_{xn}^{(3)} = \alpha_3 \widetilde{A}_{xn}^{(1)} \tag{4.20}$$

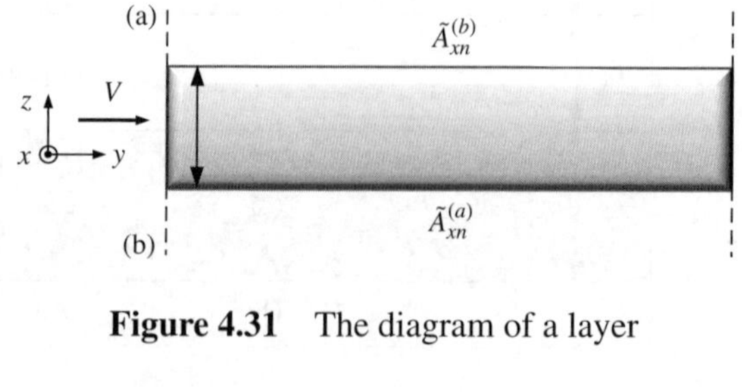

Figure 4.31 The diagram of a layer

$$\widetilde{H}_{yn}^{(1)} = \frac{\gamma}{\mu_0}\widetilde{A}_{xn}^{(1)} \tag{4.21}$$

$$\widetilde{H}_{yn}^{(2)} = \alpha_2\widetilde{A}_{xn}^{(1)} \tag{4.22}$$

$$\widetilde{H}_{yn}^{(3)} = \alpha_4\widetilde{A}_{xn}^{(1)} \tag{4.23}$$

$$\widetilde{H}_{yn}^{(4)} = -\frac{\gamma}{\mu_0}\widetilde{A}_{xn}^{(4)} \tag{4.24}$$

where Equations (4.10)–(4.13) are the boundary conditions in Figure 4.30, and α_1–α_4 are the transfer constants for different layers, defined:

$$\alpha_1 = \frac{\mu_{\mathrm{c}}}{\gamma_{\mathrm{c}}}\frac{\gamma_0}{\mu_0}\sinh(\gamma_{\mathrm{c}}T_{\mathrm{c}}) + \cosh(\gamma_{\mathrm{c}}T_{\mathrm{c}}) \tag{4.25}$$

$$\alpha_2 = \frac{\gamma_{\mathrm{c}}}{\mu_{\mathrm{c}}}\sinh(\gamma_{\mathrm{c}}T_{\mathrm{c}}) + \frac{\gamma_0}{\mu_0}\cosh(\gamma_{\mathrm{c}}T_{\mathrm{c}}) \tag{4.26}$$

$$\alpha_3 = \frac{\gamma_{\mathrm{c}}}{\mu_{\mathrm{c}}}\frac{\mu_0}{\gamma_0}\left[\sinh\left(\gamma_0T_{\mathrm{a}}\right)\sinh(\gamma_{\mathrm{c}}T_{\mathrm{c}})\right] + \frac{\mu_{\mathrm{c}}}{\gamma_{\mathrm{c}}}\frac{\gamma_0}{\mu_0}\left[\sinh\left(\gamma_{\mathrm{c}}T_{\mathrm{c}}\right)\cosh(\gamma_0T_{\mathrm{a}})\right]$$
$$+\sinh(\gamma_0T_{\mathrm{a}})\cosh(\gamma_{\mathrm{c}}T_{\mathrm{c}}) + \cosh(\gamma_0T_{\mathrm{a}})\cosh(\gamma_{\mathrm{c}}T_{\mathrm{c}}) \tag{4.27}$$

$$\alpha_4 = \frac{\mu_{\mathrm{c}}}{\gamma_{\mathrm{c}}}\left(\frac{\gamma_0}{\mu_0}\right)^2\left[\sinh\left(\gamma_0T_{\mathrm{a}}\right)\sinh(\gamma_{\mathrm{c}}T_{\mathrm{c}})\right] + \frac{\gamma_0}{\mu_0}\left[\sinh\left(\gamma_0T_{\mathrm{a}}\right)\cosh(\gamma_{\mathrm{c}}T_{\mathrm{c}})\right]$$
$$+\frac{\gamma_{\mathrm{c}}}{\mu_{\mathrm{c}}}\left[\sinh\left(\gamma_{\mathrm{c}}T_{\mathrm{c}}\right)\cosh(\gamma_0T_{\mathrm{a}})\right] + \frac{\gamma_0}{\mu_0}\left[\cosh\left(\gamma_{\mathrm{c}}T_{\mathrm{c}}\right)\cosh(\gamma_0T_{\mathrm{a}})\right] \tag{4.28}$$

where the subscript c represents the parameters of the coil, and a and 0 the parameters of the air gap.

4.3.4 Analysis of Vector Magnetic Potential

Equation (4.12) can be rewritten as

$$\nabla^2A_x + \mu_{\mathrm{m}}\frac{\partial M_z}{\partial y} = \mu_{\mathrm{m}}\sigma_{\mathrm{m}}\left\{\frac{\partial A_x}{\partial t} + U_y\frac{\partial A_x}{\partial y}\right\} \tag{4.29}$$

where the subscript m represents the magnetic parameters.

The partial differential equation can then be used to obtain the general solution of the vector magnetic potential of the magnet

$$\widetilde{A}_{xn}(z) = A_{xn}^{(b)}\frac{\sinh(\gamma z)}{\sinh(\gamma T_{\mathrm{m}})} + A_{xn}^{(a)}\frac{\sinh\left[\gamma\left(T_{\mathrm{m}}-z\right)\right]}{\sinh(\gamma T_{\mathrm{m}})} - j\widetilde{A}_{xn} \tag{4.30}$$

After substituting the boundary conditions of the magnet into the equation and rearranging, the vector magnetic potential of the magnet is

$$\widetilde{A}_{xn}^{(1)} = \mathrm{j}\left\{\frac{1-\cosh(\gamma T_{\mathrm{m}}) - \left(\frac{\mu_0}{\mu_{\mathrm{m}}}\right)\sinh(\gamma T_{\mathrm{m}})}{\left[\cosh(\gamma T_m) + \left(\frac{\mu_0}{\mu_{\mathrm{m}}}\right)\sinh(\gamma T_{\mathrm{m}})\right]\alpha_3 + \left[\frac{\cos h^2(\gamma T_{\mathrm{m}})}{\sinh(\gamma T_{\mathrm{m}})} + \left(\frac{\mu_0}{\mu_{\mathrm{m}}}\right)\cosh(\gamma T_{\mathrm{m}}) - \frac{1}{\sinh(\gamma T_{\mathrm{m}})}\right]\alpha_4}\right\} \tag{4.31}$$

The characteristic of the hyperbola can be used to simplify $\widetilde{A}_{xn}^{(1)}$ as

$$\widetilde{A}_{xn}^{(1)} = j\left(\frac{-T_{\mathrm{m}}}{T_{\mathrm{a}} + T_{\mathrm{c}}}\right)\widetilde{A}_{xn} \tag{4.32}$$

The vector magnetic potential of the coil can be obtained by substituting the parameters of the remanent flux density B_{r}, the inner radius R_i, the outer radius R_{o}, and the number of poles, to obtain:

$$\widetilde{A}_{xn(\mathrm{coil})} = j\left(\frac{-T_{\mathrm{m}}}{T_{\mathrm{a}} + T_{\mathrm{c}}}\right)\left[\frac{2B_{\mathrm{r}}(R_{\mathrm{o}} + R_{\mathrm{i}})}{Pn^2\pi}\right]e^{j\frac{n\pi}{2}} \tag{4.33}$$

4.3.5 Analytical Solutions for Power Generation

In this study we designed an in-plane rotary electromagnetic microgenerator according to procedures described in Section 4.2. The proposed design was fabricated using LTCC technology to co-fire the Ag microcoils on the ceramic substrate with various microcoil shapes (e.g. square, circular, and sector-shaped; Figure 4.32). Investigating the relationship between the different microcoil shapes and power generation allowed us to determine the best microcoil geometry.

According to the basic concept of electromagnetics, the magnetic flux $\widetilde{\Phi}$ through a closed surface S surrounded by the coil can be defined as

$$\widetilde{\Phi} = \int_s \overrightarrow{B}\, dS \tag{4.34}$$

where $\overrightarrow{B}$ is the magnetic field.

According to Stokes' theorem, Equation (4.34) can be rewritten:

$$\widetilde{\Phi} = \oint_S \left(\nabla \times \widetilde{A}_{xn(\mathrm{coil})}\right) dS = \oint_C \widetilde{A}_{xn(\mathrm{coil})} dl \tag{4.35}$$

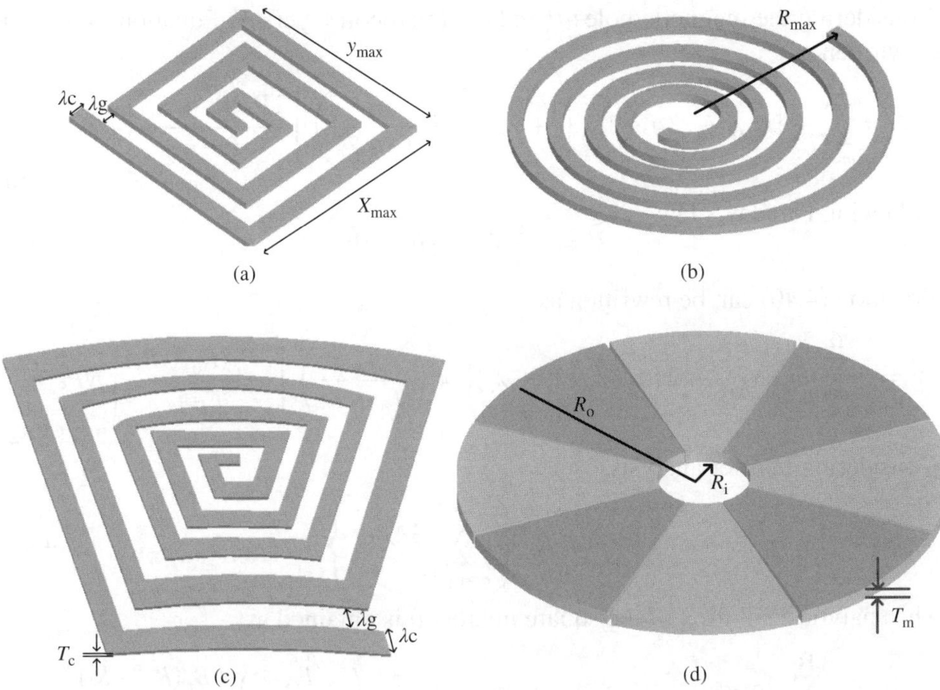

Figure 4.32 The diagram of (a) square-shaped, (b) circle-shaped, and (c) sector-shaped microcoils, and (d) magnet

4.3.5.1 Square Microcoil

The magnetic flux $\widetilde{\Phi}$ of the square microcoil can be written as

$$\widetilde{\Phi} = \int_0^{l_x} \int_0^{l_y} \widetilde{A}_{xn(\text{coil})} dxdy = -l_x l_y \left(\frac{T_m}{T_a + T_c} \right) \left[\frac{2B_r B (R_o + R_i)}{Pn^2 \pi} \right] e^{j\frac{n\pi}{2}} \tag{4.36}$$

where

$$l_x = \left[x_{max} - 2(q-1)(\lambda_g + \lambda_c) \right] \tag{4.37}$$

and

$$l_y = \left[y_{max} - 2(q-1)(\lambda_g + \lambda_c) \right] \tag{4.38}$$

and where x_{max} and y_{max} are the maximum length of one square microcoil pole in the x- and y-directions, respectively. l_x and l_y are the total lengths of one square microcoil pole in the x- and y-directions, respectively. The gap between neighboring coils is λ_g (spacing), the coil width is λ_c (line width), and q is the number of microcoil turns. Because the shape of the microcoil is square,

$$x_{max} = y_{max} \quad \text{and} \quad l_x = l_y \tag{4.39}$$

Considering the magnetic poles P and the microcoil layers N, Equation (4.36) can be rewritten as

$$\widetilde{\Phi} = \sum_{m=1}^{\Omega} - \left[x_{\max} - 2(q-1)(\lambda_g + \lambda_c)\right]^2 \pi \left(\frac{T_{\mathrm{m}}}{T_{\mathrm{a}} + T_{\mathrm{c}}}\right) \left[\frac{2B_{\mathrm{r}}(R_{\mathrm{o}} + R_{\mathrm{i}})}{Pn^2\pi}\right] NPe^{j\frac{n\pi}{2}} \tag{4.40}$$

Following Faraday's law,

$$\widetilde{V} = -\frac{d}{dt}\widetilde{\Phi} = -jn\omega_{\mathrm{m}}\widetilde{\Phi} \tag{4.41}$$

Equation (4.40) can be rewritten as

$$\widetilde{V} = -\sum_{m=1}^{\Omega} j\omega_{\mathrm{m}}\left[x_{\max} - 2(q-1)(\lambda_g + \lambda_c)\right]^2 \pi \left(\frac{T_{\mathrm{m}}}{T_a + T_{\mathrm{c}}}\right) \left[\frac{2B_{\mathrm{r}}(R_{\mathrm{o}} + R_{\mathrm{i}})}{Pn\pi}\right] NPe^{j\frac{n\pi}{2}} \tag{4.42}$$

Considering Fourier analysis,

$$V(t) = R_e\left\{\sum_{n=1,\mathrm{odd}}^{\infty} \widetilde{V}e^{j\omega_{\mathrm{m}}t}\right\} \tag{4.43}$$

The analytical solution of the square microcoil is obtained as

$$V(t) = \sum_{m=1}^{\Omega} NPW_{\mathrm{m}}\left[x_{\max} - 2(q-1)(\lambda_g + \lambda_c)\right]^2 \pi \left(\frac{T_{\mathrm{m}}}{T_{\mathrm{a}} + T_{\mathrm{c}}}\right) \left[\frac{B_{\mathrm{r}}(R_{\mathrm{o}} + R_{\mathrm{i}})}{2}\right] \times \left[\sum_{n=1,\mathrm{odd}}^{\infty} \frac{4}{n\pi} \sin(\omega_{\mathrm{m}}t)\right] \tag{4.44}$$

4.3.5.2 Circular Microcoil

The magnetic flux $\widetilde{\Phi}$ of the circular microcoil is written as

$$\widetilde{\Phi} = \int_0^r \int_0^{2\Omega} \widetilde{A}_{xn(\mathrm{coil})} r\, dr\, d\theta = -r^2\pi \left(\frac{T_{\mathrm{m}}}{T_{\mathrm{a}} + T_{\mathrm{c}}}\right) \left[\frac{2B_{\mathrm{r}}(R_{\mathrm{o}} + R_{\mathrm{i}})}{Pn^2\pi}\right] e^{j\frac{n\pi}{2}} \tag{4.45}$$

where r is the total radius of one circular microcoil pole, defined:

$$\mathrm{r} = \left[r_{\max} - (q-1)(\lambda_g + \lambda_c)\right] \tag{4.46}$$

and $r_{\max}$ is the maximum radius of one circular microcoil pole.

Similarly, the analytical solution of the circular microcoil is obtained as

$$V(t) = \sum_{m=1}^{\Omega} NPW_{\mathrm{m}}\left[r_{\max} - (q-1)(\lambda_g + \lambda_c)\right]^2 \pi \left(\frac{T_{\mathrm{m}}}{T_{\mathrm{a}} + T_{\mathrm{c}}}\right) \left[\frac{B_{\mathrm{r}}(R_{\mathrm{o}} + R_{\mathrm{i}})}{2}\right] \times \left[\sum_{n=1,\mathrm{odd}}^{\infty} \frac{4}{n\pi} \sin(\omega_{\mathrm{m}}t)\right] \tag{4.47}$$

4.3.5.3 Sector-Shaped Microcoil

The magnetic flux $\widetilde{\Phi}$ of the sector-shaped microcoil is written as

$$\widetilde{\Phi} = \int_r \int_\theta \widetilde{A}_{xn(\mathrm{coil})} r \, dr \, d\theta = -\left(\frac{r_\mathrm{o}{}^2\pi - r_\mathrm{i}{}^2\pi}{8}\right)\left(\frac{T_\mathrm{m}}{T_\mathrm{a} + T_\mathrm{c}}\right)\left[\frac{2B_\mathrm{r}(R_\mathrm{o} + R_\mathrm{i})}{Pn^2\pi}\right] e^{j\frac{n\pi}{2}} \tag{4.48}$$

where r_o and r_i are the total outer radius and internal radius of one sector-shaped microcoil pole, respectively, defined:

$$r_\mathrm{o} = \left[r_{o(\max)} - (q-1)(\lambda_g + \lambda_c)\right] \tag{4.49}$$

and

$$r_\mathrm{i} = \left[r_{\mathrm{i}(\min)} + (q-1)(\lambda_g + \lambda_c)\right] \tag{4.50}$$

where $r_{o(\max)}$ and $r_{i(\min)}$ are the maximum outer radius and the minimum internal radius of one sector-shaped microcoil pole, respectively.

Similarly, the analytical solution of the sector-shaped shaped microcoil is obtained as:

$$V(t) = \sum_{m=1}^{\Omega} NPW_\mathrm{m}\left(\frac{r_\mathrm{o}{}^2 - r_\mathrm{i}{}^2}{8}\right)\pi\left(\frac{T_\mathrm{m}}{T_\mathrm{a} + T_\mathrm{c}}\right)\left[\frac{B_\mathrm{r}(R_\mathrm{o} + R_\mathrm{i})}{2}\right]\left[\sum_{n=1,\mathrm{odd}}^{\infty} \frac{4}{n\pi} \sin(\omega_\mathrm{m} t)\right] \tag{4.51}$$

4.4 Fabrication

Figure 4.33 shows the isometric views of the microgenerator. The figure illustrates the concept of this microgenerator, which consists of a permanent magnet and planar LTCC Ag microcoils. The rotor is a permanent NdFeB magnet with an outer diameter of 9 mm and an inner diameter of 4 mm, whereas the stator is a planar microcoil with the same dimension as that of the rotor. A sector-shaped area of NdFeB is used to cover the microcoil. To fabricate the microcoil in this area with as many turns as possible, three configurations of microcoil with eight poles are designed: (1) sector-shaped; (2) circular-shaped; and (3) square-shaped, as shown in Figure 4.34. The dimensions of both the rotor and stator are only 9 mm in outer diameter and 0.7 mm in thickness to allow a compacted and integrated system.

4.4.1 LTCC Process

Because it is easy to fabricate microcoils with vias, cavities, and embedded inductors, the microcoil is designed and fabricated using LTCC. In the green state, the green sheets are processed to define microcoil patterns. The individual sheet is then stacked and laminated under heat and pressure to obtain the desired microcoil structure. In this study, the planar microcoil is fabricated with the lamination of two individual layers of green sheets, each of which is individually processed. The two layers of LTCC planar microcoils with 100 μm in the line width and spacing are designed.

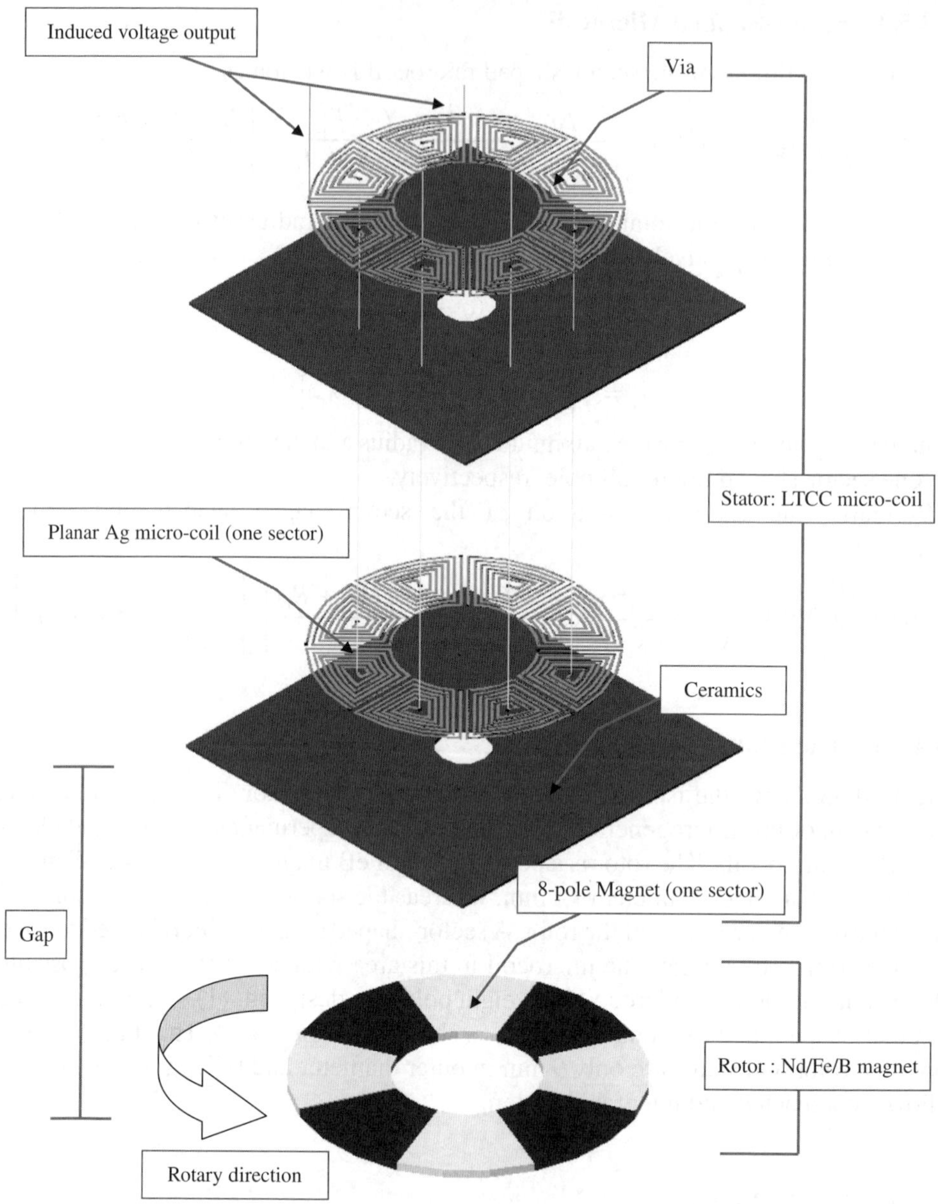

Figure 4.33 Eelectromagnetic microgenerator in three dimensions [61]

The process of manufacturing a LTCC planar microcoil is as follows. First, a tape of ceramics is used to trim the edges and cut them as a green sheet. It is set on the frame as shown in Figure 4.35a. The ceramics is composed of punched holes acting as vias to conduct between the layers above and below (Figure 4.35b). The vias are filled with Ag paste as a conductor metal (Figure 4.35c). The Ag metal microcoil patterns

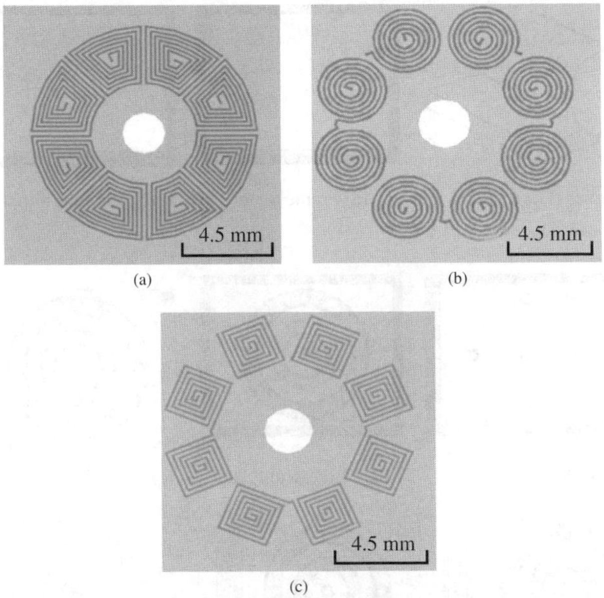

Figure 4.34 Schematic illustration of (a) sector-shaped, (b) circle-shaped, and (c) square-shaped microcoils [61]

are printed on ceramics under the alignment system (Figure 4.35d). Alignment keys are punched on the ceramics, and the metal frame is removed (Figure 4.35e). These ceramics are stacked in turn and bonded in a vacuum surrounding (Figure 4.35f). The sample is package sealed before being uniformly bonded (Figure 4.35g); it is then laminated at a high pressure of 850 psi (Figure 4.35h). After lamination, it is co-fired at a specific temperature (800–900 °C) for a certain time (Figure 4.35i). Finally, the sample is post-printed on the surface (Figure 4.35j) and the microcoil is tested (Figure 4.35k).

4.4.2 Magnet Process

The manufacturing technology of a sintering NdFeB magnet is used to fabricate eight-polar magnets. The compact yoke is custom designed for an eight-pole magnet 9 mm in diameter. The sintering of magnetic materials involves raw powder material molding, sintering, heat treatment, and product anisotropic magnetizing. A specially designed yoke was used to magnetize this sintered NdFeB magnet with a vertical-direction anisotropic magnetization, with which the power output of the microgenerator can be enhanced. Because the miniaturization of a magnet is limited by current yoke fabrication technology, the dimension of the magnet must be carefully considered in this type of design. The multipolar magnet is composed of both S-pole and N-pole magnetic segments. Segment-shaped magnets are easier to fabricate than

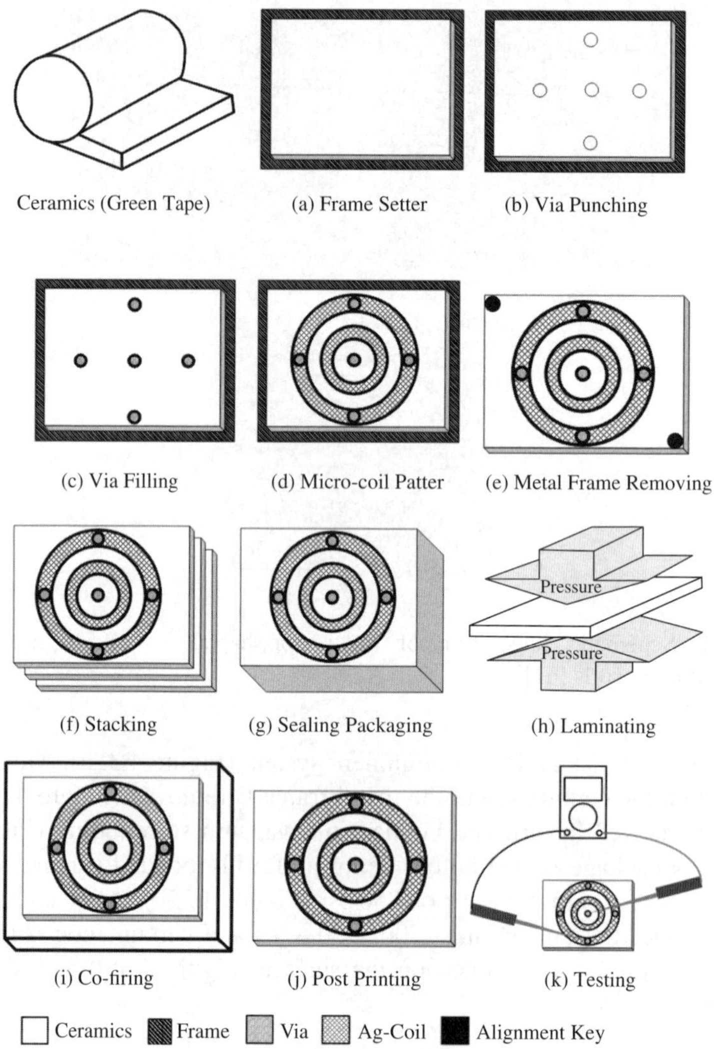

Figure 4.35 Schematic LTCC coil process [61]

contiguous annular magnets. The yield rate of the magnetic segment is higher than that of a contiguous annular magnet. Moreover, the equipment to create a uniform field to magnetize these magnets anisotropically would cost more than the yoke.

4.4.3 Measurement Set-up

The planar microcoil and planar magnet were assembled and tested using the self-built drive mechanism for power measurement, as shown in Figure 4.36. Figure 4.36a

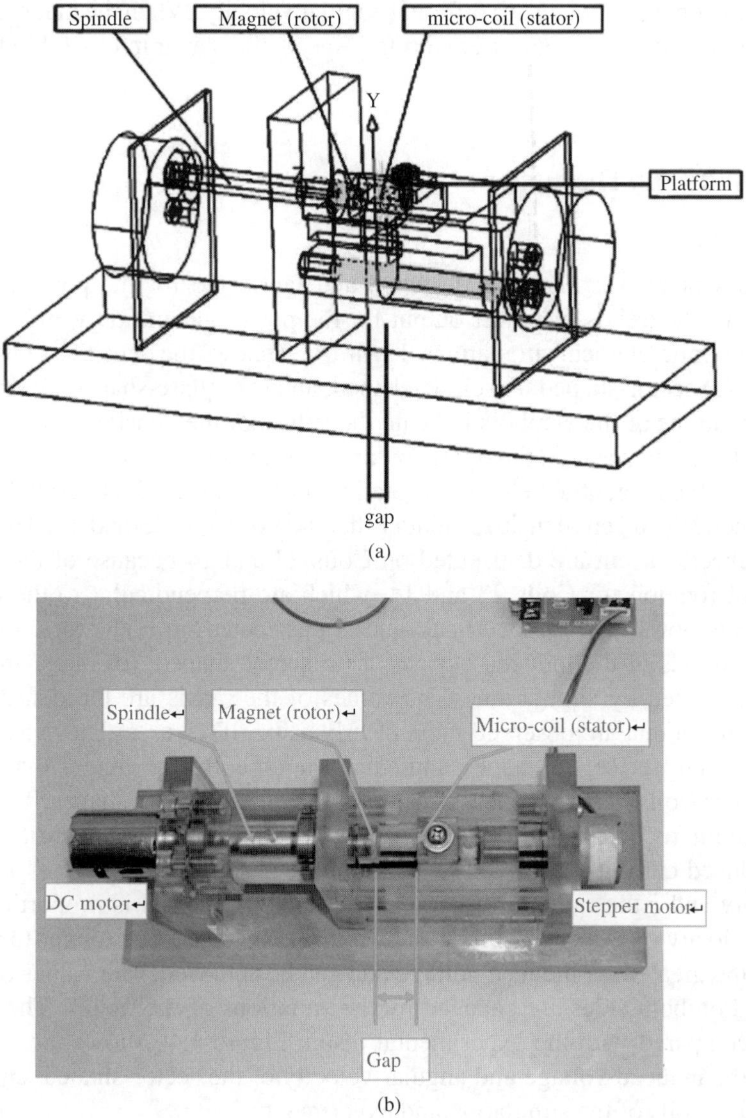

Figure 4.36 Self-designed drive of (a) CAD design chart and (b) mechanism system [61]

depicts the measurement set-up, and Figure 4.36b is a photograph of the mechanism. The self-built drive mechanism included a stepper motor, a rotary mechanism, a static platform, and a spindle.

First, the magnet is fixed on a rotary mechanism, and the microcoil is attached to a static platform. The gap between the microcoils and magnet can be controlled by a stepper motor. Second, the rotary mechanism is driven by the power supply to make the multipolar magnet rotate at a different angular velocity. When the magnet is rotated relative to the microcoil at an external rotary speed, the planar microcoil induces voltage and current.

4.5 Results and Discussion

4.5.1 Design

Three types of microcoils were designed and fabricated for the power generation experiment. The trends of power output for the power generation were obtained by performing finite-element simulations. Figure 4.37 shows the densities of the induced current of (a) sector-shaped, (b) circle-shaped, and (c) square-shaped by simulations. The distributions of the symbols indicate the values of the densities and the induced current. The arrowheads on the microcoil are the directions of the induced current. In Figure 4.37, the tangential velocity of rotation for Coils 12 and 34 parallel to the coil leads to induced current densities smaller than that of Coils 23 and 14. The densities of the induced current are distributed on Coils 23 and 14 because of the tangential velocity of rotation for Coils 23 and 14, which are perpendicular to the microcoil. Figure 4.37 shows that the electrical field is distributed on both sides. Figure 4.38 shows the trends of the induced current of (a) sector-shaped, (b) circle-shaped, and (c) square-shaped by simulations. The values of the z-axis are the densities of the induced current that are observed more clearly using 3D figures. The densities of the induced current for sector-shaped and square microcoils are greater than those for circular microcoils in four corners because of the differences in shapes. The four corners of the microcoils with the greatest deformations lead to large density variations in the induced current. The velocity of the microcoil is $\tilde{\omega} \times \tilde{r}$ where $\tilde{\omega}$ is the angular velocity and $\tilde{r}$ the radius of rotation. The induced voltage is proportional to the angular velocity of the microcoil, as shown in Figure 4.39. The longer the radius of rotation, the greater the induced voltage that can be achieved. The values of the electrical field on both sides are changed by the variations of the radius. The simulated result is compared with the experimental result. Figure 4.40 shows the relationship between the induced voltage and angular velocity of the sector-shaped, circular, and square microcoils of the simulation and experiment.

The results of the experiment and simulation reveal that the microgenerator with a sector-shaped microcoil generates a maximum effective value of the induced voltage of 205.7 and 232.7 mV separately at 13 325 rpm. The lifetime of this structure was not measured; that is an area for future study. The planar microcoil and planar magnet are assembled and tested using a self-drive mechanism for powering measurement. The gap between the microcoil and the magnet can be controlled using a stepper motor and feedback control program. Although the gap is set at 1 μm, the actual gap might vary between one and several microns during the spinning test. The values between the induced voltage and the angular velocity in this study were determined as shown in Figure 4.40. For an induced microsource, the ratio of energy-to-size is an

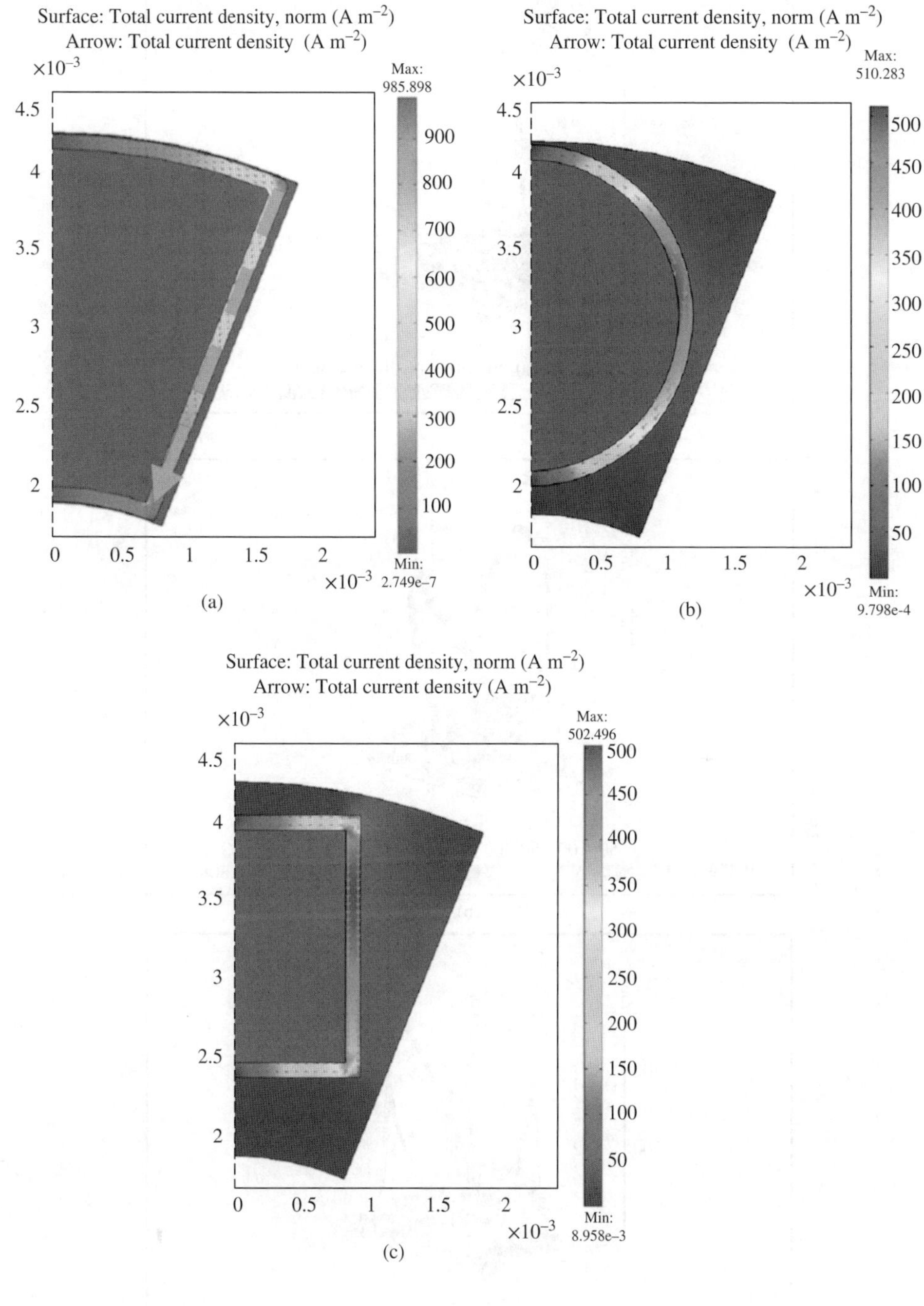

Figure 4.37 Densities of the simulated induced current of (a) sector-shaped, (b) circle-shaped, and (c) square-shaped microcoils

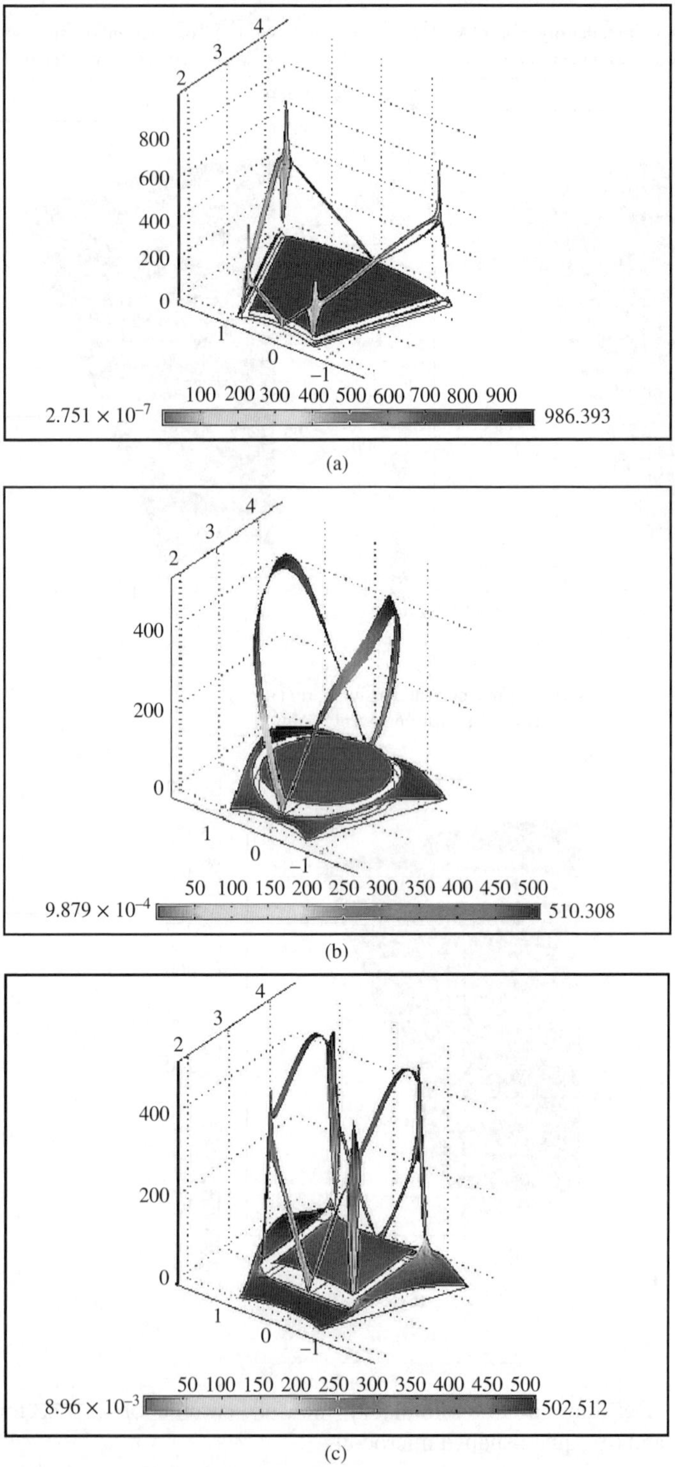

Figure 4.38 Trends of the simulated induced current of (a) sector-shaped, (b) circle-shaped, and (c) square-shaped microcoils

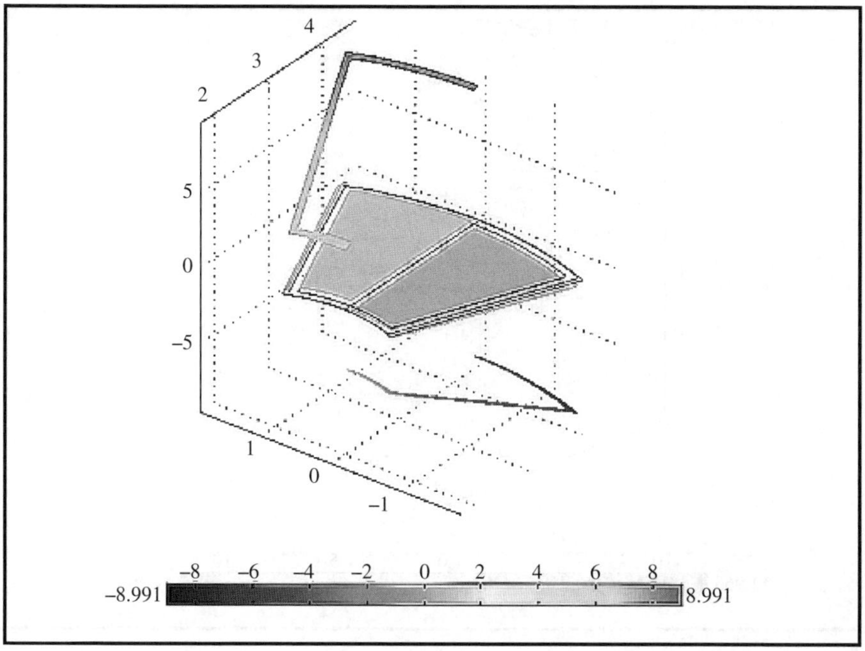

(a)

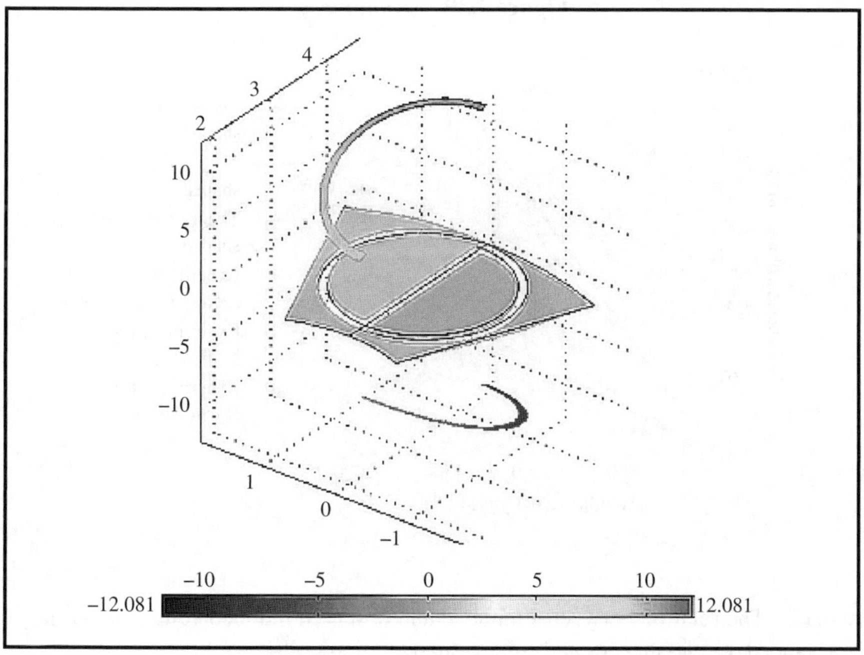

(b)

Figure 4.39 Distributions of the simulated electrical field of (a) sector-shaped, (b) circle-shaped, and (c) square-shaped microcoils

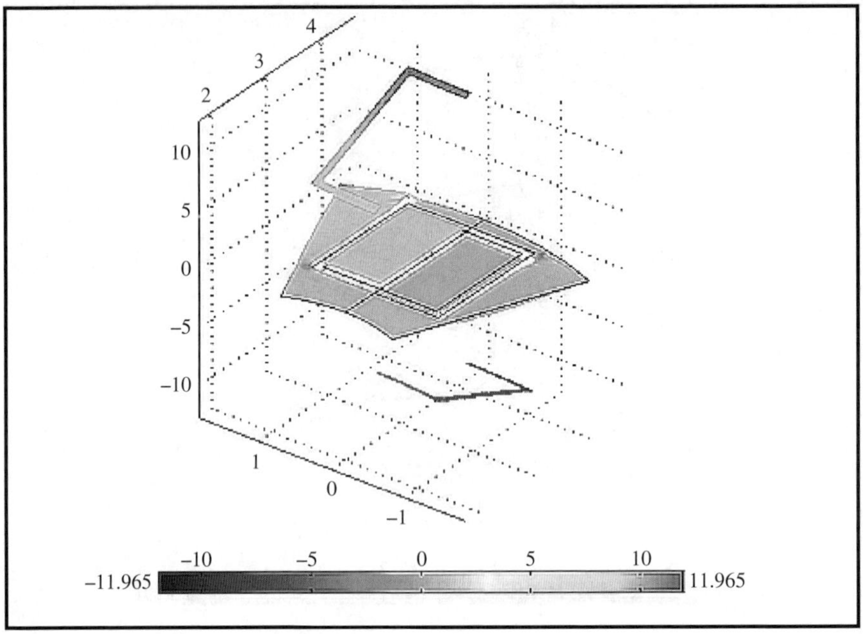

(c)

Figure 4.39 (*continued*)

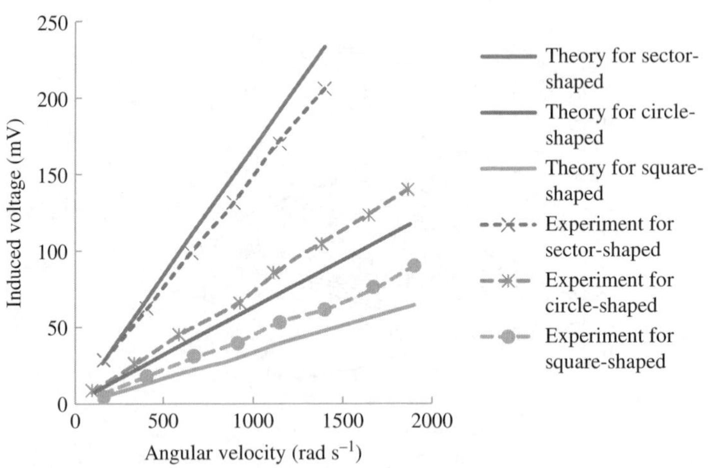

Figure 4.40 The relation between simulated and observed induced voltage and rotary speed of sector-shaped, circle-shaped, and square-shaped microcoils

essential criterion for evaluating the performance of generators. The power density (PD) expressed in units of mW cm^{-3} is the power output divided by the entire volume of the generator. Table 4.8 lists the values of the experiment and simulation. The sector-shaped microcoil is the most favorable pattern compared to the other two. The experiment and simulation data show that the microgenerator with a sector-shaped microcoil exhibits a maximum power output of 1.89 and 2.53 mW separately, as

Table 4.8 The values for the simulation and experiment (volume of microgenerator: 0.761 cm^3)

	Sector-shaped		Circle-shaped		Square-shaped	
	Simulation	Experiment	Simulation	Experiment	Simulation	Experiment
Maximum rotary speed (rpm)	13 325		13 317		13 367	
Entire volume of the microgenerator (cm^3)	0.761					
Induced voltage (mV)	232.7	205.7	86.7	105.8	47.5	61.8
Power output (mW)	2.53	1.89	0.57	0.85	0.23	0.39
Power density (mW cm^{-3})	3.32	2.48	0.75	1.12	0.30	0.51

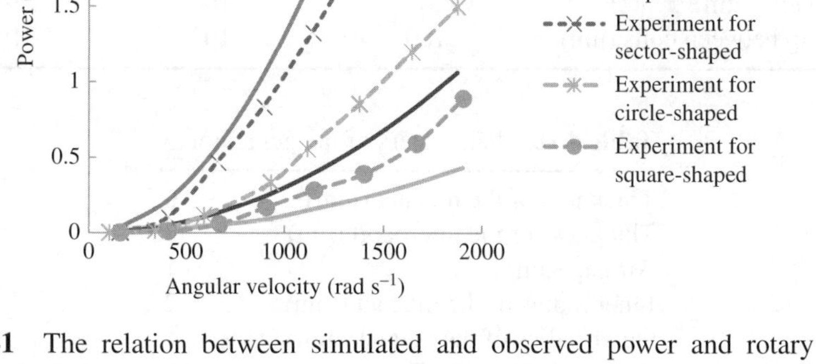

Figure 4.41 The relation between simulated and observed power and rotary speed of sector-shaped, circle-shaped, and square-shaped microcoils

shown in Figure 4.41. The sector-shaped microcoil exhibits the highest power output compared with the circular and square microcoil. Experimental measurements show a trend with finite-element simulations.

4.5.2 *Analytical Solutions*

The square, circular, and sector-shaped microcoils were designed and fabricated with eight poles and two laminated layers. The square, circular, and sector-shaped microcoils were printed with four, five, and five coils, respectively, in one pole. This study uses the theory of electromagnetism and multiple aspects of coupling to study the mathematical model of the proposed in-plane rotary electromagnetic microgenerator. The analytical solutions for the three types of microcoil were obtained using Equations (4.44), (4.47), and (4.51). The power generation is influenced by many variables, including the rotational speed of the rotor, the number of layers, the number of poles, the gap between neighbor coils, the coil width, microcoil circles, the coil thickness, and the parameters of the magnet. The analytical solutions in this study consider these variables to calculate the power generation. Table 4.9 lists the properties of the three microcoils, and Table 4.10 lists the other parameters of the microgenerator. The multilayer plane microcoil and multipolar plane permanent magnet are composed together and tested using the self-designed drive mechanism. The permanent magnet induces an EMF when it moves relative to a microcoil. Substituting the parameters of the microgenerator into Equations (4.44), (4.47), and (4.51) produces the induced voltages of the analytical solutions for the three different-shaped planar microcoils (Figure 4.42). The analytical solutions of the square, circular, and sector-shaped microcoils are represented by sinusoidal waveforms, and the simulations are matched by observations. Figure 4.42 shows that

Table 4.9 Parameters of the planar microcoil

	Square-shaped	Circle-shaped	Sector-shaped
Layers	2	2	2
Poles	8	8	8
Turns	4	5	5
Coil width (μm)	100	100	100
The gap between coils (μm)	100	100	100

Table 4.10 Parameters of the generator

Thickness of the magnet (mm)	0.7
Thickness of the microcoil (μm)	30
Air gap (μm)	1
Inner radius of the magnet (mm)	2
Outer radius of the magnet (μm)	4.5
Residual induction (T)	1.4

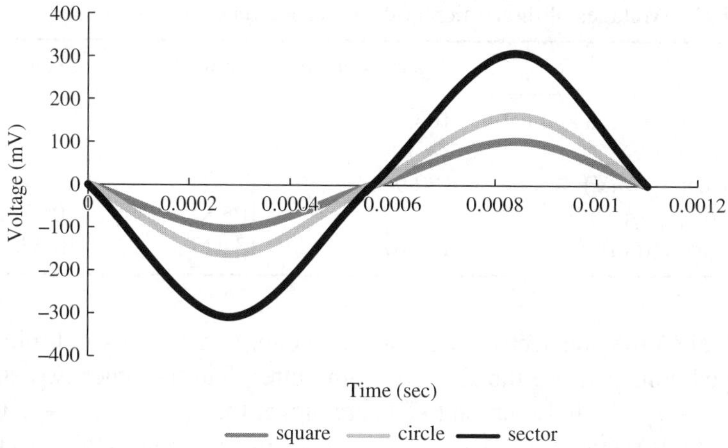

Figure 4.42 The simulated induced voltages of sector-shaped, circle-shaped, and square-shaped planar microcoils

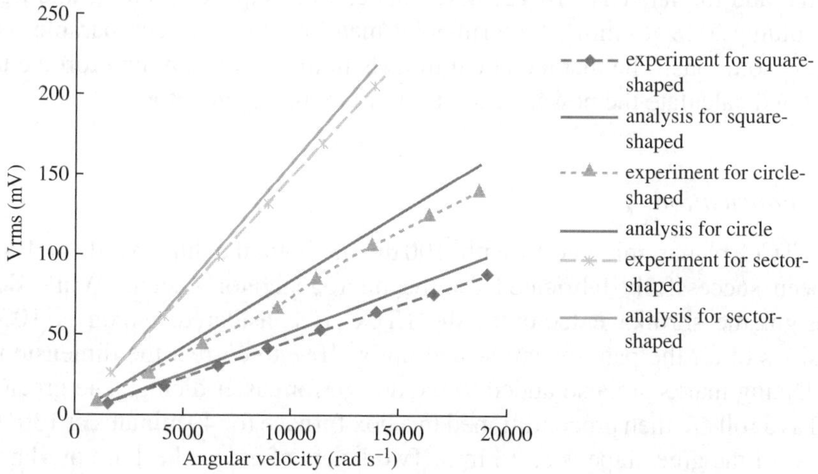

Figure 4.43 The relationship between the simulated and observed induced voltage and rotary speed of sector-shaped, circle-shaped, and square-shaped microcoils

the sector-shaped microcoils have the best induced voltage, as they fully use the area to distribute the coils. The effective value of the induced voltage V_{rms} was measured by a voltmeter, and the analytical solutions of the root mean square were computed as shown in Figure 4.43.

The comparison between V_{rms} of analytical solutions and V_{rms} of experiments is shown in Table 4.11. The maximum effective values of the induced voltage of the square, circular, and sector-shaped microcoils are 61.8, 105.8, and 205.7 mV at rotary speeds of 13 367, 13 317, and 13 325 rpm, respectively. The analytical solutions of the root mean square of the square, circular, and sector-shaped microcoils are 72.9,

Table 4.11 Voltages of three microcoils with rotational speeds

	Square-shaped	Circle-shaped	Sector-shaped
Induced voltage of analytical solutions (mV)	103.1	162.2	308.5
Simulated V_{rms} (mV)	72.9	114.7	218.1
Observed V_{rms} (mV)	61.8	105.8	205.7
Rotary speed (rpm)	13 367	13 317	13 325

114.7, and 218.1 mV, respectively. Room to accommodate coils is limited, but the sector-shaped coils can use the limited room better than the other two shapes. The values of $(r_o^2 - r_i^2)/8$ in Equation (4.51) are bigger than $[x_{max} - 2(q-1)(\lambda_g + \lambda_c)]^2$ in Equation (4.44) and $[r_{max} - (q-1)(\lambda_g + \lambda_c)]^2$ in Equation (4.46). Consequently, the results of the experimental and analytical model showed that the sector-shaped microcoils have the best induced voltage. Although the experimental and analytical results both show 13 325 rpm, they show different induced voltages: the former is 205.7 mV and the latter is 218.127 mV. This can be achieved using a microgenerator of volume $9 \times 9 \times 1\,mm^3$. Experimental measurements are comparable with the analytical solutions. The mathematical models in this study can therefore be used to forecast and calculate the power generation of the microgenerator.

4.5.3 Fabrication

Three LTCC planar microcoils with 100 μm in both the line width and spacing have been successfully fabricated for the microgenerator system. While devising the design, the shrinkage factor of the LTCC is considered, which is 10%. The dimensions of all the patterns are scaled up by 10% to correct the dimensions after firing. Dicing marks are also added to the design for easier dicing. The green sheets packed as a roll are then preconditioned in a box furnace for 40 minutes at 130 °C. The thickness of the green tape is *c.* 0.1 mm. Two layers of green sheet are pre-laminated at 850 psi and 75 °C for 15 minutes. The stack is then inspected under a microscope to ensure that all the vias are aligned. The laminating fixture plate is then sealed in a vacuum bag and heated at 800 psi and 850 °C for 15 minutes. This is followed by sintering the stack in a furnace programmed to follow the temperature-time curve for the firing process. Experimental photos of the LTCC microcoils are shown in Figure 4.44 of (a) sector-shaped, (b) circle-shaped, and (c) square-shaped patterns of microcoils. The number of the poles n was designed as 8. The dual-layered LTCC planar microcoils with 100 μm in line width are fabricated. The planar microcoil with one pole is shown in Figure 4.45. There are three types of planar microcoils: sector-shaped (Figure 4.45a), circular (Figure 4.45b), and square (Figure 4.45c). The electrical resistance in LTCC Ag is listed in Table 4.12a. The comparison between bulk Ag and bulk Cu microcoils is listed in Tables 4.12b and 4.12c.

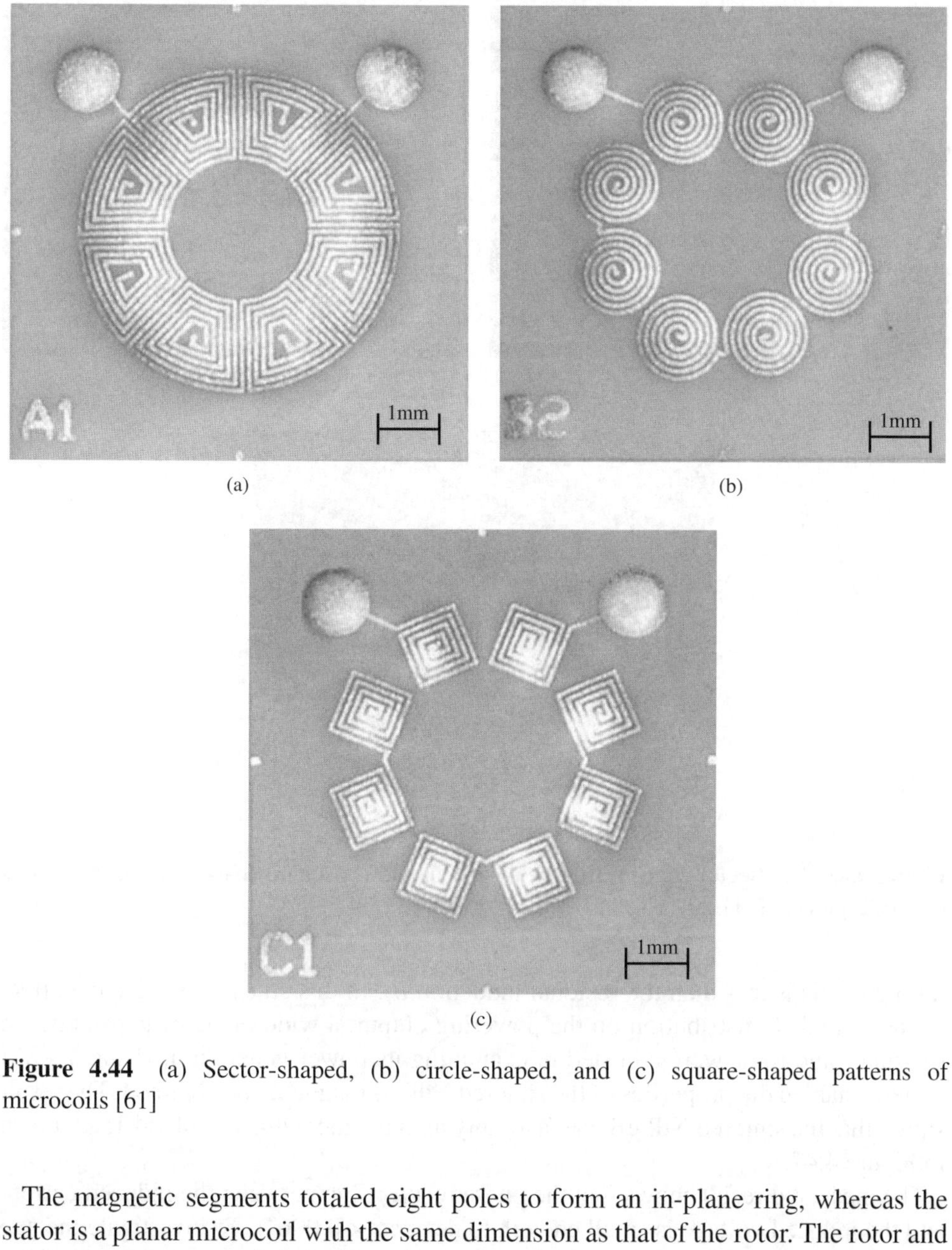

Figure 4.44 (a) Sector-shaped, (b) circle-shaped, and (c) square-shaped patterns of microcoils [61]

The magnetic segments totaled eight poles to form an in-plane ring, whereas the stator is a planar microcoil with the same dimension as that of the rotor. The rotor and stator both had an outer diameter of 9 mm and an inner diameter of 4 mm, achieving the goal of a compact and integrated system. Figure 4.46 shows (a) the multipolar permanent magnet and (b) the single-pole permanent magnet. The true spatial magnetic field B distribution in the air gap caused by the flux leakage of an open-circuit condition should be considered. For an open magnetic circuit, the B-field strength diminishes if the magnet surface is further away from it. The actual magnetic field

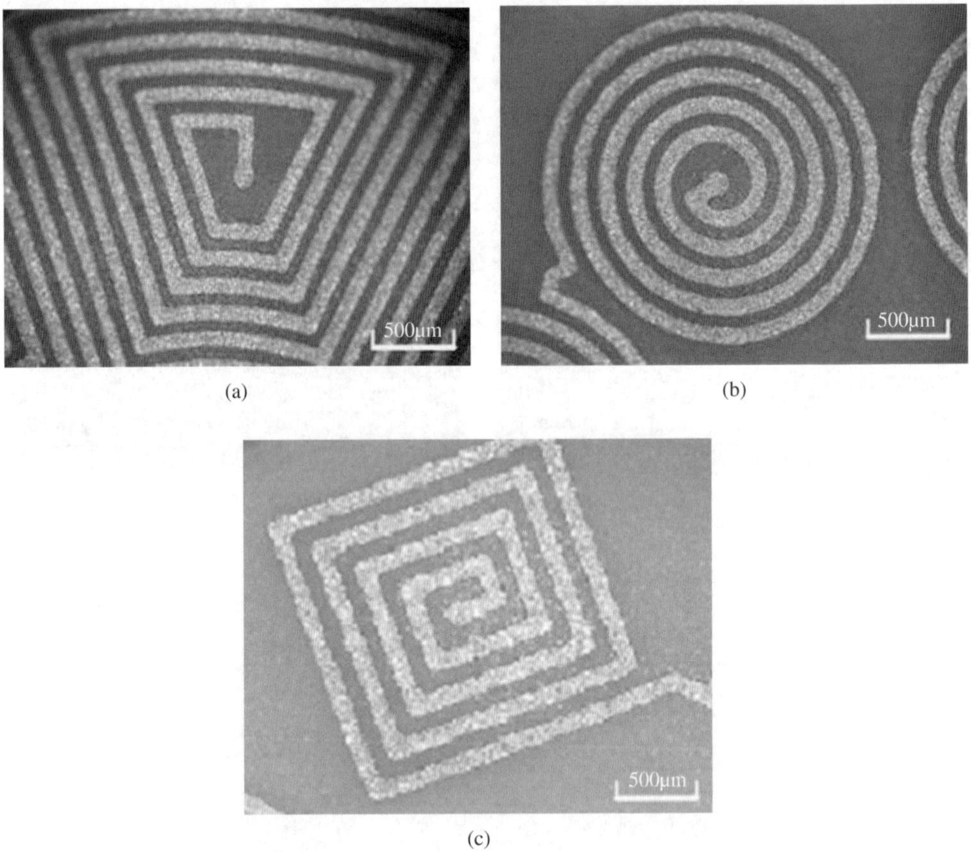

Figure 4.45 (a) Sector-shaped, (b) circle-shaped, and (c) square-shaped planar microcoils by LTCC process [61]

through coils is less than the residual induction B_r. In this study, to realize the effect of the spatial B_r distribution on the powering output, a wide range of gaps between the stator and rotor were adjusted in examining its power generation. Using a VSM system enabled the properties of the sintered NdFeB magnet to be obtained. The result shows that the sintered NdFeB is a favorably hard magnet with a B_r of 1.4 T, as shown in Figure 4.47.

The measured result shows that the phase relationship between the induced current and the induced voltage is small enough to neglect ($< 0.006°$). To investigate the gap effect on the generation, the gap between the microcoil and the magnet is adjustable from 500 to 1 μm. The spindle is driven to rotate the multipolar magnet. As the multipolar magnet is rotated relative to the microcoil, an electric current is induced. A sinusoidal signal waveform of voltage is induced in the planar microcoil and observed using an oscilloscope.

Table 4.12 Electrical properties

(a) LTCC Ag planar microcoil	
Electrical conductivity (S m^{-1})	2.95×10^4
Sector-shaped resistance (Ω)	21.38
Circle-shaped resistance (Ω)	13.15
Square-shaped resistance (Ω)	9.63
Resistivity (Ωm)	3.41×10^{-5}
(b) Calculated bulk Ag properties	
Electrical conductivity (S m^{-1})	6.19×10^7
Sector-shaped resistance (Ω)	9.82×10^{-3}
Circle-shaped resistance (Ω)	8.37×10^{-3}
Square-shaped resistance (Ω)	6.91×10^{-3}
Resistivity (Ωm)	1.62×10^{-8}
(c) Calculated bulk Cu properties	
Electrical conductivity (S m^{-1})	5.81×10^7
Sector-shaped resistance (Ω)	10.04×10^{-3}
Circle-shaped resistance (Ω)	8.57×10^{-3}
Square-shaped resistance (Ω)	7.07×10^{-3}
Resistivity (Ωm)	1.72×10^{-8}

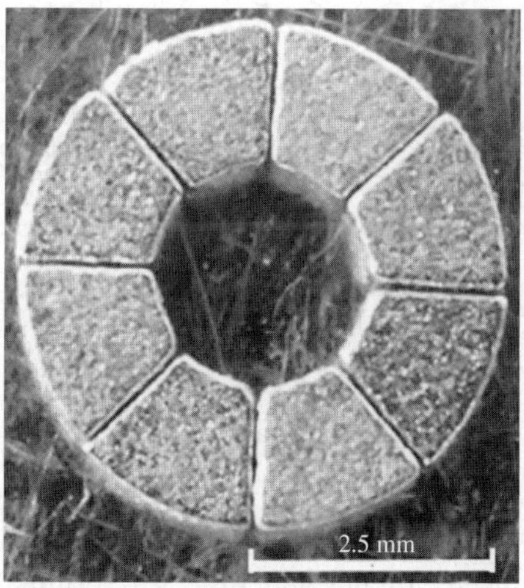

(a)

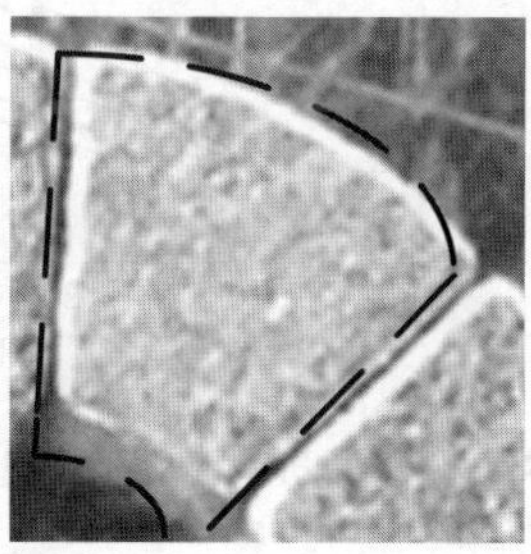

(b)

Figure 4.46 (a) Multipolar and (b) single-pole permanent magnets

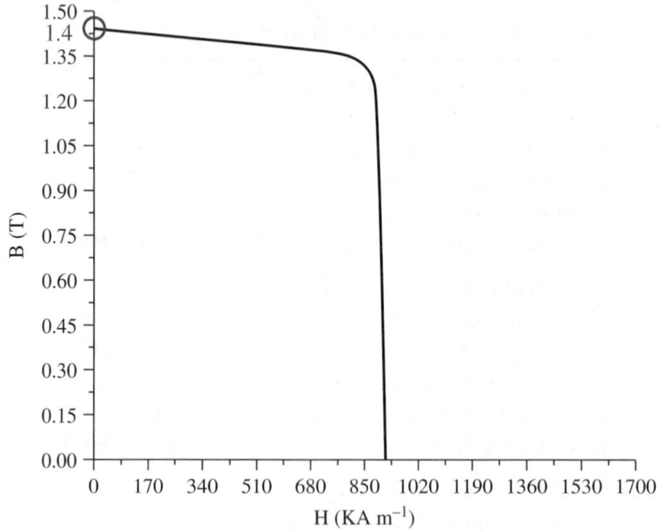

Figure 4.47 Measurements by a vibrating sample magnetometer

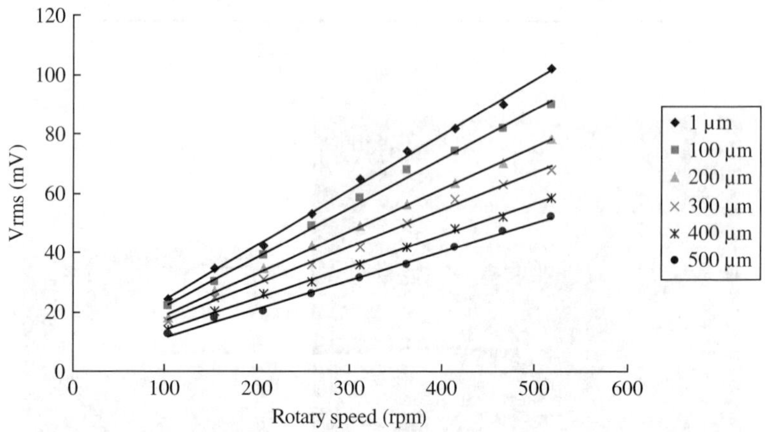

Figure 4.48 Relation between the induced voltage and rotary speed for different-sized gaps in a sector-shaped microcoil [61]

The relationship between the induced voltage and time was originally a sinusoidal waveform. The effective value of the induced voltage V_{rms} is plotted in Figures 4.48–4.50. The relationship between the induced voltage and rotary speed (rpm) with different gaps is discussed. The experimental data show that the induced voltage is proportional to the frequency. As expected, the faster the rotary speed of the magnet, the greater the induced voltage that can be achieved. Furthermore, to explain the effect of the gap between the microcoil and magnet on the

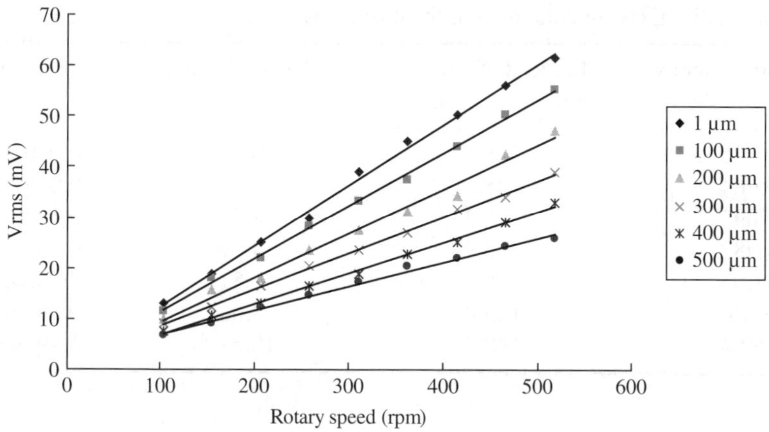

Figure 4.49 Relation between the induced voltage and rotary speed for different-sized gaps in a circle-shaped microcoil [61]

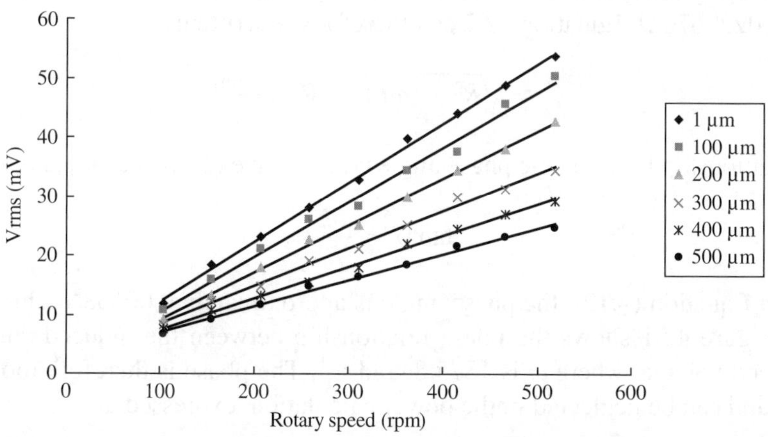

Figure 4.50 Relation between the induced voltage and rotary speed for different-sized gaps in a square-shaped microcoil [61]

power-generating process, samples with different gaps were tested and realized. Each line in Figures 4.48–4.50 shows the comparison between samples with different gaps. The smaller the gap between the microcoil and magnet, the larger the induced voltage.

In this study, the effects of resistance and inductance are listed below to explain the phase problem. Table 4.13 shows the circular coil data.

Impedance Z (Ω) is defined:

$$Z = \frac{V_{\text{rms}}}{I_{\text{rms}}} = \sqrt{R^2 + (\omega L)^2} \tag{4.52}$$

Table 4.13 Experimental data of the circle-shaped coil

Angle velocity, ω (rad s^{-1})	Induced effective voltage V_{rms} (mV)	Induced effective current I_{rms} (mA)	Power (mW)
162.106	8.1	0.09	0.000729
409.664	25.8	0.55	0.01419
659.892	45.2	2.54	0.114808
895.982	66.9	4.99	0.333831
1141.03	85.4	6.51	0.555954
1395.34	105.8	8.06	0.852748
1875.84	140.2	10.67	1.495934

where V_{rms} and I_{rms} are the effective values of voltage and current, respectively; R is resistance (13.14 Ω measured using a Keithley 2400 source meter with a four-wire measure); ω is angular velocity (1875.84 rad s^{-1}); and L is inductance (1.22439 μH measured using an LCR meter). According to Equation (4.10), the inductive reactance ωL is 0.00229676 Ω. Equation 4.52 can therefore be written: . . . .

$$Z \approx \sqrt{R^2 + (\omega L)^2} \approx \mathrm{R} \approx \frac{V_{rms}}{I_{rms}} \tag{4.53}$$

The relationship between the phase angle ψ, inductive reactance ωL, and resistance R is:

$$\tan \psi = \frac{\omega L}{R} \tag{4.54}$$

Based on Equation (4.12), the phase angle is approximately 0.00638°, which is negligible. Figure 4.51 shows the phase relationship between the induced current and the induced voltage, where ω is 1875.84 rad s^{-1}. The phase is therefore too small to observe and can be neglected in the power calculation, expressed as

$$P = {I_{rms}}^2 R = I_{rms} V_{rms} \tag{4.55}$$

Figure 4.52 shows the relationship between the induced power and frequency at a gap of 1 μm of sector-shaped, circle-shaped, and square-shaped micro-coils. Each power output in Figure 4.52 is the multiplication of the induced voltage and induced current. Table 4.14 lists the induced voltage and induced current of the microcoils. The experimental data show that the microgenerator with a sector-shaped microcoil exhibits a maximum power output of 1.89 mW. The sector-shaped microcoil is the most favorable pattern, compared to the other two. The experimental result reveals that this microgenerator generates the maximum effective value of the induced voltage, 205.7 mV, at 13 325 rpm and at a gap of 1 μm.

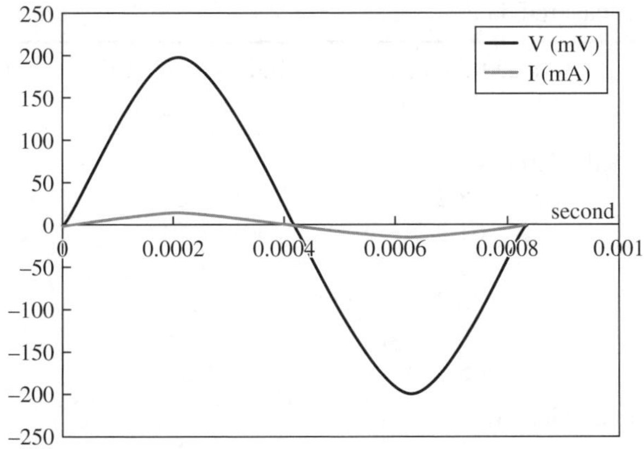

Figure 4.51 The phase relation between induced current and induced voltage [61]

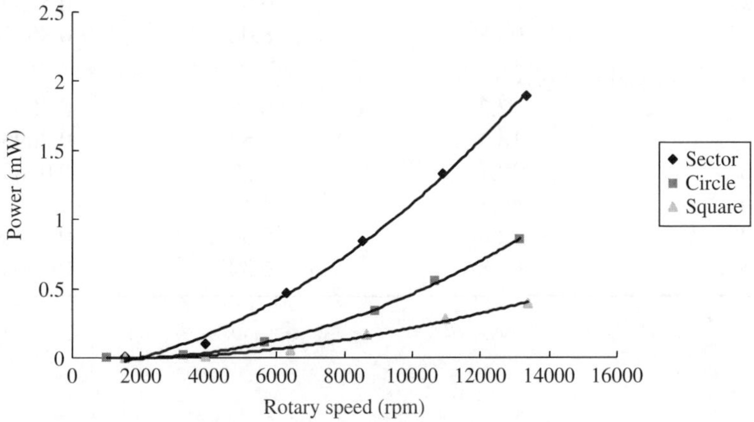

Figure 4.52 Relation between power and rotary speed of sector-shaped, circle-shaped, and square-shaped microcoils [61]

The different shapes of microcoils are tested for power generation. A sector-shaped microcoil exhibits the highest power output, compared to the circular and square microcoils. The given explanation is based on Ampere's right-hand rule. When the magnet moving in a tangential direction is perpendicular to the microcoil wire, a maximum electric current can be induced. Fabricating microcoils is easier using the LTCC method with acute-angled microcoil patterns, such as sector-shaped and square coils, compared to the traditional LIGA and LIGA-like processes and the filament-winding method.

Table 4.14 Measured data

Rotary speed (rpm)	Induced effective voltage V_{rms} (mV)	Induced effective current I_{rms} (mA)	Power (mW)
(a) Sector-shaped microcoil			
1549	28.2	0.2	0.00564
3914	61.9	1.69	0.104611
6305	99.4	4.75	0.47215
8560	130.7	6.42	0.839094
10901	170.9	7.78	1.329602
13325	205.7	9.2	1.89244
(b) Circle-shaped microcoil			
1015	8.1	0.09	0.000729
3266	25.8	0.55	0.01419
5655	45.2	2.54	0.114808
8915	66.9	4.99	0.333831
10669	85.4	6.51	0.555954
13317	105.8	8.06	0.852748
(c) Square-shaped micro-coil			
1555	7.3	0.13	0.000949
3906	18.4	0.52	0.009568
6396	30.1	1.87	0.056287
8670	40.1	4.19	0.168019
10995	52.8	5.36	0.283008
13367	61.8	6.32	0.390576

References

1. Espinosa, D.C.R., Bernardes, A.M. and Tenório, J.A.S. (2004) An overview on the current processes for the recycling of batteries. *Journal of Power Sources*, **135** (1–2), 311–319.
2. Hande, A. and Stuart, T.A. (2004) A selective equalizer for NiMH batteries. *Journal of Power Sources*, **138** (1–2), 327–339.
3. Yang, S. and Knickle, H. (2002) Design and analysis of aluminum/air battery system for electric vehicles. *Journal of Power Sources*, **112** (1), 162–173.
4. Fergus, J.W. (2010) Recent developments in cathode materials for lithium ion batteries. *Journal of Power Sources*, **195** (4), 939–954.
5. Bernardes, A.M., Espinosa, D.C.R. and Tenório, J.A.S. (2004) Recycling of batteries: a review of current processes and technologies. *Journal of Power Sources*, **130** (1–2), 291–298.
6. Jones, F.G., Beeby, S.P. and White, N.M. (2001) Towards a piezoelectric vibration-powered micro-generator. *IEEE Proceedings: Science Measurement and Technology*, **148** (2), 68–72.
7. Swallow, L.M., Luo, J.K., Siores, E. *et al.* (2008) A piezoelectric fibre composite based energy harvesting device for potential wearable applications. *Smart Materials and Structures*, **17** (2), 1–7.
8. Fang, H.B., Liu, J.Q., Xu, Z.Y. *et al.* (2005) A MEMS-based piezoelectric power generator for low frequency vibration energy harvesting. *Chinese Physics Letter*, **23** (3), 732–734.

9. Fang, H.B., Liu, J.Q., Xu, Z.Y. *et al.* (2006) Fabrication and performance of MEMS-based piezoelectric power generator for vibration energy harvesting. *Microelectronics Journal*, **37** (11), 1280–1284.
10. Rowe, D.M., Morgan, D.V. and Kiely, J.H. (1991) Low cost miniature thermoelectric generator. *Electronics Letter*, **27** (25), 2332–2334.
11. Gould, C.A., Shammas, N.Y.A., Grainger, S. and Taylor, I. (2008) Thermoelectric technology: micro-electrical and power generation properties. 43rd International Universities Power Engineering Conference, 2008 (UPEC 2008), pp. 1–5.
12. Tashiro, R., Katayama, N., Ishizuka, Y. *et al.* (2000) Development of an electrostatic generator that harnesses the motion of a living body. *JSME International Journal Series C: Mechanical Systems Machine Elements and Manufacturing*, **43** (4), 916–922.
13. Meninger, S., Mur-Miranda, J.O., Amirtharajah, R. *et al.* (2001) Vibration-to-electric energy conversion. *IEEE Transactions on Very Large Scale Integration (VLSI) Systems*, **9** (1), 64–76.
14. Roundy, S., Wright, P.K. and Pister, K.S. (2002) Micro electrostatic vibration to electricity converters. ASME International Mechanical Engineering Congress and Congress and Exposition, pp. 1–10.
15. Wang, X., Shang, J., Luo, Z. *et al.* (2012) Reviews of power systems and environmental energy conversion for unmanned underwater vehicles. *Renewable and Sustainable Energy Reviews*, **16** (4), 1958–1970.
16. Arnold, D.P. (2007) Review of microscale magnetic power generation. *IEEE Transactions on Magnetics*, **43** (11), 3940–3951.
17. Kulkarni, S., Koukharenko, E., Torah, R. *et al.* (2008) Design, fabrication and test of integrated micro-scale vibration-based electromagnetic generator. *Sensors and Actuators A: Physical*, **145–146**, 336–342.
18. Wang, P.H., Dai, X.H., Fang, D.M. and Zhao, X.L. (2007) Design, fabrication and performance of a new vibration-based electromagnetic micro power generator. *Microelectronics Journal*, **38** (12), 1175–1180.
19. Kulkarni, S., Roy, S., O'Donnell, T. *et al.* (2006) Vibration based electromagnetic micropower generator on silicon. *Journal of Applied Physics*, **99** (8), 08P511–08P511-3.
20. Sari, I., Balkan, T. and Kulah, H. (2008) An electromagnetic micro power generator for wideband environmental vibrations. *Sensors and Actuators A: Physical*, **145–146**, 405–413.
21. Serre, C., Rodriguez, A.P., Fondevilla, N. *et al.* (2008) Design and implementation of mechanical resonators for optimized inertial electromagnetic microgenerators. *Microsystem Technologies*, **14** (4–5), 653–658.
22. Serre, C., Rodriguez, A.P., Fondevilla, N. *et al.* (2007) Vibrational energy scavenging with Si technology electromagnetic inertial microgenerators. *Microsystem Technologies*, **13** (11–12), 1655–1661.
23. Beeby, S.P., Torah, R.N., Tudor, M.J. *et al.* (2007) A micro electromagnetic generator for vibration energy harvesting. *Journal of Micromechanics and Microengineering*, **17**, 1257–1265.
24. Williams, C.B., Shearwood, C., Harradine, M.A. *et al.* (2001) Development of an electromagnetic micro-generator. *IEE Proceedings-Circuits, Devices and Systems*, **148** (6), 337–342.
25. Chen, P.C., Lin, H.Y., Chang, S.B. and Huang, Y. C. (2010) The torque control of human power assisted electric bikes. 2010 International Conference on System Science and Engineering, pp. 373–378.
26. Lomonova, E.A., Vandenput, A.J.A., Rubacek, J. *et al.* (2002) Development of an improved electrically assisted bicycle. IEEE Industry Applications Society 37th Annual Meeting, Vol. **1**, pp. 384–389.
27. Valerius, C., Krupar, J. and Schwarz, W. (2005) Electronic power management for bicycles. European Conference on Power Electronics and Applications, pp. 1–10.
28. Pan, C.T. and Wu, T.T. (2007) Development of a rotary electromagnetic microgenerator. *Journal of Micromechanics and Microengineering*, **17**, 120–128.

29. Wang, Y.J., Chen, C.D. and Sung, C.K. (2010) Design of a frequency-adjusting device for harvesting energy from a rotating wheel. *Sensors and Actuators A: Physical*, **159** (2), 196–203.
30. Arnold, D.P., Das, S., Park, J.W. *et al.* (2006) Microfabricated high-speed axial-flux multiwatt permanent-magnet generators – part II: design, fabrication, and testing. *Journal of Microelectromechanical Systems*, **15** (5), 1351–1363.
31. Iizuka, A., Takato, M., Kaneko, M. *et al.* (2012) Millimeter scale MEMS air turbine generator by winding wire and multilayer magnetic ceramic circuit. *Modern Mechanical Engineering*, **2** (2), 41–46.
32. Liao, A.D., Chao, P.C.P., Chen, J.T. *et al.* (2009) A miniaturized electromagnetic generator with planar coils and its energy harvest circuit. *IEEE Transactions on Magnetics*, **45** (10), 4621–4627.
33. Arnold, D.P., Herrault, F., Zana, I. *et al.* (2006) Design optimization of an 8 W, microscale, axial-flux, permanent-magnet magneto. *Journal of Micromechanics and Microengineering*, **16**, S290–S296.
34. Raisigel, H., Cugat, O., Delamare, J. *et al.* (2005) Magnetic planar micro generator. Transducers 2005, Solid-State Sensors, Actuators and Microsystems, Vol. **1**, pp. 757–761.
35. Raisigel, H., Cugat, O. and Delamare, J. (2006) Permanent magnet planar micro-magnetos. *Sensors and Actuators A: Physical*, **130–131**, 438–444.
36. Das, S., Arnold, D.P., Zana, I. *et al.* (2005) Multi-watt electricpower from a microfabricated permanent-magnet generator. 18th IEEE International Conference on Micro Electro Mechanical Systems, pp. 287–290.
37. Arnold, D.P., Joung, Y.H., Zana, I. *et al.* (2005) High-speed characterization and mechanical modeling of microscale, axial-flux, permanent-magnet generators. The 13th International Conference on Solid-Static Sensors, Actuators and Microsystems, pp. 701–704.
38. Holmes, A.S., Hong, G. and Pullen, K.R. (2005) Axial-flux permanent magnet machines for micropower generation. *Journal of Microelectromechanical Systems*, **14** (1), 54–63.
39. Hosseini, S.M., Mirsalim, M.A. and Mirzaei, M. (2008) Design, prototyping, and analysis of a low cost axial-flux coreless permanent-magnet generator. *IEEE Transactions on Magnetics*, **44** (1), 75–80.
40. Herrault, F., Ji, C.H., Shafer, R.H. *et al.* (2007) Ultraminiaturized milliwatt-scale permanent magnet generators. The 14th International Conference on Solid-State Sensors, Actuators and Microsystems, pp. 899–902.
41. Zwyssig, C. and Kolar, J.W. (2006) Design considerations and experimental results of a 100W, 500,000 rpm electrical generator. *Journal of Micromechanics and Microengineering*, **16**, 297–302.
42. Herrault, F., Arnold, D.P., Zana, I. *et al.* (2008) High temperature operation of multi-watt, axial-flux, permanent-magnet microgenerators. *Sensors and Actuators A: Physical*, **148** (1), 299–305.
43. Krähenbühl, D., Zwyssig, C., Weser, H. and Kolar, J.W. (2009) Theoretical and experimental results of a mesoscale electric power generation system from pressurized gas flow. *Journal of Micromechanics and Microengineering*, **19**, 1–6.
44. Cordero, R., Rivera, A., Neuman, M. *et al.* (2012) Micro-rotational electromagnetic generator for high speed applications. Micro Electro Mechanical Systems (MEMS), 2012 IEEE 25th International Conference, pp. 1257–1260.
45. Boland, J., Chao, Y.H., Suzuki, Y. and Tai, Y.C. (2003) Micro electret power generator. *Micro Electro Mechanical Systems*, **2**, 538–541.
46. Pan, C.T. and Wu, T.T. (2007) Simulation and fabrication of magnetic rotary microgenerator with multipolar Nd/Fe/B magnet. *Microelectronics Reliability*, **47** (12), 2129–2134.
47. Epstein, A.H. (2003) Millimeter-scale, MEMS gas turbine engines. Proceedings of ASME Turbo Expo 2003 International Joint Power Generation Conference, Atlanta, GA, pp. 669–696.
48. Toh, T.T., Mitcheson, P.D., Holmes, A.S. and Yeatman, E.M. (2008) A continuously rotating energy harvester with maximum power point tracking. *Journal of Micromechanics and Microengineering*, **18**.

49. Chao, P.C.P., Shao, C.I., Lu, C.X. and Sung, C.K. (2011) A new energy harvest system with a hula-hoop transformer, micro-generator and interface energy-harvesting circuit. *Microsystems Technologies*, **17** (5–7), 1025–1036.
50. Willke, T.L. and Gearhart, S.S. (1997) LIGA micromachined planar transmission lines and filters. *IEEE Transactions on Advanced Packaging*, **45** (10), 1681–1688.
51. Cheng, P.J., Shen, S.C., Pan, C.T. *et al.* (2006) In-plane microgenerator with high energy product magnet. *Journal of the Chinese Society of Mechanical Engineers*, **27**, 185–190.
52. Wi, S.H., Kim, J.S., Kang, N.K. *et al.* (2007) Package-level integrated LTCC antenna for RF package application. *IEEE Transactions on Advanced Packaging*, **30** (1), 132–141.
53. Ponchak, G.E. (2000) The use of metal filled via holes for improving isolation in, LTCC RF and wireless multichip packages. *IEEE Transactions on Advanced Packaging*, **23** (1), 88–99.
54. Jang, J.H., Ishitobi, N., Kim, B.C. and Jeong, S.G. (2006) Effects of microstructural, defects in multilayer LTCC stripline. *IEEE Transactions on Advanced Packaging*, **29**, 314–319.
55. Sarkar, S., Pinel, S., Kidera, N. and Laskar, J. (2007) Analysis and application of 3-D LTCC directional filter design for multiband millimeter-wave integrated module. *IEEE Transactions on Advanced Packaging*, **30** (1), 124–131.
56. Zhang, K.L., Chou, S.K. and Ang, S.S. (2005) Development of a low-temperature cofired ceramic solid propellant microthruster. *Journal of Micromechanics and Microengineering*, **15**, 944–952.
57. Golonka, L.J. (2006) Technology and applications of low temperature cofired ceramic (LTCC) based sensors and microsystems. *Bulletin of the Polish Academy of Sciences Technical Sciences*, **54**, 221–231.
58. Sebastian, M.T. and Jantunen, H. (2008) Low loss dielectric materials for LTCC applications: a review. *International Materials Reviews*, **53** (10), 57–90.
59. Tager, A.E., Bray, J. and Roy, L. (2003) High-Q LTCC resonators for millimeter wave applications. *Microwave Symposium Digest*, **3**, 2257–2260.
60. Thelemann, T., Thust, H. and Hintz, M. (2002) Using LTCC for microsystems. *Microelectronics International*, **19** (3), 19–23.
61. Pan, C.T. and Chen, Y.J. (2008) Application of low temperature co-fire ceramics on in-plane micro-generator. *Sensors and Actuators A: Physical*, **144** (1), 144–153.
62. Herrault, F., Yen, B.C., Ji, C.H. *et al.* (2010) Fabrication and performance of silicon-embedded permanent-magnet microgenerators. *Journal of Microelectromechanical Systems*, **19** (1), 4–13.
63. O'Donnell, T., Saha, C., Beeby, S. and Tudor, J. (2007) Scaling effects for electromagnetic vibrational power generators. *Microsystems Technologies*, **13** (11–12), 1637–1645.
64. Mateu, L., Villavieja, C. and Mol, F. (2007) Physics-based time-domain model of a magnetic induction microgenerator. *IEEE Transactions on Magnetics*, **43** (3), 992–1001.
65. Sreedharan, V. and Zhang, D. (2003) Finite element modeling of cellular responses of gap junction connected osteocytes under extremely low-frequency electromagnetic fields. Bioengineering Conference, pp. 160–161.
66. Smistrup, K., Hansen, O., Bruus, H. and Hansen, M.F. (2005) Magnetic separation in microfluidic systems using microfabricated electromagnets-experiments and simulations. *Journal of Magnetism and Magnetic Materials*, **293** (1), 597–604.
67. Pan, C.T., Chen, Y.J. and Shen, S.C. (2009) Simulation and analysis of electromagnetic in-plane microgenerator. *Journal of Micro/Nanolithography, MEMS and MOEMS*, **8** (3), 031304-1–031304-8.
68. Sari, I., Balkan, T., and Kulah, H. (2008) A micro power generator with planar coils on parylene cantilevers. In *Proceedings of PRIME* (PhD Research in Microelectronics and Electronics), IEEE, pp. 133–136.
69. Sari, I., Balkan, T. and Kulah, H. (2010) An electromagnetic micro power generator for low-frequency environmental vibrations based on the frequency upconversion technique. *Journal of Microelectromechanical Systems*, **19** (1), 14–27.

70. Zhu, D., Roberts, S., Tudor, M.J. and Beeby, S.P. (2010) Design and experimental characterization of a tunable vibration-based electromagnetic micro-generator. *Sensors and Actuators A: Physical*, **158** (2), 284–293.
71. Lin, G.J., Chen, M. and Song, D. (2009) Research of micro-power generator based on the dielectric electro active polymer. International Conference on Energy and Environment Technology, pp. 782–786.
72. Das, S., Arnold, D.P., Zana, I. *et al.* (2006) Microfabricated high-speed axial-flux multiwatt permanent-magnet generators-part I: modeling. *Journal of Microelectromechanical Systems*, **15** (5), 1330–1350.
73. Cheng, D.K. and Wesley, A. (1989) *Field and Wave Electromagnetic*, Addison-Wesley, Menlo Park, CA.
74. Pan, C.T. and Chen, Y.J. (2008) Application of low temperature co-fire ceramics on in-plane micro-generator. *Sensors and Actuators A: Physical*, **144**, 144–153.
75. Herrault, F., Yen, B.C., Ji, C.-H. *et al.* (2010) Fabrication and performance of silicon-embedded permanent-magnet microgenerators. *Journal of Microelectromechanical Systems*, **19**, 4–13.
76. O'Donnell, T., Saha, C., Beeby, S. and Tudor, J. (2007) Scaling effects for electromagnetic vibrational power generators. *Microsystems Technologies*, **13**, 1637–1645.
77. Mateu, L., Villavieja, C. and Mol, F. (2007) Physics-based time-domain model of a magnetic induction microgenerator. *IEEE Transactions on Magnetics*, **43**, 992–1001.
78. Smistrup, K., Hansen, O., Bruus, H. and Hansen, M.F. (2005) Magnetic separation in microfluidic systems using microfabricated electromagnets – experiments and simulations. *Journal of Magnetism and Magnetic Materials*, **293**, 597–604.
79. Pan, C.T., Chen, Y.J. and Shen, S.C. (2009) Simulation and analysis of electromagnetic in-plane microgenerator. *Journal of Micro/Nanolithography, MEMS and MOEMS*, **8**, 031304-1–031304-8.
80. Sari, I., Balkan, T. and Kulah, H. (2010) An electromagnetic micro power generator for low-frequency environmental vibrations based on the frequency upconversion technique. *Journal of Microelectroelectromechanical Systems*, **19**, 14–27.
81. Zhu, D., Roberts, S., Tudor, M.J. and Beeby, S.P. (2010) Design and experimental characterization of a tunable vibration-based electromagnetic micro-generator. *Sensors and Actuators A: Physical*, **158**, 284–293.

5

Design and Fabrication of Electrospun PVDF Piezo-Energy Harvesters

5.1 Introduction

One-dimensional nanostructures such as nanotubes, nanowires, and nanofibers have attracted tremendous research interest for various potential applications. Among them, the flexible nature of nanofibers could be advantageous in structures such as embedded zero-dimensional nanomaterials for sensing and actuation applications. Electrospinning is a versatile method to produce nanofibers of different kinds of materials with diameters ranging from a few micrometers down to a few nanometers. The method can be applied for synthetic and natural polymers, polymer alloys, polymer composites, and metals and ceramics [1]. Typical applications of electrospinning include filtrations, membranes, wound dressing, bio-scaffolds, and medical implants [2]. In recent years, researchers have further explored the possibilities of using electrospinning for fabrication of micro- and nanodevices such as field effect transistors [3], gas sensors [4], optical sensors [5], and DNA functional chips [6].

In a typical electrospinning process, ultra-thin polymer fibers are formed when a droplet of a viscoelastic polymer solution is subjected to a high-voltage electrostatic field [7]. After the jet is emitted from the polymer cone, it bends into a complex path and changes its shape when an electrical force stretches the fiber [8]. The stretching

Design and Fabrication of Self-Powered Micro-Harvesters: Rotating and Vibrating Micro-Power Systems, First Edition.
C. T. Pan, Y. M. Hwang, Liwei Lin and Ying-Chung Chen.

and whipping process causes the formation of thin polymer fibers, which are randomly deposited on a counter electrode serving as a collector.

Since the nanofiber formation in an electrospinning process relies on the chaotic whipping of the liquid jet, electrospinning is unstable in nature. Highly aligned nanofibers are therefore difficult to achieve using this method. Several researchers have investigated different techniques to achieve alignment of electrospun nanofibers, using for example a rotating collector [9–12] or modified electrical field [13, 14]. At the same time, numerous studies have contributed to the fundamental physics and chemistry of electrospinning [15] as well as modeling of the process [16] for further improvement and better control of the process.

In this chapter, the approach of controllable electrospinning based on near-field electrospinning (NFES) is presented as shown in Figure 5.1. The electrode-to-collector distance, which is typically of the order 10 cm in conventional electrospinning process, is dramatically reduced to *c.* 0.5 mm in order to utilize the stable liquid jet region. At such a shortened distance, electrospinning can be operated at a voltage of *c.* 1 kV, in comparison to several tens of kilovolts in conventional electrospinning. Using this concept, NFES presents a simple yet powerful method for direct-write deposition of nanofibers with a high level of controllability.

A second generation of NFES, called continuous NFES, has been developed to further extend its capability to construct large-scale patterns and to simplify its operation procedure for possible process automation. The main differences between these two generations are the use of a syringe needle for continuous polymer solution supply and the probe tip for process initiation as shown in Figure 5.2. With these modifications, continuous NFES maintains the continuous characteristic of conventional electrospinning while the superior controllability is preserved. The concept and initial

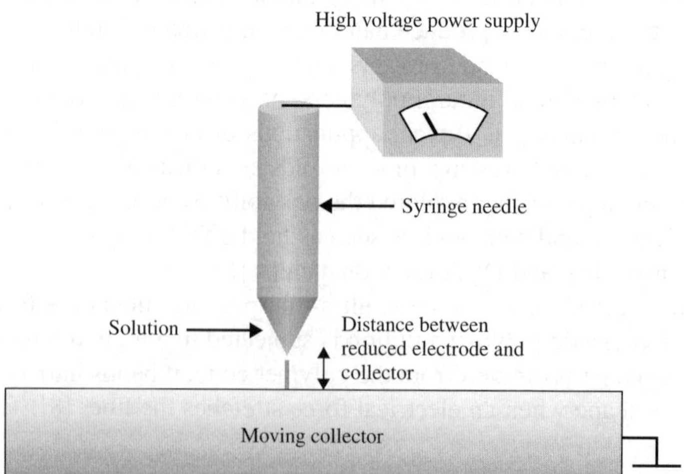

Figure 5.1 Schematic diagram of NFES [26]

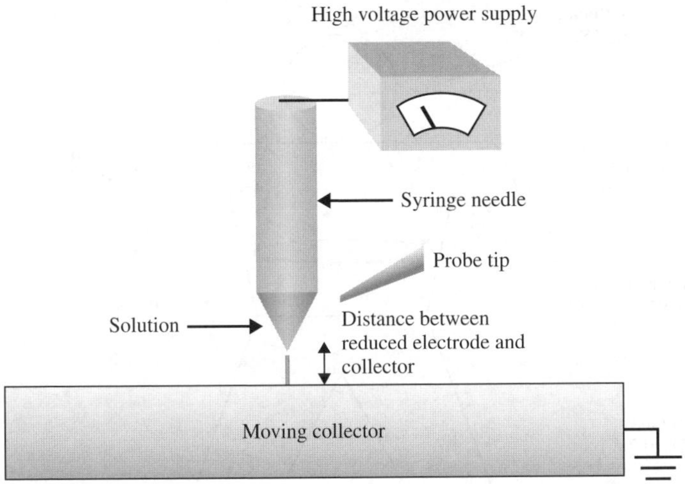

Figure 5.2 Schematic illustration of continuous NFES [26]

demonstrations of NFES suggest that NFES can be extended to new applications previously unachievable by conventional electrospinning.

NFES was used to make piezoelectric nanogenerators for possible applications of energy harvesting. Electrical power supply generated by scavenging mechanical movements are of great interest in systems of all dimensions [17]. Specifically, a self-powering system that harvests its operating energy directly from the environment is an attractive proposition for sensing, personal electronics, and defense technologies [18]. For example, buildings often have a constant low-frequency vibration (100–200 Hz) and energy could be scavenged to power stand-alone wireless transducers such as temperature and humidity sensors [19]. When a car is running, the rotation of wheels could be the power source for wireless tire pressure sensors [20]. A walking human being could provide energy from various locations such as the heels [21], elbows [22], and clothes [23]. One key element in these energy scavenging schemes is an efficient mechanical-to-electrical energy converter and the utilization of piezoelectric materials has been the major tactic in many research projects.

This chapter focuses on the scavenging of mechanical energy by utilizing electrospun piezoelectric nanofibers, which have great potential in applications beyond energy scavenging including piezoelectric sensing and actuation. Of the various applications, energy conversion is both fundamentally and scientifically challenging. Figure 5.3 shows a classical mechanical power distribution diagram from body-driven sources redrawn from Starner [24]. The total power of each action are included in parentheses while the estimations of possible scavenged power outputs are illustrated using various energy harvesters. Note that these numbers are very optimistic; most of the estimated energy outputs have never been demonstrated experimentally. For example, although footfalls have a theoretical energy scavenge output of 8.3 W,

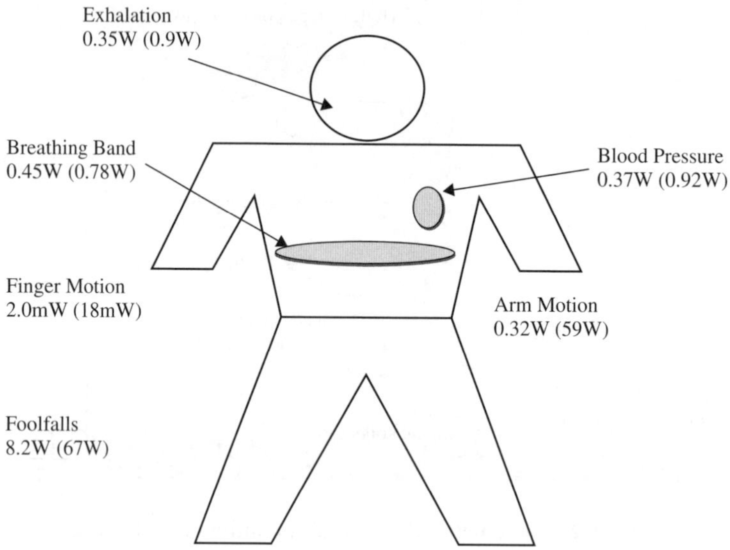

Figure 5.3 Power from body-driven movements. Total power from each action is included in parentheses [24]

a previous project at MIT more than 10 years ago was able to demonstrate 80 mW peak and 1 mW average power outputs, well below the analytical prediction [25, 26].

We begin with manufacturing protocols, material properties, and optimizations and then focus our discussions on single nanofiber nanogenerators. With the aid of NFES, polyvinylidene fluoride (PVDF) nanofibers can be direct-written and simultaneously polarized to obtain the piezoelectric crystalline structure as shown in Figure 5.4. This principle and the successful experimental demonstration of PVDF nanogenerator could be the basis for new self-powered nanodevices that harvest electricity from the environment for applications such as implantable biomedical devices, wireless sensors, and portable electronics.

The chapter is organized in six sections to describe the research and development efforts. Section 5.1 briefly introduces the background of electrospinning and nanogenerators, the development of NFES and the use of electrospun nanofibers for nanogenerators. Section 5.2 focuses on the fundamentals of the electrospinning process and describes relevant work on the alignment and assembly of electrospun nanofibers. In Section 5.3, the details of the NFES process are presented, including how to fabricate nanofibers in a direct and controllable manner. The dip-pen-like approach has limited deposition length for nanofibers and the fiber thickness is inevitably non-uniform because the polymer droplet is consumed during the process. Continuous NFES is therefore proposed and demonstrated in Section 5.4, having both the continuous nature of conventional electrospinning and the superior location control capability of NFES to produce orderly nanofiber patterns over

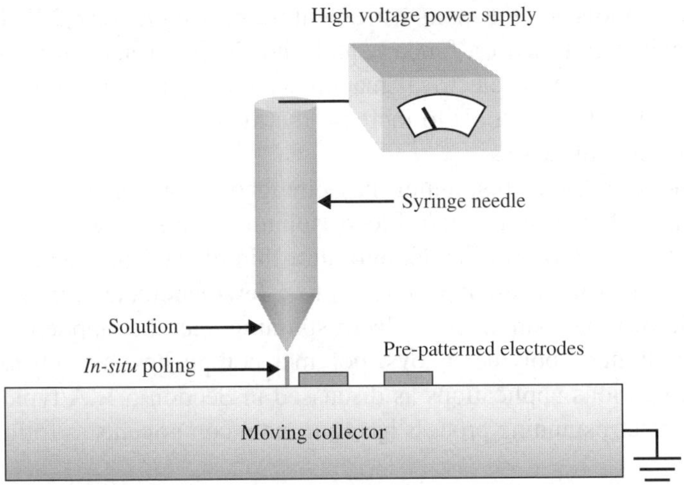

Figure 5.4 Schematic illustration of the direct-write piezoelectric nanogenerator [26]

large areas. In Section 5.5, the development of PVDF piezoelectric nanogenerator via continuous NFES is presented, including fabrication process development, experiments, measurements, and validations of piezoelectric response. Section 5.6 discusses the investigations of the material, structure and operation of piezoelectric nanogenerators with future prospects. A case study of the electrospun piezoelectric devices based on a large array of electrospun PVDF nanogenerator on a flexible substrate is described in Section 5.7. Sections 5.8 and 5.9 provide a conclusion to the chapter and discuss future research directions, including other possible applications of NFES and additional improvements of the performance and functionality of piezoelectric nanogenerators.

5.2 Fundamentals of Electrospinning Technology

5.2.1 *Introduction to Electrospinning*

Conventional tools for spinning and weaving to make yarn and fabrics may date back to the earliest relics of human habitations. For example, linen fabrics dated back to 5000 BC have been discovered in Egypt; woolen textiles have been found in Scandinavia and Switzerland from the early Bronze Age; cotton has been spun and woven in India since 3000 BC; and silk has been woven in China since at least 1000 BC. By the fourteenth century, splendid fabrics have been woven on the hand looms of the Mediterranean countries in practically all the basic structures known to modern artisans, and there has been no change in fundamental processes since that time although methods and equipment have been radically altered. Mechanical spinning using traditional tools was the main instrument for making fabrics [27]. Today, industrialized machineries based on fundamental knowledge of ancient history are used to make fabrics; advanced electrical machinery tools and control algorithms have replaced the old-fashioned tools to increase the manufacturing speed and lower the manufacturing cost of fabrics.

On the other hand, electrospinning has emerged as an interesting material processing and manufacturing method. Electrospinning is a simple but highly versatile method to process solutions or melts into ultra-thin fibers. The diameters of electrospun fibers range from a few micrometers to a few nanometers; they are therefore commonly known as nanofibers. Electrospinning can be applied for synthetic and natural polymers, polymer alloys, polymer composites, as well as metals and ceramics with various applications as discussed in Section 5.1. A typical set-up for conventional electrospinning process has four main components as follows.

1. A syringe pump to maintain a constant flow rate of the polymer solution.
2. A dispense tip that is connected to a high-voltage source to serve as a cathode, through which the polymer is squirted. Although it is often referred to as a needle, nozzle, or spinneret in the literature, the hollow needle that is commonly used with a syringe is not always suitable for solution dispensing in electrospinning because of the bevels. In most electrospinning processes, the dispense tip has a flat edge.
3. A high-voltage power supply.
4. A counter electrode which serves as a collector for the electrospun fibers.

Figure 5.5 illustrates schematically a typical set-up for conventional electrospinning with a vertical feeding direction. Horizontal or angular feeding for specific processing is also common. The inner diameter of the dispense tip is usually of the order hundreds of micrometers (needle size 32 to 20 gauge). The applied voltage is of the order several tens of kilovolts. Typical electrode-to-collector distance is of the order 10 cm.

The formation of thin fibers in an electrospinning process is based on the uniaxial stretching of a viscoelastic solution by electrostatic force [28]. Such a polymer solution can be prepared using a suitable solvent. Additives can be added for certain purposes, commonly salts to increase the conductivity and/or surfactants to decrease the surface tension. In a typical electrospinning process, a polymer solution is fed through the tip by means of a syringe pump. In idle condition, the droplet that is formed at the opening of the tip has a spherical shape due to the surface tension of the solution. When a voltage is applied to the solution, the electrostatic field causes a conical deformation of the droplet in the direction of the counter electrode into a so-called Taylor cone as shown in Figure 5.6 [29–31]. At a sufficiently high voltage, the electrostatic force can overcome the surface tension of the polymer droplet and thus initiate the ejection of a thin jet. For a very short distance immediately at the tip, whose length depends on the types of polymer and solvent system, the jet goes on a more or less stable path. The jet then undergoes a stretching and whipping process in the electrical field toward the counter electrode. The solvent evaporates along the way

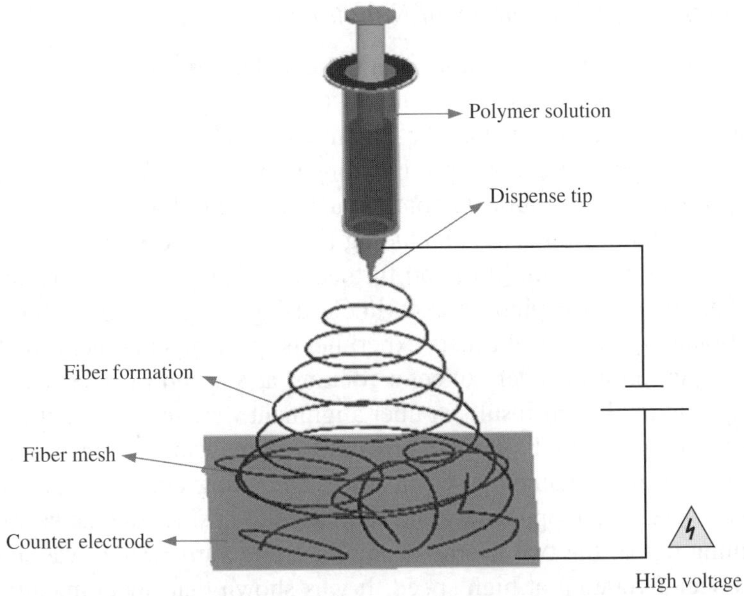

Figure 5.5 Schematic illustration of a typical laboratory set-up for conventional electrospinning with vertical solution feeding

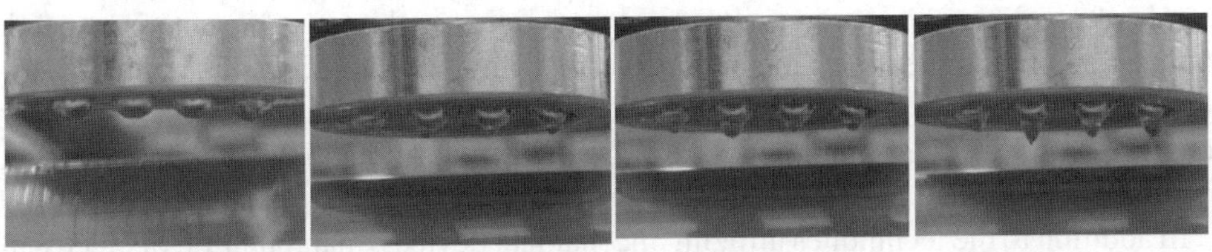

Figure 5.6 Photographs showing the meniscus at the end of the needle developing into a jet as the voltage is progressively increased

and the polymer jet solidifies. Thin solid polymer fibers are deposited on the counter electrode, also referred to as the 'collector', as randomly oriented non-woven mesh.

Nanofibers of more than 50 different types of organic polymers, including engineering plastics, biopolymers, and those with electrical, optical, or electro-optical activity, have been successfully fabricated with electrospinning [32]. A comprehensive review of polymers which have been successfully electrospun into nanofibers has been summarized in a research report [2]. More recently, the technique was developed further to the fabrication of metal oxides, ceramic, and composite nanofibers with various compositions and properties [3]. All of these nanofibers have important applications in textiles, membranes, composite reinforcement, drug-delivery-systems, tissue engineering, as well as actuators and sensors technologies.

5.2.2 *Alignment and Assembly of Nanofibers*

As described in the previous section, in a typical electrospinning process fibers are deposited randomly on the collector due to the bending instability of the charged polymer jet in an electrical field. The alignment of electrospun fibers has been a long-standing challenge. Research groups from all over the world have experimented with different methods to fabricate continuously aligned fibers. One technique is to use high-speed rotating collectors, including the use of a drum [9–12], a wheel-like disk [7, 33, 34], a wire drum [35], and further modifications of mechanical devices [36–39]. The first electrospinning experiment using a rotating collector was introduced by Boland *et al.* [40]. In their experiments, poly(glycolic acid) and collagen were electrospun on a cylinder collector rotating at a speed of 1000 and 4500 rpm, respectively. Although their results in fiber alignment were less successful, this study motivated researchers to find further improvements. For instance, Pan *et al.* [37] developed a modified electrospinning set-up using a rotating collector. Two horizontal dispense tips were placed oppositely and connected to positive and negative voltages. Fibers coming out of the two needles combined in a yarn, which was looped by a cylinder collector rotating at high speed. It was shown that fibers manufactured by this method were continuous, well-aligned, and could be deposited over a large area. The yarn was obtained from two polymer jets coming from two needles which were placed 14 cm apart. The vertical distance between the needles and collector was set at several centimeters. The collector rotated at a very high speed from hundreds to thousands of revolutions per minute. It was observed that with increasing rotation speed, the degree of alignment became better for poly(vinyl alcohol) fibers at 1.3 $m\,s^{-1}$ surface velocity and 4.3 $m\,s^{-1}$ surface velocity, respectively. The average fiber diameter of the fibers was *c.* 350 nm. This method shows good potential for manufacturing continuously aligned nanofibers using electrospinning [38].

In addition to the techniques utilizing mechanical devices, manipulation of electrical field has also been exploited for aligned fibers deposition. These techniques include the use of one or several positively charged rings as auxiliary electrodes [41–44] and well-ordered counter electrodes [13, 14]. For instance, Li *et al.* [13] demonstrated that two electrodes placed in parallel were able to collect aligned fibers in the gap. Figure 5.7 shows a schematic illustration of the set-up with the typical grounded electrode replaced by a collector consisting of two parallel silicon strips separated by a void gap. A dark-field optical micrograph of nanofibers that were electrospun from a poly(vinyl pyrrolidone) solution are directly deposited on the collector. This image demonstrates good alignment of the nanofibers with their longitudinal axes oriented perpendicular to the edges of the silicon strips. It was suggested that the fiber arrays could also be stacked layer-by-layer into a multilayered structure.

Another idea for controllable electrospinning takes advantage of the very short electrode-to-collector distance. The first experiment of electrospinning at a reduced distance was reported by Kameoka *et al.* in 2003 [45]. The method used a microfabricated scanned tip as an electrospinning source, as illustrated in Figure 5.8. As a first

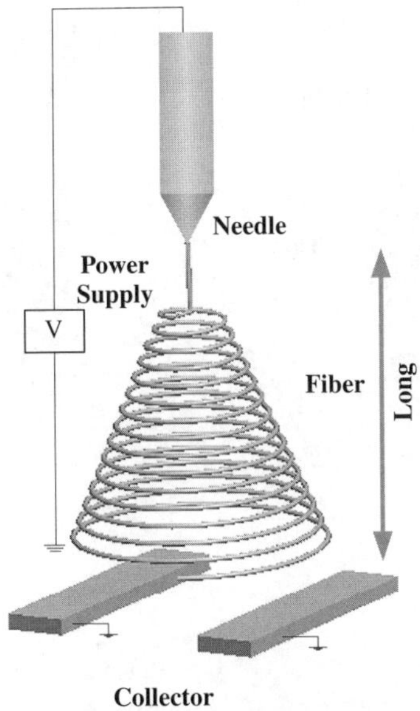

Figure 5.7 Schematic illustration of the set-up used to electrospin nanofibers to form aligned arrays. The collector was composed of two conductive substrates separated by a void gap

step, the tip was dipped into a polymer solution to gather a droplet as source material (Figure 5.8a,b). The arrow shape of the tip helped to establish a droplet located on the tip. For the electrospinning process, the tip is then placed horizontally 2 cm from the counter electrode, which was connected to an optical chopper motor to control the motion of the tip relative to the substrate (Figure 5.8c). Using a sufficiently high voltage of *c.* 4–6 kV, a polymer jet was extracted from the droplet. By moving the tip relative to a surface, oriented nanofibers can be deposited. Figure 5.8e is a scanning electron micrograph (SEM) image of a straightened electrospun polyethylene oxide (PEO) fiber with diameter of *c.* 150 nm suspended over a 3 μm wide trench.

5.3 Near-Field Electrospinning

5.3.1 Introduction and Background

Electrically driven liquid jets and the stability of electrically charged droplets have been studied for hundreds of years [46, 47]; the practical apparatus of electrospinning, in which a charged jet of polymer solution is deposited onto a collector under the influence of an electrical field, dates back to 1934 [48]. The feasibility to construct long and continuous polymeric [14, 49, 50], ceramic [51], and composite [52] nanofibers as well as nanotubes [53] with diameters less than 100 nm has been demonstrated using electrospinning. Typical applications include bio-scaffold [54],

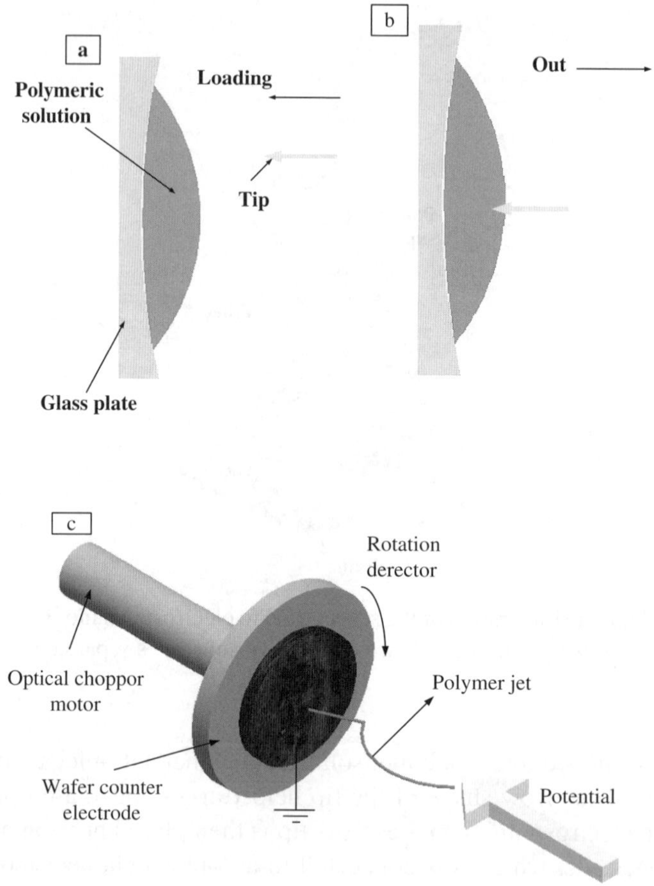

Figure 5.8 Schematic diagrams of a horizontal electrospinning process. (a) Scanning tip is dipped in a polymer solution. (b) Scanning tip is taken out of the polymer solution. (c) An electrical potential is applied to a gold wire connected to the scanning tip and the liquid jet extracted. The fiber jet is deposited on a counter electrode that is attached to an optical chopper motor for controlling nanofibers orientation. (d) Optical image of polymer liquid jet extracted from an apex of a Taylor cone established on a silicon scanning tip. (e) Scanning electron micrograph of a suspended PEO nanofiber over a 3-μm-wide trench [26]

wound dressing [55], and filtrations [56] to name a few. Researchers have further explored the possibilities of using electrospun nanofibers in fabricating micro- and nanodevices such as field-effect-transistors [3], gas sensors [4], optical sensors [5], and deposition of DNA on functional chips [6]. In these and other applications, the controllability of the electrospinning process is critical. Unfortunately, the current set-up of electrospinning is unstable in nature as it relies on the chaotic whipping of liquid jets to generate nanofibers.

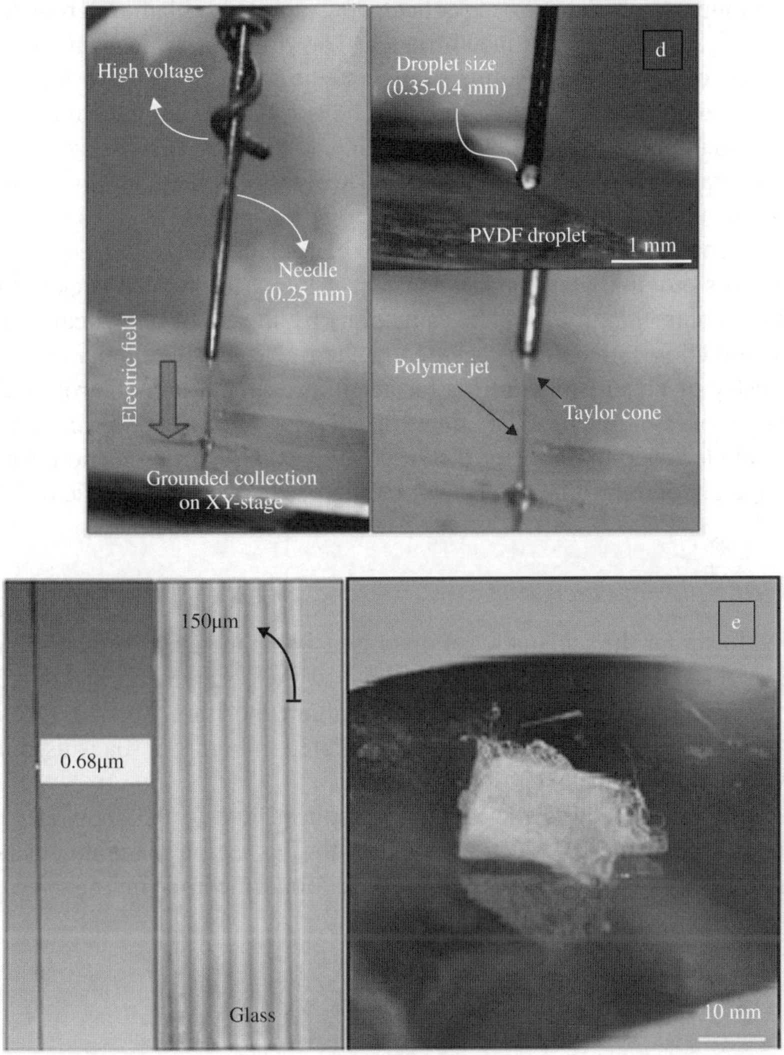

Figure 5.8 (*continued*)

Several groups in recent years have been developing orderly electrospinning processes using dynamic mechanical devices to improve the alignment and placement of electrospun nanofibers, including the use of a wheel-like reel which has been shown to position and align individual polymer nanofibers into parallel arrays [12, 35, 57]. However, highly aligned nanofibers in a large and flat area are difficult to achieve using this method. Manipulation of the electrical field has also been exploited, including: the use of an electrostatic lens element and collection target of opposite polarity, implemented to dampen bending instability for controlled deposition [42]; aligned yarns of nylon-6 nanofibers, which have been collected by rapidly oscillating a grounded frame within the jet [58]; the use of a metal frame as the collector, which can also generate parallel arrays of polyamide nanofibers [59]; and the use of electrostatic forces guiding fibers across voids in the collector [13, 14]. The combination of dynamic mechanical devices and manipulation of the electrical field can fabricate better-aligned nanofiber patterns as parallel arrays and grids [7, 60]. However, these methods do not have good control on the pitch width and cannot make complex patterns such as circles. Buckling phenomena have also been studied in which electrically charged jets impinge onto collectors moving laterally at a constant velocity to produce buckling patterns with limited controllability [61]. Investigations by means of analytical and experimental methodologies have also been conducted to study the fundamental physics and chemistry of electrospinning for further improvement and control of the electrospinning processes, such as the effects of polymer solution concentration, applied voltage, and electrode-to-collector distance [49, 62, 63]. In this section, we present the development of controllable electrospinning based on the technique of NFES.

5.3.2 Principles of Operation

Figure 5.9 illustrates the schematic set-up of NFES that merges several disparate concepts. First, the electrode-to-collector distance is in the range of 500 μm to 3 mm to utilize the stable liquid jets region for controllable deposition. Second, a solid tungsten spinneret of 25 μm in tip diameter as illustrated in Figure 5.10a is used in NFES to achieve nanofibers of sub-100 nm resolution.

Third, the applied electrostatic voltage is reduced due to the short electrode-to-collector distance while the electrical field in the tip region maintains the strength in the range of 10^7 V m^{-1} as those used in conventional electrospinning to activate the

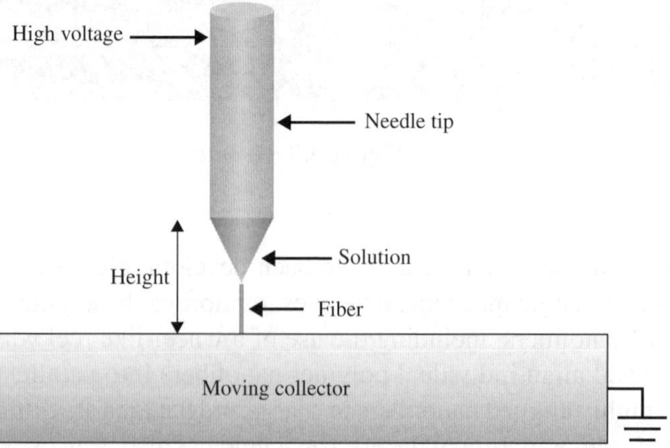

Figure 5.9 Schematic representation of NFES. The polymer solution is attached to the tip of the tungsten electrode in a manner analogous to that of a dip pen [26]

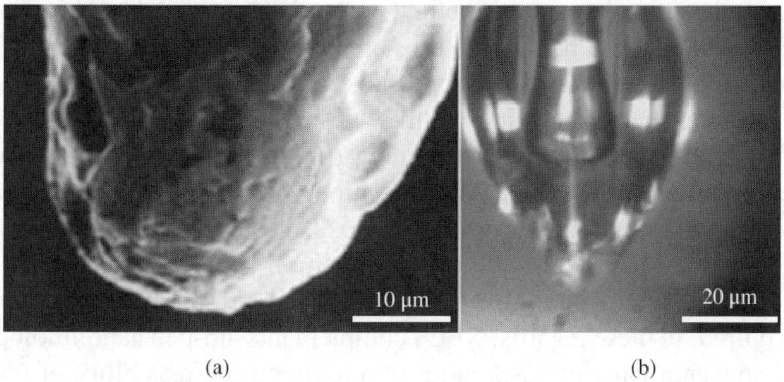

Figure 5.10 (a) SEM microphoto showing the tip region of the tungsten electrode used in the experiment with a tip diameter of 25 μm. (b) An optical photo showing a 50-μm-diameter polymer solution droplet attached to the tip of the tungsten electrode [26]

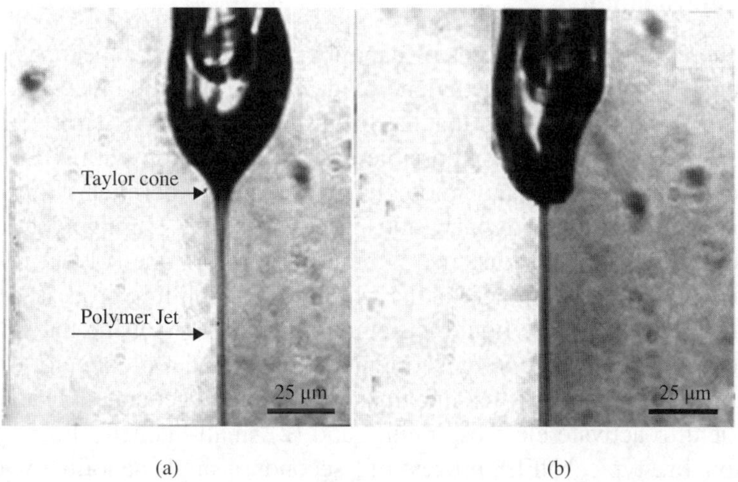

Figure 5.11 (a) A polymer jet is ejected from the apex of a Taylor cone under an applied electrical field and observed by optical microscope. (b) The size of the polymer droplet decreases as the polymer jet continues to eject polymer solution [26]

process. Forth, discrete droplets of polymer solution are supplied in a manner analogous to that of a dip pen by immersing and pulling the tungsten electrode into and out of the polymer solution. Figure 5.10b shows that a droplet of 50 μm in diameter is extracted from the polymer solution under an optical microscope for NFES. A Taylor cone is observed under an optical microscope during the process of NFES as shown in Figure 5.11a. As the process proceeds, the polymer solution on the tungsten tip is consumed and its diameter shrinks, leading to a smaller Taylor cone and thinner nanofibers as observed in Figure 5.11b.

NFES is not the only direct-write method for 'serial' material deposition in nanofabrication. For example, dip pen which uses an atomic force microscopy (AFM) tip to deliver collections of molecules to a solid substrate via capillary transport can deposit materials of better than 30 nm line-width with fine position control [64]. This is an advantage over conventional electrospinning (unless trying to deposit over a large area) in terms of continuous and fast deposition. On the other hand, inkjet printing [65] can cover a large area and quickly deposit discrete droplets with sizes in the micrometer range. The controllability of deposition location is again an advantage over conventional electrospinning, unless continuous and sub-micron line-width resolution is required. In these regards, NFES complements dip-pen nanolithology (DPN), inkjet and conventional electrospinning by providing the feasibility of controllable electrospinning for sub-100 nm nanofabrication.

5.3.3 Process and Experiment

5.3.3.1 Controllability

Although there are various choices of nanofiber material and collector systems, we focus on the deposition of polyethylene oxide (PEO, viscosity-averaged molecular weight $M_v = 300\,000$), a polymer that has been studied extensively in electrospinning onto silicon-based collectors for possible system integration with MEMS (microelectromechanical systems) and microelectronics. All experiments are conducted under room temperature and one atmospheric pressure. During electrospinning, adequate electro-to-collector distance is required to allow the solvent to evaporate and nanofibers to become thinner via the whipping and splitting processes [8]. NFES takes advantage of the stable liquid jet region immediately outside the spinneret for position controllable deposition by shortening the electrode-to-collector distance. At the same time, an electrode with a fine tip is introduced to generate: (1) an intensified electrical field to activate electrospinning; and (2) small-diameter liquid jets in the stable region. In a typical NFES process of 5 seconds, a single nanofiber with a length of several centimeters can be constructed. The droplet size, polymer solution concentration, applied voltage, and the electrode-to-collector distance collectively affect the morphology of the electrospun nanofibers. Overall, the near-field electrospun solid nanofibers have similar morphology to the conventional electrospinning process.

The controllability of nanofibers is demonstrated by moving the collector along a straight line to generate line-shape nanofibers. Figure 5.12 shows an example where parallel lines perpendicular to each other are constructed using a x–y stage (Nano Workcell, Adept Japan Inc.) to control the collector movement [26]. In this experiment, the electrode-to-collector distance is 1 mm; applied voltage is 1000 V; and PEO solution is 5 wt%. The close view SEM microphoto in Figure 5.13 shows that the typical diameter of the nanofiber by NFES is 300 nm. More controllability is illustrated in Figure 5.14 when the collector is moving along a straight line with speeds of 5, 10, and 15 cm s^{-1}, respectively. In these cases, the nanofiber electrospinning

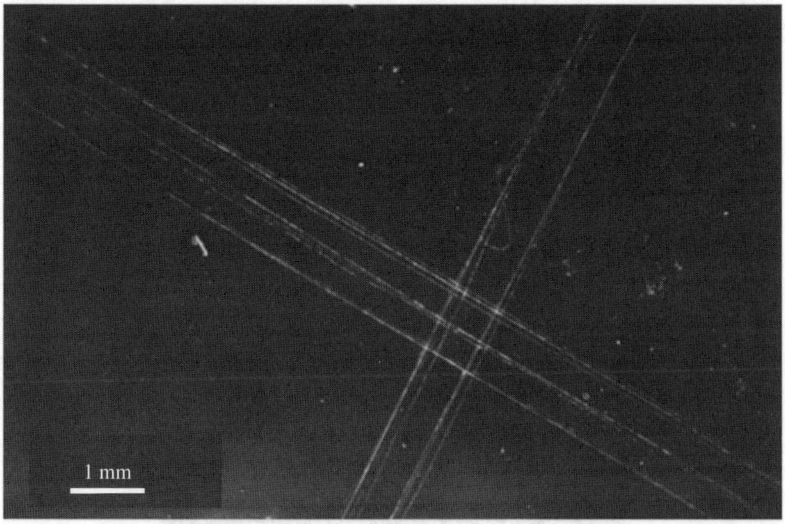

Figure 5.12 Two groups of parallel lines are constructed perpendicular to each other by NFES [26]

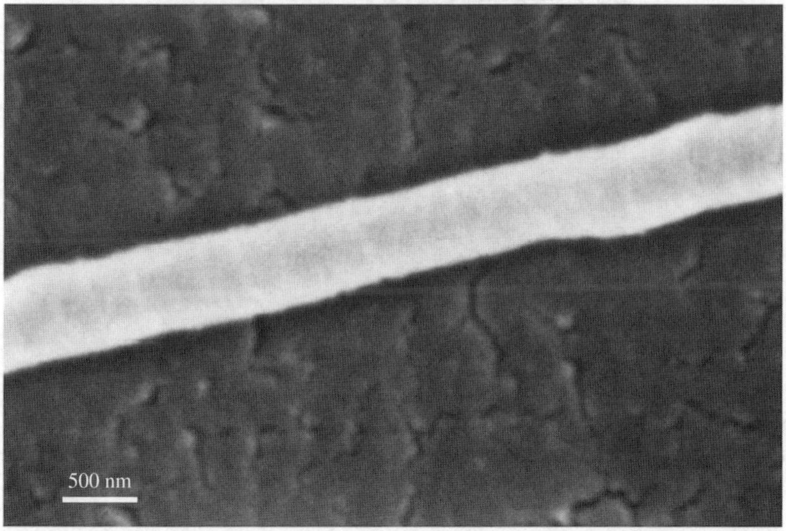

Figure 5.13 Enlarged SEM microphoto showing a nanofiber with diameter of 300 nm [26]

speed is faster than the collector moving speed. The result is that 'local spiraling' occurs as in Figure 5.14a and gradually diminishes as the collector is moving faster in Figure 5.14b,c. When the collector speed reaches 20 cm s^{-1} as shown in Figure 5.14d, straight-line-shape nanofibers can be constructed. In this case, we control the x–y stage to deposit two straight nanofibers 25 μm from each other.

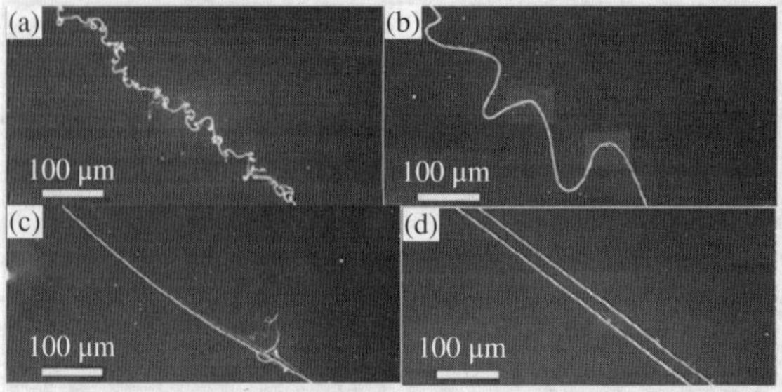

Figure 5.14 (a) NFES result when the collector has a moving speed of 5 cm s^{-1} showing local spiraling. (b) Collector moving speed of 10 cm s^{-1}. (c) Collector moving speed of 15 cm s^{-1}. (d) Collector moving speed of 20 cm s^{-1}, allowing straight lines to be constructed. The two nanofibers are separated by 25 μm under the control of the x–y stage [26]

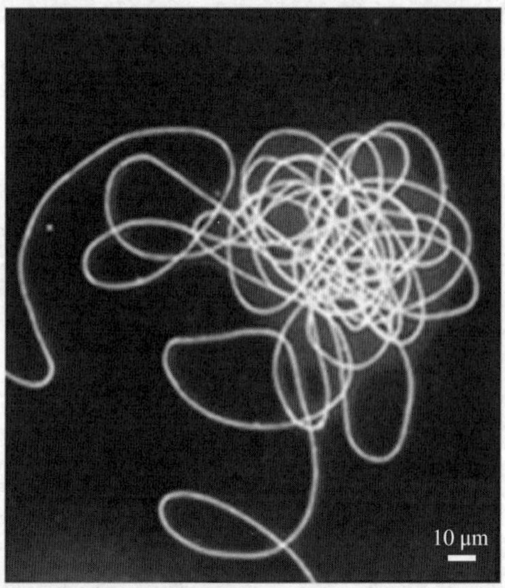

Figure 5.15 When the collector is stationary for 0.5 s while making a 90-degree turn, the local spiraling effect is observed as the result of self-expelling as nanofibers are electrically charged. The spread is of the order 50 μm in diameter

5.3.3.2 Deposition of Various Patterns

The location of the nanofiber formation under NFES is investigated by keeping the spinneret and collector at a fixed distance. It is observed that in a period of less than 1 second without moving the collector, the NFES deposition is concentrated within a short radial distance of 50 μm as seen in Figure 5.15. The local spiraling is the result of the self-expelling of nanofibers as they are electrically charged. This demonstrates the feasibility of location control of NFES on a conductive collector as opposed to conventional electrospinning which results in widespread and random deposition. Figure 5.16 shows a 2-second NFES deposition result on a stationary, 2-μm-thick insulating oxide-coated silicon collector. Although the collector does not move, concentric and elliptical patterns of nanofibers are constructed as shown and cover an area with base length of more than 300 μm. The close view SEM photo in Figure 5.17 shows that the typical distance between adjacent nanofibers of *c*. 1.5 μm is relatively uniform over the whole pattern with no external control.

Figures 5.18 and 5.19 further illustrate the controllability of NFES under manual operations. Figure 5.18 shows that an L-shaped nanofiber is generated under one single NFES process. A straight line is plotted to construct the top line of the L-shape structure. The collector then stops to change to the downward direction and stops again before moving leftwards. It is shown during each stop of less than 0.5 second the

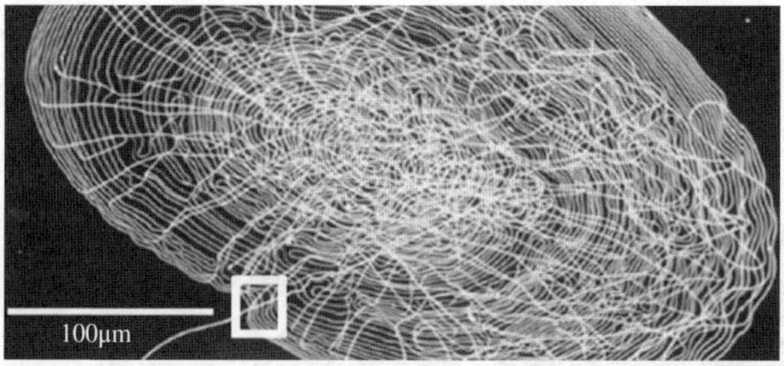

Figure 5.16 When the substrate is covered with an electrical isolation layer of silicon dioxide, the local spiraling is enhanced as the concentric/elliptical rings are formed in an area of *c*. 300 μm in diameter [26]

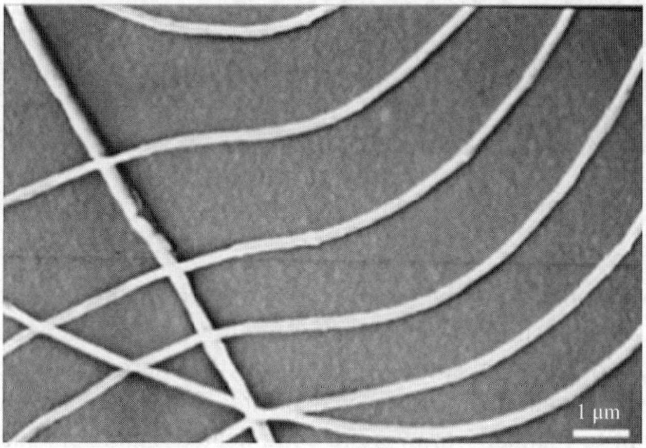

Figure 5.17 Close view of Figure 5.16, showing that the distance between the adjacent nanofibers is *c*. 1.5 μm [26]

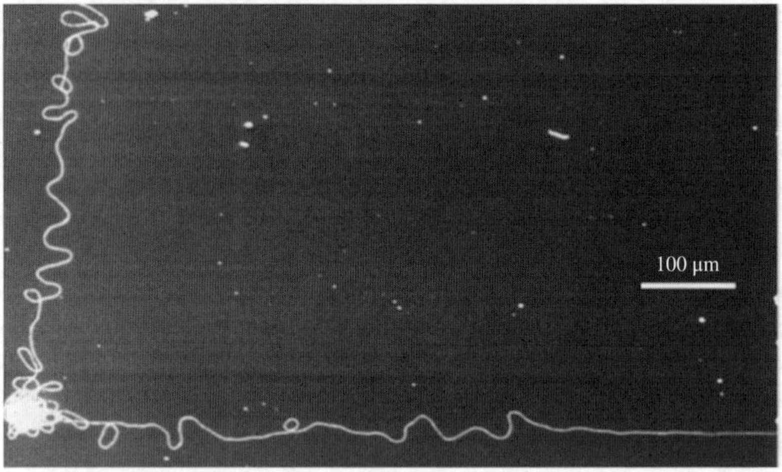

Figure 5.18 A L-shaped symbol is plotted manually by NFES. Excessive nanofiber depositions are observed at the two corners when the collector stopped shortly to adjust moving direction

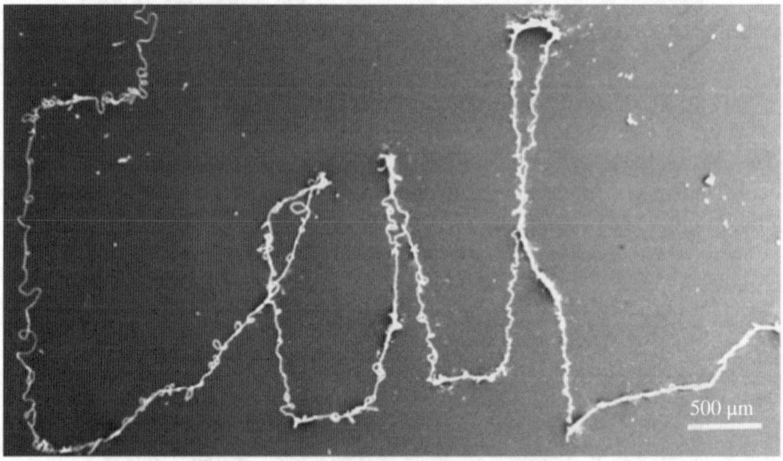

Figure 5.19 The three-letter word 'Cal' is manually drawn over a period of 3 s [26]

continuous deposition of nanofibers causes accumulation at the corners. We purposely slow down the moving speed of the collector to draw the middle and bottom portion of the L-shape, and the 'local spiraling' occurs as the charged nanofibers expel each other when excessive nanofibers are to be deposited at the same location. Figure 5.19 shows an attempt to manually draw more complicated characters such as the letters 'Cal' in one single NFES process. Each character is *c*. $1 \times 2\,\text{mm}^2$ in size and it takes *c*. 1 s to manually write one character. The writing speed is not fast enough to have a clean shape under the manual control.

We conclude that location and pattern control of NFES are achievable to deposit nanofibers without spiraling effects when the relative moving speed between the spinneret and collector is comparable to the electrospinning speed, that is, determined by various factors including the viscosity, conductivity, and surface tension of the polymer solution; applied electrical field; tip diameter of the spinneret; the size of the droplet; and ambient parameters including temperature, humidity, and air velocity. Figure 5.20 shows experiments on the minimum required voltages versus electro-to-collector distances to activate NFES under various polymer concentrations and types of collectors when the polymer droplets have a nominal size of 50 μm in diameter. The minimum applied voltage increases when either the electrode-to-collector distance or the polymer solution concentration increases.

An electrical field simulation was performed using COMSOL for both a 500-μm-thick silicon wafer and a 500-μm-thick silicon wafer coated with a 2-μm-thick layer of silicon dioxide. Simulation results indicate that the magnitudes of the electrical field for the collectors with and without the oxide insulating layer are similar under the same applied bias, as the thin oxide layer does not affect the electrical field strength significantly. The maximum electrical fields for both cases are 1.097 and $1.089 \times 10^7\ \mathrm{V\,m^{-1}}$, respectively.

Furthermore, it is observed that nanofibers collected on a silicon collector are nominally bigger than those on an oxide-coated collector under the same deposition conditions, since nanofibers travel a longer distance on average (Figure 5.16) and have more time for the solvent to evaporate. Statistical experimental results show that when the electrode-to-collector distance is 500 μm with 3 wt% PEO, the diameters of nanofiber on the silicon and oxide-coated silicon collector are in the range of 150–300 nm and 50–200 nm, respectively.

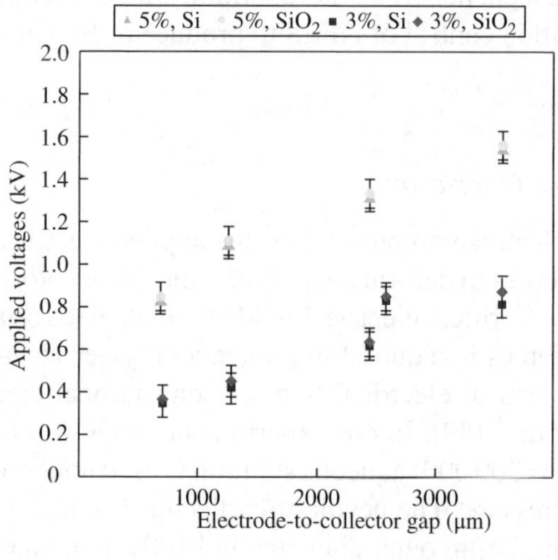

Figure 5.20 Experimental results of the minimum required voltages versus electrode-to-collector distance to activate the NFES process under various polymer concentrations on either silicon or silicon oxide collectors [26]

5.3.4 *Summary*

NFES is a simple yet powerful method for direct-write deposition of nanofibers with unprecedented controllability at resolutions comparable to those achieved with much more expensive and sophisticated lithography tools. It should be especially useful for the heterogeneous integration of nanoscale materials to devices prepared by conventional lithographic and manufacturing methods, such as microelectronics and MEMS structures. Furthermore, NFES could play a key role in building larger-area ordered non-woven nanofibers for various applications.

5.4 Continuous NFES

5.4.1 *Introduction and Background*

Conventional electrospinning processes producing randomly deposited nanofibers have been used in various applications including filtration [56], texturing [66], composite reinforcement [67, 68], and tissue scaffolds [69]. The disorderly fashion of such deposited nanofibers has limited potential, however. For example, well-controlled architectures rather than random configurations could give rise to improvements in the aforementioned examples as well as other applications. Electrospinning using short needle-to-collector distance, including scanning tip electrospinning [45] and NFES, have shown easier and more predictable location control for the deposition of nanofibers. However, the polymer droplet approach in these processes limits the total length of nanofiber deposition, and the thickness/diameter of the fibers is inevitably non-uniform because the polymer droplet is gradually consumed during the process. In this section, the principle and methodology of continuous NFES is proposed and demonstrated, which benefits from the continuity of conventional electrospinning and the superior location control of NFES to produce orderly nanofiber patterns over large areas.

5.4.2 *Principles of Operation*

In the conventional electrospinning process, the applied electrical field generates sufficient electrostatic forces to deform the polymer meniscus into a conical shape called a 'Taylor cone' [29]. A critical electrical field at which electrostatic forces overcome the surface tension forces is required to induce a polymer jet from the tip of Taylor cone. Typically, the critical electrical field of conventional electrospinning process is of the order 10^5 V m^{-1} [49]. In our experimental demonstration, 7 wt% polyethylene oxide (PEO, $Mv = 300\,000$) aqueous solution is used under room temperature and one atmosphere of pressure. The needle-to-collector distance is fixed at 500 μm and a syringe needle with 200 μm outer diameter and 100 μm in inner diameter is utilized as the spinneret to continuously supply polymer solution.

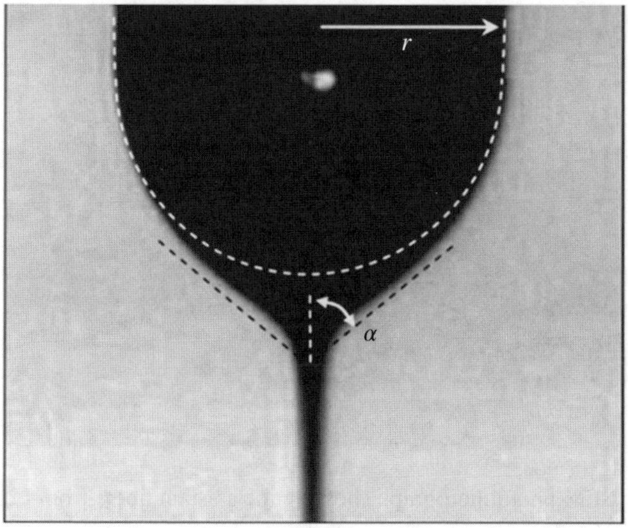

Figure 5.21 Optical image showing the syringe needle ($r = 100\,\mu m$), droplet and cone ($\alpha = 49°$) of conventional electrospinning in a near-field set-up. The white dashed curve represents the fluid surface just before jet initiation and the cone part is the fluid surface right at jet initiation. A polymer jet starts to eject from the droplet while 1.5 kV is applied to the syringe needle and the needle-to-collector distance is 500 μm [70]

Figure 5.21 shows that when the applied voltage on the syringe needle reaches a critical value of 1.5 kV for jet initiation and the cone has a semi-vertical angle of around 49°, the corresponding critical electrical field ($3 \times 10^6\ V\,m^{-1}$) is 1 order of magnitude larger than that of conventional electrospinning because the surface tension of the droplet increases as syringe needle diameter decreases. The polymer jet below the tip of the cone is *c.* 25 μm in diameter, similar to that of the conventional electrospinning process in which the needle-to-collector distance is *c.* 5–50 cm. However, since the needle-to-collector distance is only 500 μm in the near-field set-up, the solvent in the polymer jet does not have enough time to fully evaporate. As a result, as-spun fibers are 3–6 μm in diameter and remain in liquid form shortly after deposition as shown in Figure 5.22. Consequently, fibers will often merge with those that have just been deposited onto the collector.

In order to have electrospun nanofibers of diameter in the sub-100 nm range, the applied electrical field for conventional electrospinning can be increased to cause greater stretching of the polymer solution. The controllability of nanofiber deposition is however severely affected by bending instability [8]. The key strategy for producing sub-100 nm nanofibers via NFES, on the other hand, is to reduce the size of the polymer jet emerging from the cone by reducing the applied electrical field below the critical value.

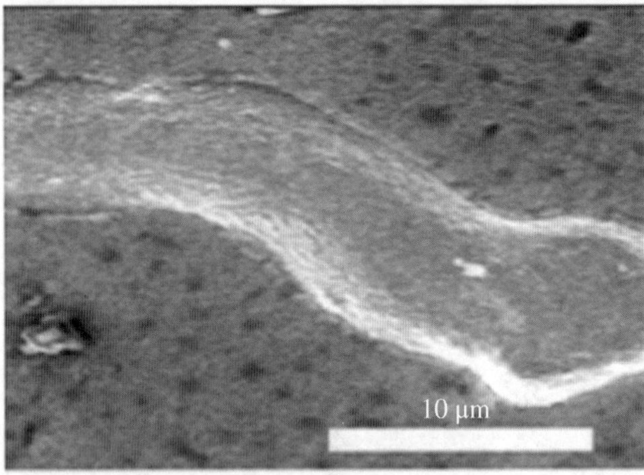

Figure 5.22 SEM photomicrograph showing the as-spun fibers from Figure 5.21 [70]

Theoretically, this can be accomplished by simultaneously reducing the applied voltage and polymer solution supply rate after the onset of electrospinning is achieved (Figure 5.21). Practically, this is very difficult to control as a high supply rate would result in excess polymer plummeting to the collector, disrupting the delicate electrospinning process; a low supply rate would cause the pendant droplet to shrink quickly and terminate the electrospinning process. To circumvent the limitations of these approaches, we have developed a simple and reliable technique to achieve continuous NFES by means of an initial mechanical drawing process that aids electrical force in overcoming surface tension to initiate the ejection of a polymer jet.

This process starts by applying a subcritical voltage to deform the polymer meniscus without inducing electrospinning. Mechanical drawing is applied by using a tungsten probe with 1 μm tip diameter to poke inside the meniscus. The probe is then rapidly pulled away from the polymer droplet to activate the continuous electrospinning process as illustrated in Figure 5.23a–c. The initial location of the probe tip is not crucial since the polymer jet will automatically move downwards and align itself with the applied electrical field as shown in Figure 5.23c,d.

Instead of mechanical drawing, the subcritical voltage is the key control parameter in determining the size of deposited nanofibers as depicted in Figure 5.23b where the fiber jet is relatively thick during the mechanical drawing process; Figure 5.23c,d shows the thinner fiber jet after the mechanical probe has been removed.

In the case depicted in Figure 5.24, the applied voltage is 600 V (corresponding to an electrical field of 1.2×10^6 V m^{-1}). The polymer jet diameter immediately outside of the cone is *c*. 3 μm, and the deposited nanofibers on the collector have diameters of *c*. 50 nm as shown in Figure 5.25. The semi-vertical angle of the cone is 75°. A polymer feed rate of *c*. 0.1 μL h^{-1} can maintain the droplet shape for continuous and uniform deposition of nanofibers. With the aid of this initial mechanical drawing process,

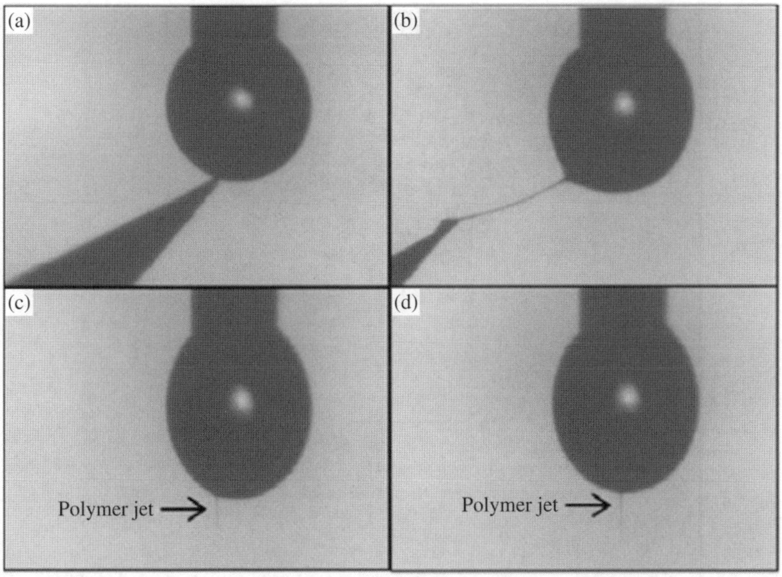

Figure 5.23 (a) Optical image showing a tungsten probe tip poking inside the polymer meniscus. (b) Optical image showing mechanical drawing of a fiber from polymer droplet. (c) Optical image showing the initiation of the electrospinning process. (d) Optical image showing how the polymer jet automatically moves downwards due to the applied electrical field [70]

a continuous supply of polymer solution is maintained for NFES while the size of polymer jet emerging from the cone is drastically reduced to achieve sub-100 nm deposition of nanofibers. Moreover, this approach avoids the cumbersome issue of consistently extruding a polymer droplet of the same size through a syringe needle of small diameter.

5.4.3 Controllability and Continuity

To demonstrate the controllability and continuity of this process, a variety of orderly patterns are deposited onto silicon collectors using a programmable x–y stage (XPS controller driver and Newport Inc.). Experimentally, the syringe needle is fixed while the stage translates at a speed of 120 mm s^{-1}. Figure 5.26 shows the smooth writing of the text 'Cal' in an area of 1×0.5 mm^2 where the diameter of the fiber is 150 nm. The smoothness of the pattern was achieved by setting the translational speed of the stage as close as possible to the deposition rate. A slow translational speed would cause the nanofiber to form spiraling patterns as shown in the region preceding the letter 'C' in Figure 5.26, while a high translational speed would not afford the nanofiber enough time to anchor itself at all points on the intended pattern. A circular design, for instance, would appear as a polygonal nanofiber pattern.

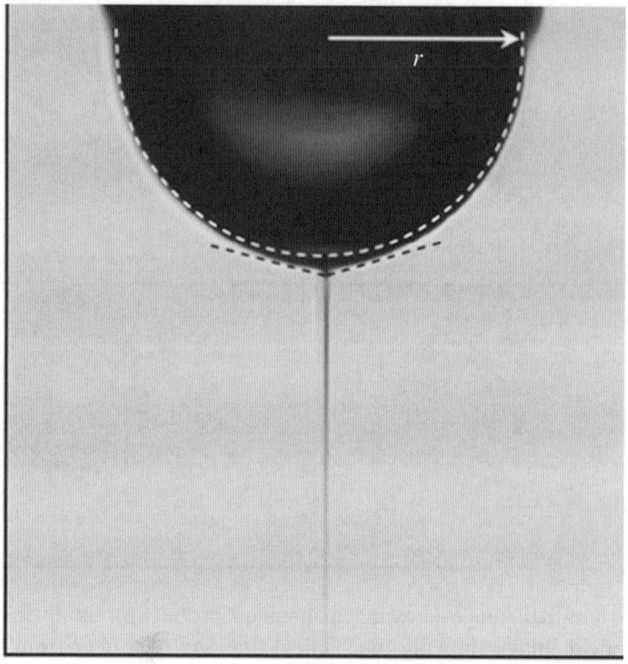

Figure 5.24 Optical image showing the syringe needle ($r = 100\,\mu m$), droplet and cone ($\alpha = 75°$) of continuous NFES. The process is initiated with the aid of a probe tip at a voltage of 600 V while the needle-to-collector distance is 500 μm [70]

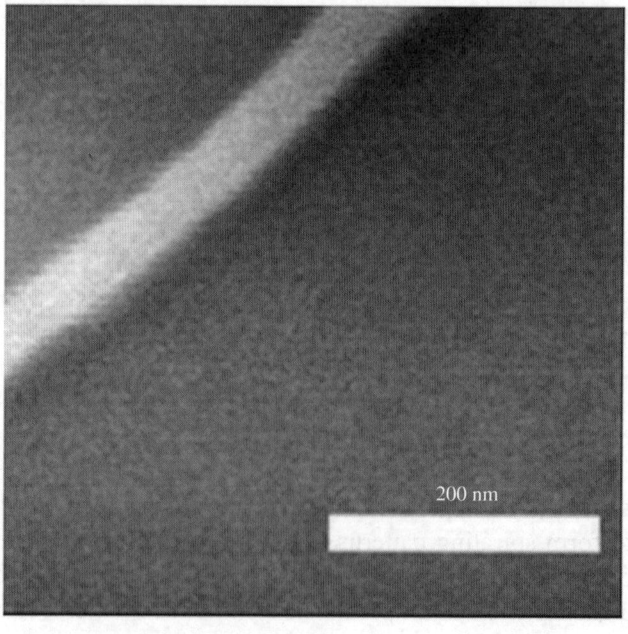

Figure 5.25 SEM photomicrograph showing the as-spun fibers [70]

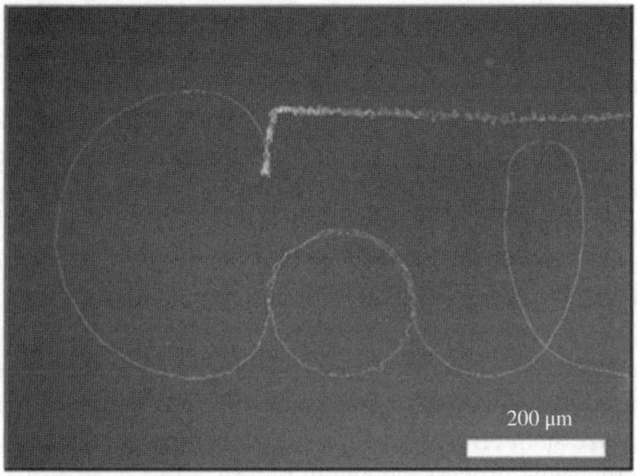

Figure 5.26 The three-letter word 'Cal' is drawn on a silicon chip by a programmable *x–y* stage [70]

Figure 5.27 A single nanofiber deposited on a silicon chip in a designed trajectory covers a $4 \times 4\,\mathrm{cm}^2$ area in 15 min via continuous NFES. In the foreground is a US quarter [70]

Additionally, the size of the pattern of Figure 5.26 does not represent the smallest design that can be drawn using continuous NFES. Since the speed of the *x–y* stage must match the electrospinning deposition rate for smoothness, this criterion must be met regardless of the intricacy of the pattern; however, smaller arc trajectories demand extremely large radial acceleration, and the extent to which this acceleration can be executed depends on the capabilities of the *x–y* stage. Figure 5.27 illustrates experimental results of the continuous NFES of a single nanofiber deposited on a silicon chip in a designed trajectory over a $4 \times 4\,cm^2$ large area. The process can continue running as long as polymer solution is supplied from the syringe needle. In this case, the deposition period is 15 minutes for a total length of 108 m and the nanofiber has a diameter of 709 ± 131 nm. The diameter and standard deviation are calculated based on 100 data points measured directly from SEM. The optical image in Figure 5.28 shows that the nanofiber is deposited on the silicon chip with a fiber pitch of 49.88 ± 3.98 μm (from 50 data points), while the designed trajectory has a fiber pitch of 50 μm. We believe this uneven spacing is mainly due to the oscillation of the cone on the polymer droplet, which is caused by the interference of the fiber charges with the electrical field, mechanical buckling during impact between polymer jet and collector [61], and external perturbations [8]. Further investigation is required in order to minimize such errors on nanofiber deposition. A triangular pattern is also deposited using the same process, shown in Figure 5.29.

5.4.4 Process Characterization

Similar to the conventional electrospinning process, the diameter of nanofibers can be adjusted by controlling various operating parameters. The following characterization of fiber diameters and standard deviation are calculated based on 50 data points measured directly from SEM in each experiment. When the PEO concentration is 7 wt% and the needle-to-collector distance is fixed at 500 μm, the average fiber diameter can be varied in the range of 38 ± 4 to 2314 ± 622 nm under different

Figure 5.28 Optical image of a grid pattern with controlled 50 μm spacing of the area highlighted by the red box in Figure 5.27 [70]

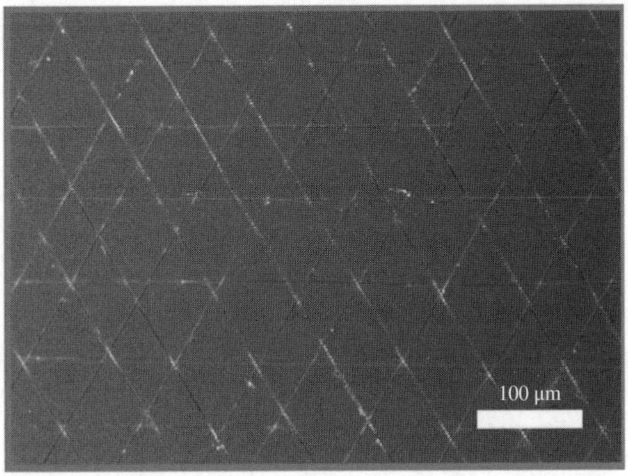

Figure 5.29 SEM photomicrograph of a crosshatch pattern of the area highlighted by the green box in Figure 5.27 [70]

applied voltages as shown in Figure 5.30. The proportional relationship between the fiber diameter and applied voltage confirms the previous observation that the cone size and polymer jet diameter both increase with applied electrical field. Higher voltage therefore results in thicker fibers in continuous NFES, which is contrary to the trends observed in conventional electrospinning [72–74]. In the conventional electrospinning process, higher bias voltages imply stronger Columbic repulsion force in the polymer jet to further stretch the nanofibers during the whipping process. In the continuous NFES process, this effect is diminished due to short needle-to-collector distance such that the initial polymer jet diameter dominates the final nanofiber diameter rather than bending instability.

The polymer concentration also affects the morphology of the electrospun nanofibers. If the PEO solution concentration is lower than *c.* 3 wt%, the continuous NFES process cannot be activated due to its insufficient viscoelasticity. At such low polymer concentrations, the elongational viscosity is insufficient to suppress capillary break-up. When the polymer concentration is increased nanofibers are obtained due to an increase in viscosity, thus counteracting the stretching via electrical force. As shown in Figure 5.31, the average fiber diameter can range from 37 ± 3 to 119 ± 6 nm by increasing the polymer concentration when the needle-to-collector distance is fixed at 1000 μm under a bias voltage of 500 V.

The needle-to-collector distance is another key factor that affects the diameters of deposited nanofibers. As shown in Figure 5.32, larger needle-to-collector distance results in thinner fibers due to a longer period of stretching as polymer jets travel towards the collector. The average fiber diameter ranges from 49 ± 7 to 74 ± 9 nm when the applied voltage is 800 V and the concentration of PEO solution is 7 wt%.

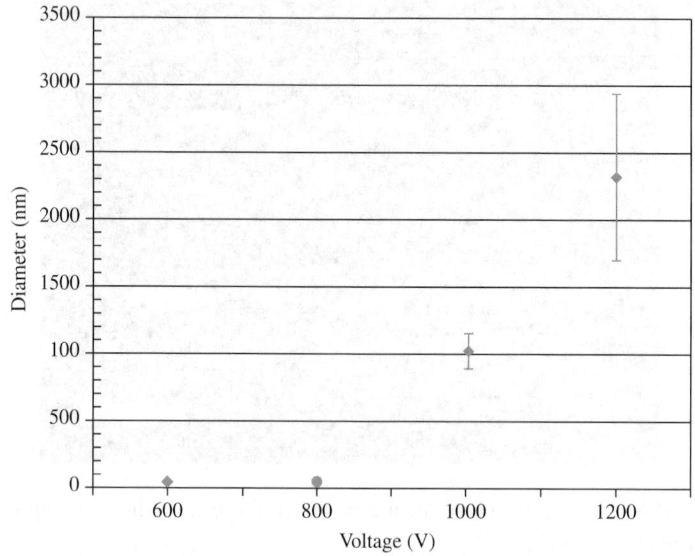

Figure 5.30 Plots showing the dependence of nanofiber diameters on applied voltage. PEO concentration is maintained at 7 wt% and needle-to-collector distance at 500 μm. Experiments were conducted under room temperature and one atmosphere pressure [70]

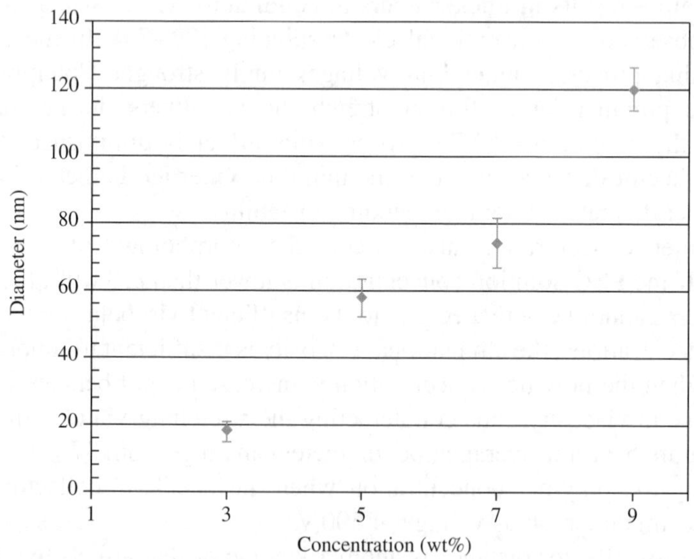

Figure 5.31 Plots showing the dependence of nanofiber diameters on polymer concentration. Applied voltage is maintained at 500 V and needle-to-collector distance at 1000 μm. Experiments were conducted under room temperature and one atmosphere pressure [70]

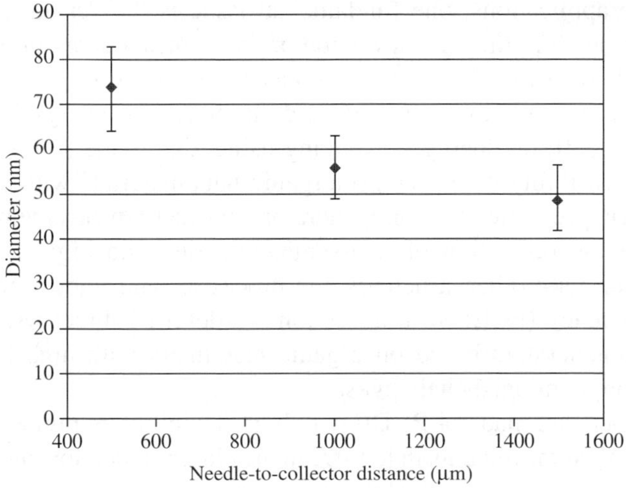

Figure 5.32 Plots showing the dependence of nanofiber diameters on needle-to-collector distance. PEO concentration is maintained at 7 wt% and applied voltage at 800 V. Experiments were conducted under room temperature and one atmosphere pressure [70]

5.4.5 Summary

The concept and demonstration of continuous NFES allows new applications that were previously impossible by conventional electrospinning. We believe that the ability to deposit polymeric materials at specific locations in specific patterns opens up opportunities for heterogeneous integration of a variety of materials; continuous NFES is a necessary technology for cost-effective manufacturing. There are many other areas to be explored within the utilization of continuous NFES. For example, continuous NFES has the potential to create orderly patterns through automated processes to enable bioengineers to create scaffolds *in vitro* that are highly representative of real extracellular matrices (ECMs). Such scaffolds could potentially extend current tissue engineering techniques to develop previously unachievable organ types and increase the viability of these tissues.

5.5 Direct-Write Piezoelectric Nanogenerator

5.5.1 Introduction and Background

Mechanical energy scavenging from ambient environments is an attractive renewable source of power for various applications. Examples range from large-scale power generators which convert mechanical actuation found in nature, such as waterfalls, wind, and ocean waves, into electricity [75, 76] to small-scale energy harvesters which scavenge energy from mechanical vibrations in buildings, automotives, appliances, and human movement [18, 77]. In pursuing these mechanical energy harvesters, especially for small-scale applications, one fundamental issue is the design and selection of structural materials for efficient conversion of mechanical energy into electricity.

Recent work in the field of nanomaterials has provided new opportunities and directions in engineering effective materials and structures for energy harvesters. Prior research [23, 78–80] on energy scavenging using zinc oxide (ZnO) nanowires has illustrated the feasibility of utilizing inorganic nanomaterials with semiconducting and piezoelectric properties for nanogenerators. Further advancements for nanogenerators include the utilization of non-semiconducting and organic nanomaterials, direct-integration with other structures and processes, and improvements in energy conversion efficiency (ECE). In this section we describe direct-write piezoelectric polymeric nanogenerators based on organic nanofibers with high ECE to address some of the aforementioned challenges.

These nanofibers are made of PVDF with high flexibility, minimizing resistance to external mechanical movements in low-frequency large-deflection energy scavenging applications.

5.5.2 Polyvinylidene Fluoride

PVDF is a semi-crystalline polymer performing good chemical resistance, mechanical strength, and flexibility and is well known for its ferroelectric and piezoelectric properties. PVDF has four major crystalline phases – α, β, γ, and δ – which are distinguished by the conformation of the C-C bonding along the chain backbone. Among these, α-phase is the most energetic stable phase and therefore the most common. It is usually observed in commercial PVDF powders and as-manufactured films. The chemical structure of the α-phase is illustrated in Figure 5.33a. In this phase, the chains

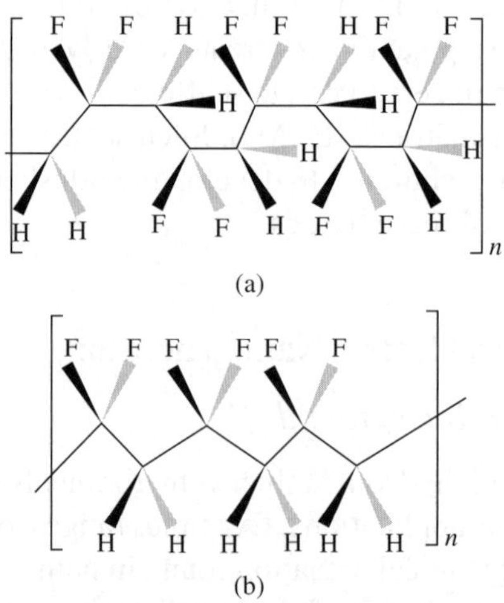

Figure 5.33 Chemical structure of PVDF in its (a) α-phase and (b) β-phase. Dipoles are oriented randomly in α-phase and organized in the β-phase, causing the material to exhibit piezoelectricity

stack with the respective polarizations in alternating directions (trans gauche + trans gauche-conformation, tg + tg-), such that the energy level is at the minimum. This phase does not exhibit piezoelectricity, however. The β-phase is known to be responsible for the piezoelectric response of PVDF. In the β-phase, the conformation is all-trans (tttt), such that the dipoles are oriented in the same direction (Figure 5.33b). This causes the piezoelectric properties of the material.

Many research groups have studied the formation of the β-phase structure of PVDF by different processing methods: crystallization under high pressure [26]; poling under a high electrical field [81]; polymer stretching [82, 83]; annealing [84]; or blending [85]. Lately, it has been reported that the β-phase can be formed during the electrospinning process [86–95]. The suggested mechanism is based on the strong electrical field in an electrospinning process which forces the dipoles to organize in the same direction, as illustrated in Figure 5.34. This introduces the essential concept of a PVDF nanogenerator.

5.5.3 Theoretical Studies for Realization of Electrospun PVDF Nanofibers

5.5.3.1 Review of Electrospinning of PVDF

As introduced in the previous section, a number of studies have investigated the fabrication of PVDF nanofibers via conventional electrospinning including processing, characterization of fiber morphology, and crystalline structures processing [86–94]. Although NFES differs from conventional electrospinning in many aspects, a good

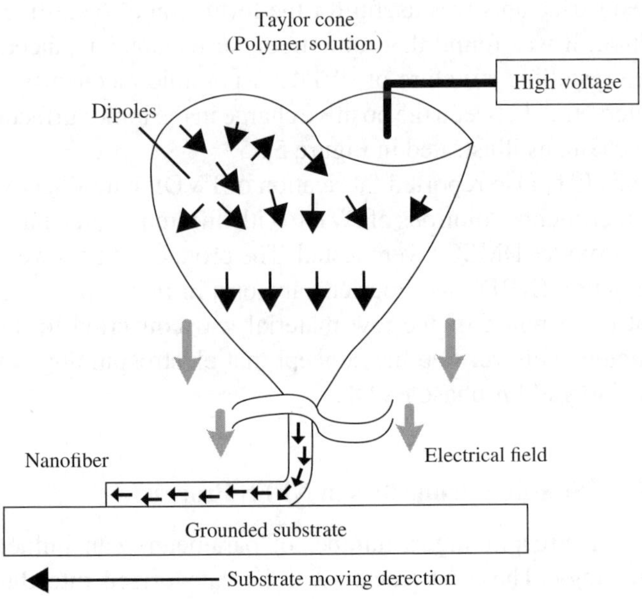

Figure 5.34 Schematic of the *in situ* poling process during electrospinning [26]

understanding of electrospinning of PVDF is still valuable. For instance, an essential point is the possibility of β-phase formation during electrospinning. This section provides a review of present studies on electrospinning of PVDF to gain better insights into process fundamentals and, in particular, factors influencing the resulting fiber characteristics.

The first extensive study on electrospinning of PVDF was presented by Nasir *et al.* [90]. Their work investigated the influence of electrospinning parameters, for example polymer concentration, applied voltage, feeding rate, and electrode-to-collector distance, on the characteristics of the resulting nanofibers, for example diameter, morphology, and crystalline structure. In their study, PVDF solutions were prepared in N,N-dimethylacetamide (DMAC) at different concentrations ranging from 5 to 26 wt%. The solution was used in a conventional electrospinning set-up with voltage of 7.5–15 kV, tip-to-collector distance of 8–15 cm, and flow rate of 1–11 μm min^{-1}. Results from the process followed typical electrospinning behavior. For example, fiber formation was promoted by higher polymer concentration or solution viscosity. The fiber diameter increased with increasing concentration and applied voltage. The distance from tip to collector had no significant effect on the resulting fibers and their diameter once the fibers were successfully deposited. The feeding rate of the polymer solution showed an effect up to a rate of 5 L min^{-1}. Fiber diameter decreased as the flow rate increased to 5 μL min^{-1} and remained constant at higher flow rates.

Nasir *et al.* [90] studied the effect of ionic fluorinated surfactants on β-phase formation of electrospun PVDF nanofibers. PVDF nanofibers were electrospun from solutions of PVDF in dimethyl formamide (DMF)/acetone mixtures with addition of ionic fluorosurfactants – both anionic and cationic. It was observed that the addition of ionic fluorinated surfactants was useful for the formation of bead-free homogeneous fibers. In addition, it was found that ionic fluorosurfactants enhanced the formation of the β-phase crystalline structure of PVDF. A possible mechanism was suggested based on the interaction between the positive charge in the ionic surfactant and fluorine atom in PVDF chain, as illustrated in Figure 5.35.

Ren and Dzenis [91] also reported fabrication of PVDF nanofibers via electrospinning. In their experiments, solutions of PVDF with different molecular weight in DMF and dimethyl sulfoxide (DMSO) were tested. The produced fibers were characterized with x-ray diffraction (XRD) and Fourier transform infrared spectroscopy (FTIR). It was found that the α-phase of the raw material was converted to the β-phase during electrospinning. This verified the concept that electrospinning was a simple yet effective method to yield β-phase PVDF.

5.5.3.2 Study of Solution Properties in NFES Process

In conventional electrospinning, a number of parameters can influence the resulting fiber morphology. These parameters can be categorized into three groups: (1) solution properties including viscosity, elasticity, conductivity, and surface tension;

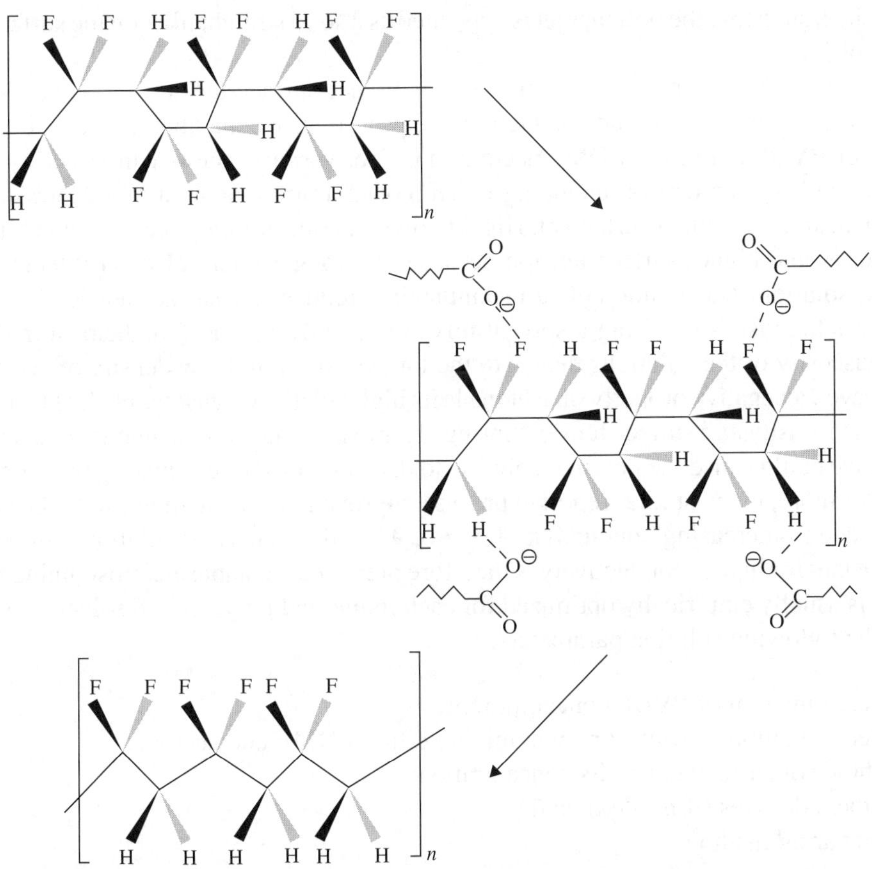

Figure 5.35 Mechanism of the transformation from α-phase to β-phase of PVDF in the presence of an anionic surfactant

(2) governing parameters including electric potential at the electrodes, flow rate of the polymer solution, and distance between the electrode and the collector; and (3) ambient parameters including solution temperature, surrounding temperature and humidity, and air velocity [95]. In NFES, fiber deposition occurs almost immediately after the jet is ejected from the tip, such that environmental factors such as temperature or air drag only play a minor role. Solution characteristics become the most important factors, discussed in the following.

A positive voltage is applied to the polymer solution while the collector is grounded, which induces an electric field E. The collector is moving with a velocity v. The electric field exerts an electric force on the charged polymer solution, so that a polymer jet is ejected from the tip and moves toward the collector. The fluid parameters of the polymer jet are: conductivity κ, dielectric constant ε, density ρ, dynamic viscosity μ, and surface tension γ. It is assumed that the collector's speed is well matched, such that the polymer jet is deposited as a solid straight fiber on the surface of the collector.

Polymer solution must have proper viscosity and surface tension for a successful electrospinning process, and conductivity helps to improve the fiber qualities. For the case of PVDF solution in DMF/acetone mixture, various factors can affect its viscosity and surface tension including PVDF concentration, amount of acetone in the solvent, and addition of surfactant. Higher polymer concentrations can result in higher viscosity and higher surface tension because of stronger intermolecular interactions in the solution. The addition of surfactant mainly reduces the surface tension but also reduces the viscosity. Using a solvent mixture of DMF and acetone helps to reduce the viscosity of the solutions thanks to the low viscosity and low density of acetone. However, a negative property of acetone is its high volatility, causing incessant changing in viscosity and surface tension during electrospinning. A high amount of acetone additive could cause the polymer solution to dry out quickly, terminating the process. Furthermore, many studies reported process improvements by adding salt to the solution, due to increasing conductivity [89, 90, 92, 96]. A polymer solution with ionic surfactant for higher conductively is therefore preferred. A stable electrospinning process is usually empirically optimized for each individual polymer and solvent system for the following solution parameters:

1. determination of PVDF concentration;
2. determination of solvent recipe: mixing ratio of DMF and acetone;
3. choice of surfactant and its concentration;
4. other additives (if needed); and
5. preparing method.

5.5.4 Electrospinning of PVDF Nanofibers

5.5.4.1 Solution Preparation

Different PVDF solutions were prepared in a three-step method by dissolving PVDF powder in a solvent mixture of DMF and acetone with the addition of different fluorochemical surfactants. NFES experiments showed that a highly homogeneous solution was required before the solution could be successfully electrospun into fibers. This could be achieved in a time-saving manner by a three-step method. The solutions prepared by this method exhibited good homogeneity that promoted continuous formation of the electrospun fibers.

Furthermore, carbon nanotubes (CNTs) are popular nanoparticles to be embedded in nanofibers to increase the conductivity as well as to store the charges. CNT can form agglomerates such that a much longer mixing time and a final sonication are required to achieve a good dispersion of CNT in the PVDF solution. It has been observed that, after sufficient mixing CNT particles were mixed in the solutions without any further stabilization. The preparation steps are described in the following.

1. PVDF solutions in DMF/acetone without CNT
   (a) PVDF in powder form is dispersed with a calculated amount of acetone for 30 minutes using a magnetic stirrer. Because acetone is highly volatile, a small amount of acetone evaporates during the process. The suspension is therefore re-weighed afterwards and acetone is added to maintain its designated weight percentage.
   (b) A suitable amount of DMF is added to the PVDF-acetone suspension. At the same time, Zonyl®UR is also added to the mixture. The mixture is stirred for at least one hour to reach a good homogeneity.
   (c) Because of the surfactant, an intensive foaming occurs during the stirring process. The solution is allowed to stand for another hour to allow the foam to escape.
2. PVDF solution with CNT (similar to step (1) with additional steps).
   (a) CNT is dispersed in DMF of proper percentage and PVDF is dispersed in acetone. The CNT–DMF suspension is sonicated for at least 1 hour to break possible agglomerates.
   (b) PVDF solution and CNT–DMF is mixed and surfactant is added.
   (c) The solution needs to stand for 1 hour before the electrospinning process.

These polymer solutions are prepared and stored at room temperature in a well-ventilated laboratory environment. To minimize the evaporation effects, containers are sealed with Parafilm.

5.5.4.2 Effect of Surfactants

It is observed that solutions of PVDF in DMF exhibited a high surface tension, making continuous NFES impossible. To reduce the surface tension, three kinds of fluorochemical surfactants were investigated: Zonyl®FSN-100, a non-ionic fluorosurfactant; Zonyl®FSP and ®FSJ, both anionic fluorosurfactant in aqueous solutions; and Zonyl®UR, an anionic phosphate fluorosurfactant that contains no solvent. According to Ren and Dzenis [91], addition of Zonyl®FSN-100 could help to eliminate the beading phenomenon for the electrospun nanofibers. In NFES however, solutions with Zonyl®FSN-100 did not work. This could be explained by its non-ionic character that did not increase the conductivity of the solutions. As a result, jet formation is difficult. Zonyl®FSP and Zonyl®FSJ are anionic fluorosurfactants. However, these surfactants include water in their as-delivered form which can cause the gel-formation of PVDF, which is unsuitable for NFES. Zonyl®UR is an anionic phosphate fluorosurfactant without solvent or water. The surface tension was significantly reduced with Zonyl®UR such that PVDF solutions were successful in the NFES process. Moreover, because the surfactant is anionic, the dissolved ions had the same effect as additional salt in the solutions which helped the conductivity of the solution and the electrospinning process.

5.5.4.3 Characterization of the Solutions

Solutions with various parameters were prepared to find the optimal recipe. The most important factor is the polymer concentration. Solutions with 15 wt% PVDF have been found to have low viscosity and could not produce continuous nanofibers. At a very high concentration of 22 wt%, PVDF was close to its equilibrium solubility and a homogeneous solution was difficult to achieve. At a concentration of 20 wt%, good homogeneity and viscosity were achieved for NFES.

Different concentrations of Zonyl®UR were examined. It was found that the better process in terms of continuity were recorded with higher concentrations of surfactants due to reduced surface tension. However, no significant changes were observed at concentrations of 8–10 wt%. The suitable concentration of Zonyl®UR was therefore set at 10 wt%. It has been reported that the correct amount of fluorochemical surfactant could facilitate β-phase formation of PVDF nanofibers [90].

Three mixing ratios of DMF to acetone have been examined, including 8:2 based on the experiences of previous NFES experiments, 6:4 and 5:5. Because of the higher concentration of PVDF than in previous experiments (20 wt% compared to 15 wt%), an increasing amount of acetone was required to promote the evaporation of the solution. The ratio of 5:5 was found to have the best electrospinning results.

5.5.4.4 Continuous NFES of PVDF

With the new solution recipe containing Zonyl®UR, PVDF solutions were successfully conducted in NFES on a silicon wafer in a continuous manner. The continuous NFES deposition was activated by using a tungsten probe to poke inside the polymer meniscus and break the surface tension. The mechanically drawn nanofiber was rapidly pulled away from the polymer droplet. The experimental set-up is shown in Figure 5.36.

Figure 5.37 shows experimental results under various parameters. For the lowest PVDF concentration of 15 wt%, discontinuous short fibers were deposited.

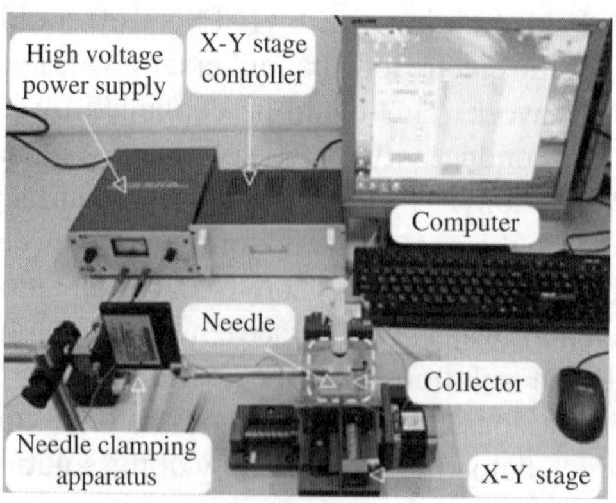

Figure 5.36 Experimental set-up of continuous NFES for PVDF nanofibers [26]

Figure 5.37a shows PVDF concentration for 22%. The addition of Zonyl®UR showed a visible effect on the fiber morphology. Figure 5.37b shows the best fiber result with the optimized recipe: 22 wt% PVDF and 10 wt% Zonyl®UR in 5:5 DMF:acetone. Straight and well-aligned fibers were deposited. The fibers were collected at a collector which was running at a speed of *c*. 10 cm s^{-1}.

A close-up look at the deposited nanofibers revealed non-uniformity in diameter (Figure 5.38), and beads or necks have been observed occasionally. One major reason could be the manual control of the collector movements. In general, uniformity of nanofibers depends strongly on the complex combination of material properties and processing parameters, which need to be further optimized.

5.5.5 Detailed Discussion of Process Parameters

5.5.5.1 Solution Viscosity and Surface Tension

Similar to any conventional electrospinning process, solution viscosity and surface tension have a great effect on the fiber morphology. These characters are determined

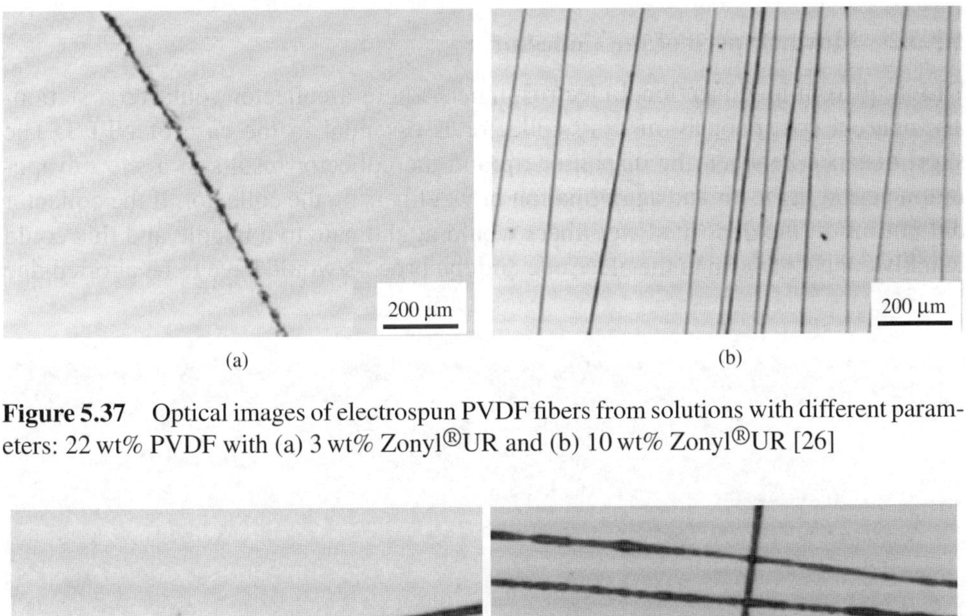

Figure 5.37 Optical images of electrospun PVDF fibers from solutions with different parameters: 22 wt% PVDF with (a) 3 wt% Zonyl®UR and (b) 10 wt% Zonyl®UR [26]

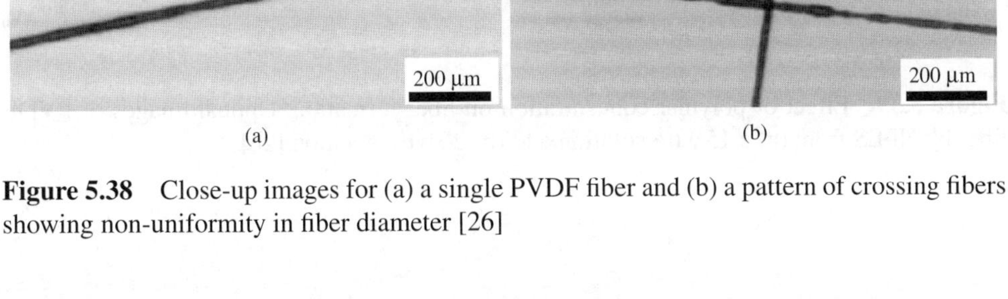

Figure 5.38 Close-up images for (a) a single PVDF fiber and (b) a pattern of crossing fibers showing non-uniformity in fiber diameter [26]

through polymer concentration and choice of solvent and surfactant; polymer concentration is the most important factor. In NFES, polymer concentration should be increased such that the solvent can evaporate in the short distance before reaching the collector. Sufficiently high viscosity also helps the polymer jet to remain stable during the NFES process.

Both viscosity and surface tension of a polymer solution will increase when the polymer concentration increases. It was observed that, similarly to conventional electrospinning, in NFES higher viscosity promoted better fiber formation. With a concentration of 15 wt% PVDF, the viscosity was not high enough such that fibers were often broken, resulting in short, discontinuous fibers Figure 5.39a. For concentration in the range of 20–22 wt%, continuous fibers could be deposited (Figure 5.39b). The looped fiber could be improved by increasing the collector speed (discussed in the following section). Higher concentration resulted in poor mixing of polymer.

The addition of Zonyl®UR at 3 wt% showed a positive effect on fiber formation under continuous NFES as shown in Figure 5.40a. With 10 wt% of Zonyl®UR, continuous and straight fibers could be successfully deposited in an orderly pattern as seen in Figure 5.40b.

5.5.5.2 Moving Speed of the Collector

Different from a conventional electrospinner, where a collector could be a stationary plate, a continuously moving collector is essential in the case of NFES. The short distance between the dispenser tip and the collector results in a short evaporation period in the air and agglutination of wet fibers on the collector. If the collector did not move, the deposited wet fibers would agglutinate to a bundle and this could quickly connect the tip to the electrode and the process would have to be aborted due

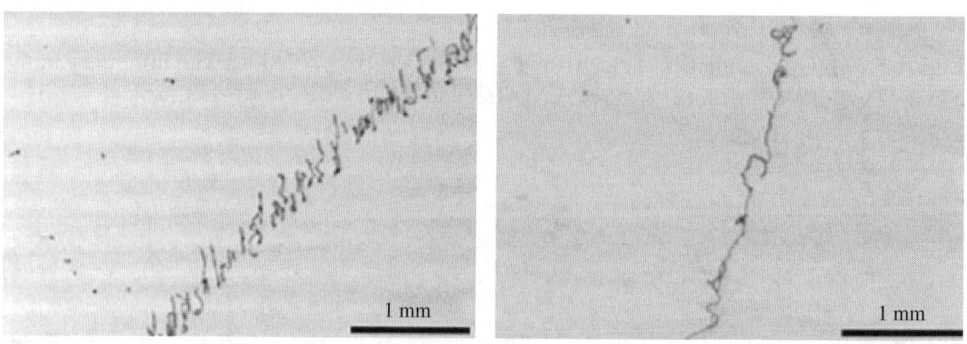

(a) (b)

Figure 5.39 Effect of polymer concentration on fiber formation. Optical images of PVDF fiber by NFES from (a) a 15 wt% solution and (b) 20 wt% solution [26]

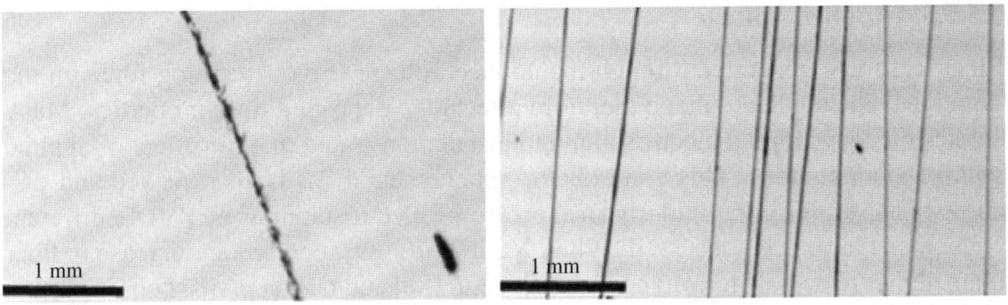

(a) (b)

Figure 5.40 Effect of fluorosurfactant on fiber formation. Optical images of PVDF fiber by NFES from a 20 wt% PVDF solution with (a) 3 wt% and (b) 10 wt% Zonyl®UR [26]

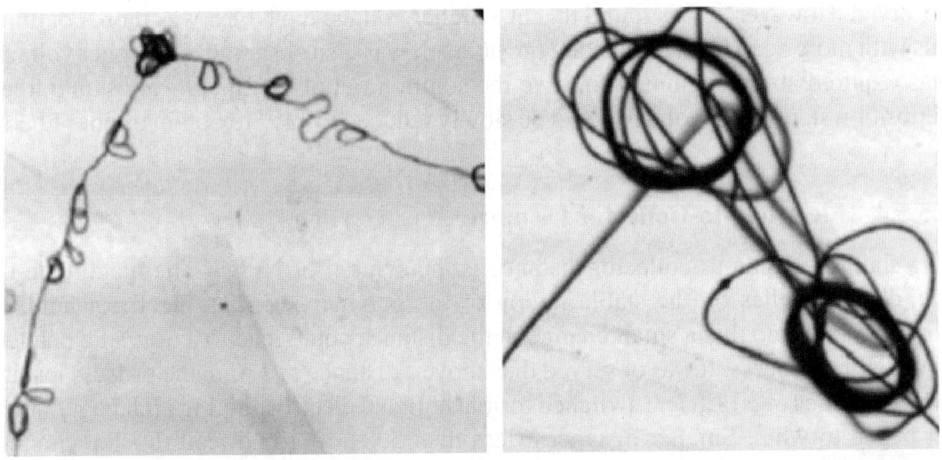

(a) (b)

Figure 5.41 Optical images of NFES PVDF fiber from a 20 wt% solution under 1 kV applied voltage to illustrate the effect of the collector moving speed on fiber straightness. (a) Spiraling occurs at low speed, *c.* 5 m s^{-1}, and (b) enlarged view [26]

to short-circuit. By continuously moving the collector, solvents can evaporate and the wet fibers become solid to avoid agglutination. The moving speed must be high enough to prevent clustering and beading of fibers to produce straight-line shapes, as demonstrated in Figure 5.41.

The movement of the collector also induces a stretching force on the deposited fibers, therefore having an effect on the fiber diameter. Assuming that the fibers adhere well on the collector surface, increasing the collector speed yields smaller fiber diameter. It was often observed that the fibers could be detached from collector because of weak adhesion, however.

5.5.5.3 Applied Voltage

Electrical force generated through the applied voltage is the stretching force on the liquid jet. The applied voltage therefore affects the deposited fiber properties. Many studies on conventional electrospinning have demonstrated that the effect of applied voltage is ambiguous. This contradictory phenomenon has been reviewed and discussed [14, 95, 97]. Particularly in the case of electrospinning of PVDF, different phenomena have been observed. While Nasir *et al.* [89] reported an increase in fiber diameter with increasing voltage, Andrew *et al.* [86] and Gao *et al.* [88] observed that higher voltage yielded thinner fibers. It is obvious that the effect of the applied voltage on the resulting fibers varies individually for each polymer-solvent system as well as other electrospinning parameters.

In this study, the suitable voltage for NFES PVDF in DMF/acetone could only be varied over the small range of 1 – 1.5 kV and no clear trend for fiber diameter could be observed. However, it was found that at a higher voltage the NFES was more continuous with more bead formation along the fibers. It was also observed that higher voltage was required at the beginning to drive the solution out of the tip. Once a stable fiber deposition started, the voltage could be slowly reduced to 1 kV to obtain thinner fibers.

5.5.5.4 Electrode-to-Collector Distance

The short electrode-to-collector distance is the symbol of NFES. The magnitude of this distance relies on the stable region of the polymer jet in an electrical field. It could be expected that a small change in the distance could lead to a dramatic change in the NFES process. It was observed that above a distance of 1 mm, the process lost its characteristics of NFES and switched to conventional electrospinning and the polymer jet began to whip. Smaller distances (less than 500 μm) could lead to short circuit. Experiments showed that at a distance of 500 μm, NFES was successfully executed and aligned straight fibers were obtained.

5.5.5.5 Diameter of the Dispenser Tip

Selection of the diameter of the tip needs to take into consideration the surface tension and the viscosity of the polymer solution. The ideal condition is that, during the idle state, a droplet of the polymer solution could form at the end of the tip by its own weight without dropping down. The inner diameter of the dispenser affects the diameter of fibers as it directly controls the diameter of the Taylor cone. A smaller tip diameter would lead to thinner fibers. However, if the diameter of the tip is too small, continuous NFES may fail. In NFES experiments with PVDF solutions, it was observed that the tip size of 30-gauge (equivalent to *c.* 150 μm inner diameter) could yield better processing and longer fibers than the tip of size 32-gauge (inner diameter *c.* 100 μm). Tip sizes larger than 30-gauge would cause the polymer solution to drop out easily and result in a short circuit, stopping the NFES process.

5.5.6 Experimental Realization of PVDF Nanogenerator

5.5.6.1 Materials

Polymer Solution

The polymer solutions were made from PVDF powders (weight-averaged molecular weight $M_w = 534\,000$) (Sigma-Aldrich), N,N-dimethylformamide (DMF) (Sigma-Aldrich), acetone (Sigma-Aldric), ZonylRUR (DuPont), and single-walled carbon nanotubes (SWCNTs) (Unidym). No reagents were purified further or modified prior to generation of the solutions. The SWCNTs (mixture of both metallic and semiconducting CNTs) had diameters of 0.8–1.2 nm and lengths of 100–1000 nm. ZonylRUR is an anionic phosphate fluorosurfactant that was provided without solvents.

Poled PVDF Thin Film

Poled PVDF thin films (Precision Acoustics) [98] were used for benchmark experimental testing. The films had dimensions of $5 \times 5\ cm^2$ with thicknesses of 28, 52, and 110 μm with sputtered metal electrodes on the top and bottom surfaces. These films were cut into smaller pieces (4–8 mm wide and 9–15 mm long) and subsequently attached to plastic substrates for experiments with glue.

Plastic Substrate

The plastic testing substrates were made of TOPAS® 8007 Cyclic Olefin Copolymer via injection molding with a thickness of *c.* 0.75 mm, width of 25–30 mm, and length of 60–70 mm. 3M™ cleanroom high-temperature electrostatic discharge (ESD) tape 1258 (0.07 mm thick) was applied on the top surface of the plastic substrate for enhanced electrostatic discharge performance. Finally, 3M™ aluminum conductive tape (0.05 mm thick) was cut and fixed to the tops of the ESD tape to serve as collectors for the NFES process and as electrodes for piezoelectricity measurements.

5.5.6.2 Fabrication of PVDF Nanogenerator

Plastic syringes with tips of 200 μm in diameter were filled with PVDF solutions. All metallic electrodes on the deposition substrate were grounded and kept at a fixed vertical separation distance of 0.5–2 mm from the needle tip. Before the onset of electrospinning, a bias voltage of 0.7–1.5 kV was applied. A probe tip was used to mechanically draw a single fiber from the droplet to initiate continuous NFES (Figure 5.42a). Experimentally, the syringe needle was stationary while the plastic substrate was moved using an x–y stage at a speed of $120\ mm\ s^{-1}$ for the controlled deposition of PVDF nanofibers (Figure 5.42b). Multiple nanofiber deposition can be achieved by repeating the direct-write process. With knowledge of the nanofiber polarities and the design of electrodes on the plastic substrate, these PVDF nanofibers can be connected in serial or parallel. An actual device is shown in Figure 5.43.

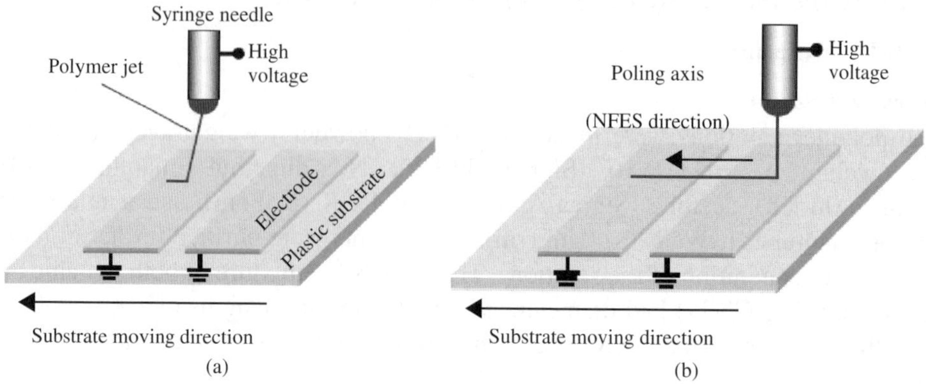

Figure 5.42 (a) Mechanically draw a single fiber from the droplet to initiate continuous NFES. (b) Fabrication of PVDF nanogenerator on a plastic substrate with pre-fabricated metallic electrodes using NFES [26]

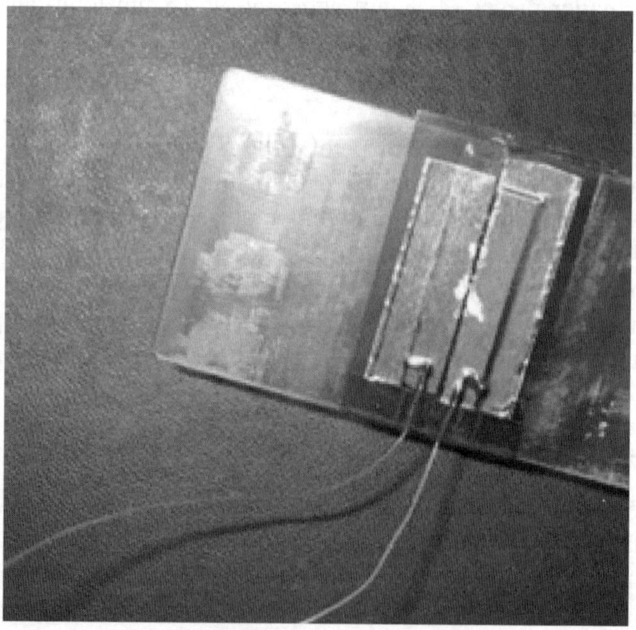

Figure 5.43 Photo of PVDF nanogenerator on a plastic substrate [26]

The diameter of deposited fibers ranged from 700 nm to 3 μm and yield was low because the plastic substrate was not conductive. Deposition of fibers were disturbed and terminated from time to time. The deposited fibers were not always uniform as seen in Figure 5.44; this could be attributed to the manual operation of the collector without a constant speed control. Nevertheless, continuous NFES was convenient to fabricate a large number of nanogenerators; superior examples were selected for detail characterizations.

With continuous NFES, it was possible to produce aligned fibers as shown in Figure 5.45. For the nanogenerator application, it was important to have the same poling direction for fibers; otherwise electrical outputs of individual fibers would cancel each other out. Manual manipulation of the collector could easily result in misalignment of fiber depositions in various forms: the formation of beads and varying distance between two fibers (Figure 5.45a); the formation of beads and non-uniform fiber diameter (Figure 5.45b); and two fibers merging together (Figure 5.45c).

5.5.6.3 Stretch of Nanofiber

Uniaxial stress is applied to the PVDF nanofibers by bending the plastic substrate, creating piezoelectric potentials that are uniformly distributed along the nanofiber

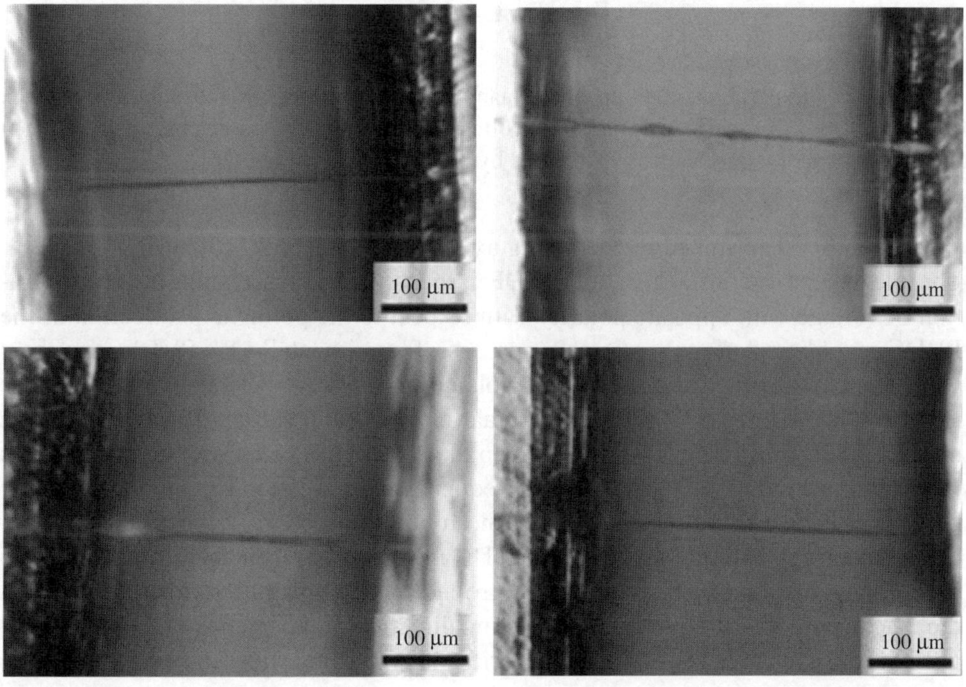

Figure 5.44 (a–d) Optical images of single PVDF fibers across a gap of 500 μm as nanogenerators. Some fibers are uniform while beads or necks are observed in some samples [26]

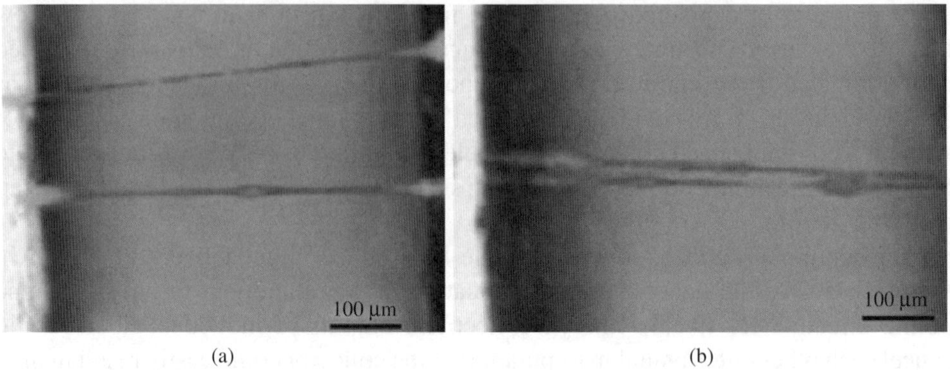

Figure 5.45 Optical images of PVDF fiber arrays across a gap of 500 μm: (a) formation of beads and different distance between two fibers and (b) formation of beads and non-uniform fiber diameter [26]

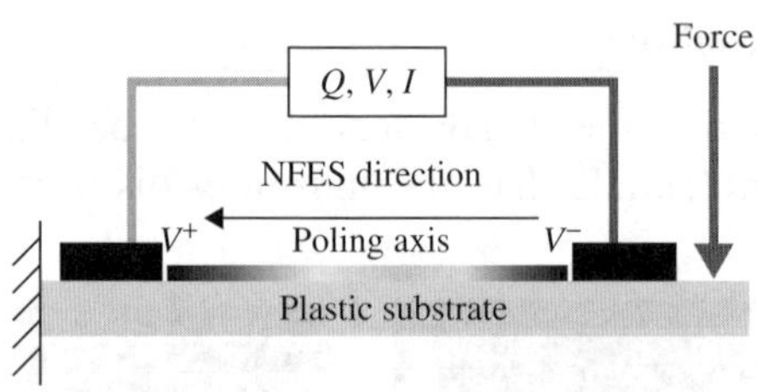

Figure 5.46 Experimental set-up of the nanogenerator with controlled polarity from the NFES process. Electrical signals including voltage, current, and charge are characterized when the nanogenerator is mechanically stretched by bending the substrate [26]

with a polarity determined by the poling axis, that is, NFES direction (Figure 5.42b). The piezoelectric constant g_{33} of PVDF is negative, which results in the generation of voltage with opposite polarity with respect to the poling axis direction of the PVDF nanofibers during the stretching process. Experimentally, the PVDF nanofibers were stretched to measure the output voltage and current between two electrodes (Figure 5.46). Figure 5.47 shows the actual experimental set-up. The loading force on the plastic substrate was exerted by an electrically powered rotating machine as illustrated in Figure 5.48. The rotating frequency and direction can be controlled by built-in switches. In most experiments, the rotating direction of the machine is not changed during the entire testing period. The plastic substrate is therefore restored to its original state at a speed which is controlled by the substrate rigidity, rather than the machine rotating frequency. The rotating direction of the machine within one testing session is changed only in the experiments where the release strain rate is the control factor.

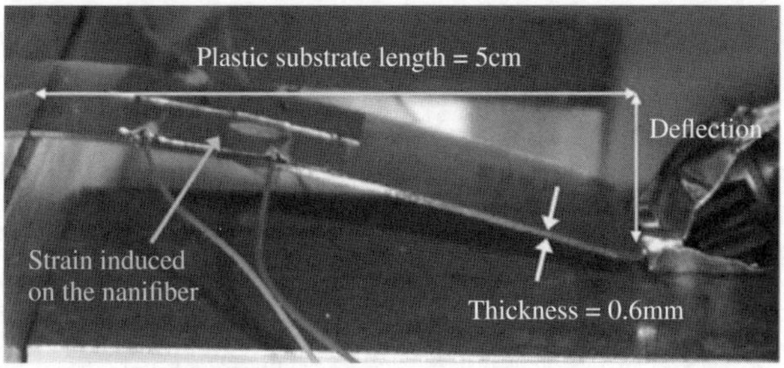

Figure 5.47 Photo of experimental strain-loading set-up [26]

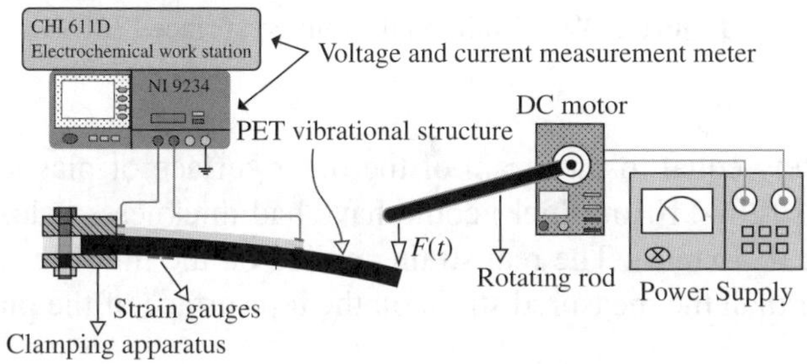

Figure 5.48 Illustration of strain-loading mechanism [26]

5.5.6.4 Measurements

Scanning Electron Microscopy

The morphology of the PVDF nanofibers was examined using a LEO 1550 SEM. A thin layer (50–100 Å) of gold-palladium was deposited on the sample surface using a Hummer Sputtering System before SEM to enhance visualization. All electrical measurements were performed prior to SEM to avoid unwanted effects such as electron charging and short circuiting of the nanofibers by the metal coatings.

Strain Measurement

A foil strain gauge (0.015 mm thick) (Omega) was glued onto the surface of the control plastic substrate to measure the axial strains. The strain ε occurring on the curved surface is defined:

$$\varepsilon = \frac{t}{2r} \tag{5.1}$$

where t is the thickness of the substrate and r is the bending radius of the substrate (Figure 5.49). Since the bending radius of the substrate in our experiment (>400 mm) and the thickness of the plastic substrate (*c*. 0.75 mm) were both much larger than the thickness of the PVDF nanofiber (500 nm to 6.5 μm), the strain of PVDF nanofiber

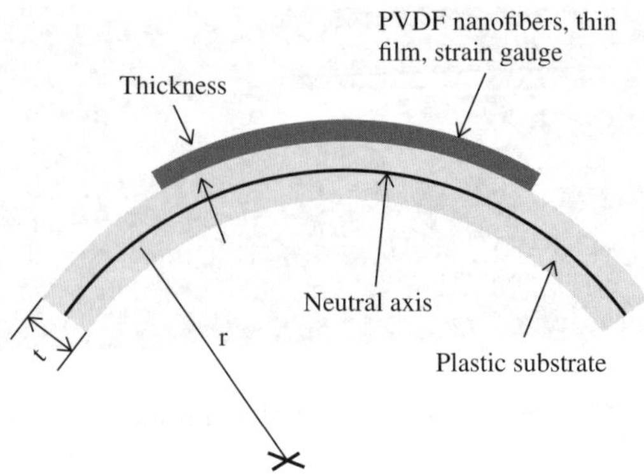

Figure 5.49 Strain on the curved surface [26]

was approximately equal to the strain of the outer surface of plastic substrate. The PVDF thin films (28–110 μm thick) could have had thicknesses close to that of the plastic substrate (0.75 mm). The real strain applied on the thin film would therefore have been larger than the measured strain on the top surface of the plastic substrate.

Electrical

All electrical measurements were performed using a Gamry Reference 600 high-precision electrometer ($I_{bias} < 5$ pA) inside a Faraday cage without connecting any external power sources to the PVDF nanogenerator. Before the electrical measurements, the two ends of the PVDF nanofiber were grounded to dissipate residual charges from the electrospinning process. The output voltages were measured with a load resistance of 10 MΩ while the output currents were measured by directly connecting the nanogenerator to the measurement system.

It was noted that the RC (resistor-capacitor circuit) time constant of the nanogenerator must be much larger than that of the measurement system or a large percentage of piezoelectric charges could leak out before being measured [99]. In our experiments, the RC time constants of the nanogenerator, voltage measurement, and current measurement systems were *c.* 2.4 s ($R_{NG} = 150$ GΩ and $C_{NG} = 160$ pF), 1.0×10^{-3} s ($R_L = 10$ MΩ and $C_L = 100$ pF), and 2.2×10^{-3} s ($R_L = 10$ MΩ and $C_L = 220$ pF) respectively, where R_{NG}, C_{NG}, R_L and C_L are the resistance and capacitance of the nanogenerator and load, respectively.

The piezoelectric charges were measured by a circuit (Figure 5.50) using the Texas Instruments OPA129 Precision Amplifier ($R_{in} = 10^{13}\Omega$, $I_{bias} < 100$ fA). The extremely small input bias current of the op-amp ensured that the charges generated by the piezoelectric response were not shadowed in our measurement system. The charge generated by the nanogenerator was transferred onto the feedback capacitance C_F, which was known and fixed. The calibration factor was fixed independently of the capacitance of the nanofiber generator. The RC time constant of the readout circuit

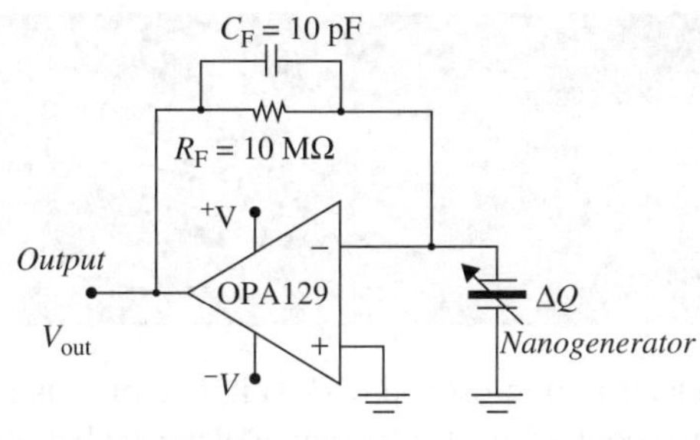

Figure 5.50 Circuit of charge amplifier [26]

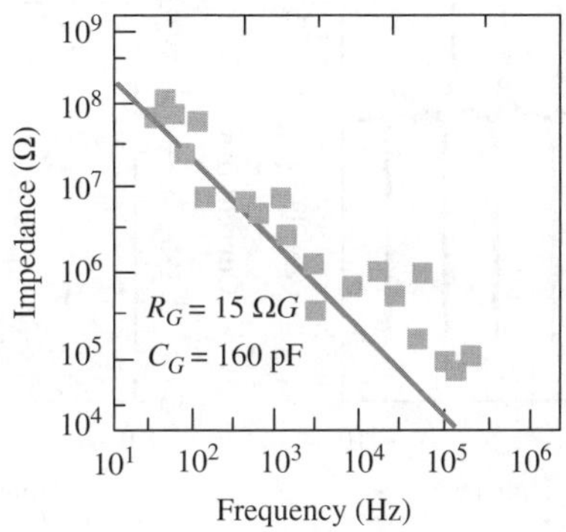

Figure 5.51 Impedance measurement (dots) and data-fitted analytical model (solid line) of the PVDF nanogenerator [26]

$R_F C_F$ (resitance times capacitance of feedback) is 10^{-4} s, which was also smaller than that of the nanogenerator.

The impedance of the nanogenerator was measured using the Electrochemical Impedance Spectroscopy (EIS) module of Gamry Reference 600. Impedance was measured by applying a sinusoidal voltage with different frequency between 10^{-2} and 10^5 Hz to the sample and measuring the current. The equivalent circuit of our nanogenerator was fit to the measured data using the Marquardt–Levenberg algorithm as shown in Figure 5.51. The resistance R_{NG} and the capacitance C_{GG} of the nanogenerator system were therefore extracted from the data fit.

5.5.6.5 Experimental Results and Discussion

The as-spun PVDF nanofibers have diameters ranging from 500 nm to 6.5 μm with variable lengths defined by the separation distance (100–600 μm) between two metallic electrodes (Figure 5.52). When an axial stress is applied by bending the plastic deposition substrate, a piezoelectric potential is generated. The piezoelectric constant

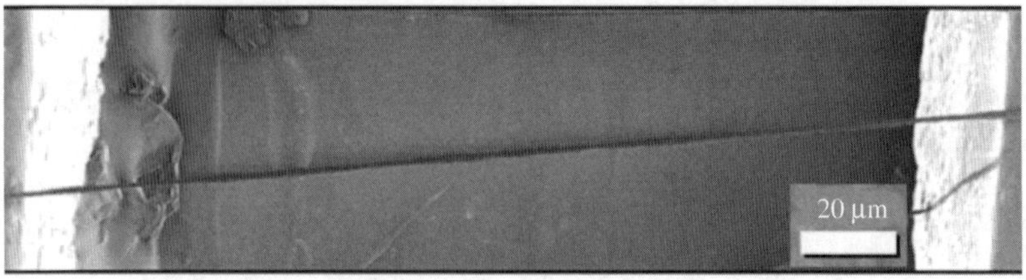

Figure 5.52 Scanning electron microscope (SEM) image of a nanogenerator comprising a single PVDF nanofiber, two contact electrodes, and a plastic substrate [26]

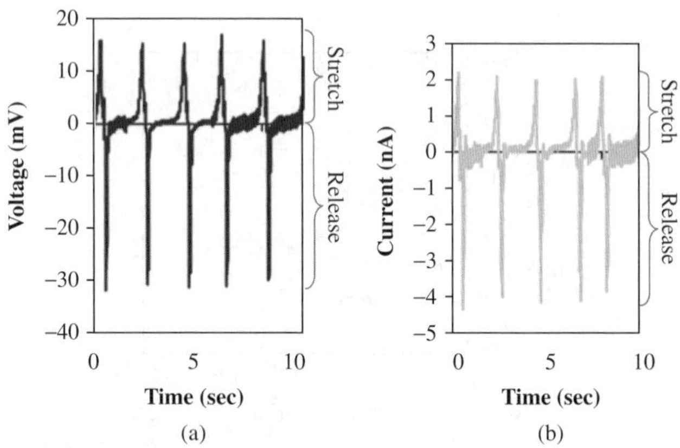

Figure 5.53 (a) Output (a) voltage and (b) current measured with respect to time under an applied strain at 2 Hz [26]

g_{33} of PVDF is negative such that stretching PVDF nanofibers along the poling axis generates a voltage with the polarity opposite to the electric field direction (the poling axis) during the electrospinning process. A positive voltage output is therefore expected by connecting the positive port of the multimeter to the distal end of the electrospun nanofiber. When the substrate is stretched and released repeatedly, voltage and current outputs can be recorded, respectively (Figure 5.53a,b). The typical electrical outputs of more than 50 tested nanogenerators were 5–30 mV and 0.5–3 nA.

As the strain is released from the nanogenerator, corresponding negative peaks can be observed in both the output voltage and current measurements. This phenomenon can be explained by examining the equivalent circuit model of the nanogenerator in Figure 5.54. The PVDF nanogenerator is modeled as a charge source in parallel with a capacitor C_{NG} and a resistor R_{NG} with estimated values of 160 pF and 15 GΩ, respectively (as measured in Figure 5.49). The equivalent circuit of a piezoelectric PVDF nanogenerator was connected to a readout circuit with an equivalent load resistor R_L and capacitor C_L. The 'positive output current' is defined as the current flow from node a (distal end of the electrospun nanofiber, Figure 5.49) to the readout circuit.

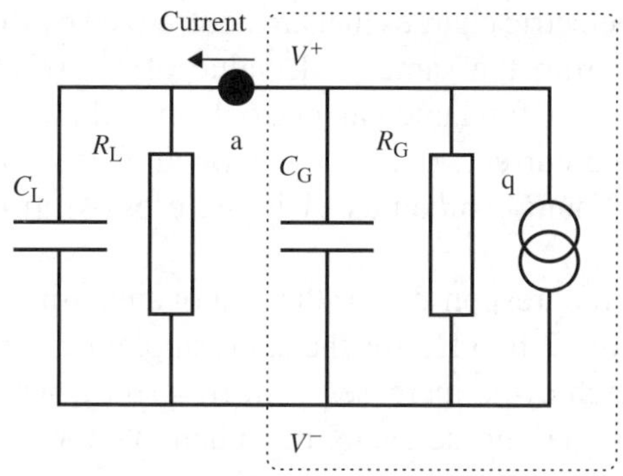

Figure 5.54 The equivalent circuit of a piezoelectric PVDF nanogenerator [26]

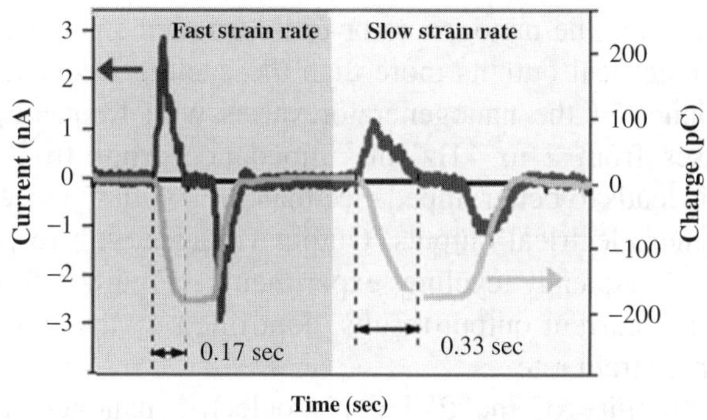

Figure 5.55 Output current (upper line) of a PVDF nanogenerator subject to different strain rates under the same applied total strain. The current output increases with strain rate. However, the total charges generated (lower line, from integral of the output current) are approximately the same under different strain rates [26]

Mechanical strain can induce piezoelectric bound charges (polarization) which results in a potential difference at the two ends of the nanogenerator. Since the nanogenerator has a large internal resistance, there is little charge leakage/loss in the nanogenerator during the short period of the stretch-hold-release experiments (Figure 5.55).

The electrical outputs are also affected by the strain rate. For a strain of 0.085% applied over 0.04 and 0.10 s, output currents of 2.74 and 1.16 nA, respectively, were experimentally measured with charges similar in both experiments (Figure 5.55). Specifically, the small discrepancy in piezoelectric charges between186 pC for fast strain rate experiments and 175 pC for slow strain rate experiments results from the RC time constant of the nanogenerator (*c.* 2.4 s) which allows a small portion of piezoelectric charge to leak through. In the fast strain rate (*c.* 0.17 s) and slow strain rate (*c.* 0.33 s) experiments, it is estimated that *c.* 6.8% and 12.8% of the piezoelectric charges leaked through the nanogenerator, respectively. This corresponds to *c.* 93.2%

and 87.2% of the piezoelectric charges that are neutralized by the free charges through the external circuitry, from the same peak value of *c*. 200 pC in both cases. The results are consistent with fundamental piezoelectric theory $I = q = d_{33}EA\acute{\varepsilon}$ [100], where I is the generated current, Q is the generated charge, d_{33} is the piezoelectric charge constant, E is Young's modulus, A is the cross-sectional area, and $\acute{\varepsilon}$ is the applied strain rate.

Figure 5.56 shows the responses of the nanogenerator under various cycling frequencies ranging from 2 to 4 Hz for the same applied external strain. The output current during the stretch cycle increases with frequency, while it remains approximately constant during the release cycle; this phenomenon that can be attributed to the applied strain rate. In the experimental set-up, the loading machine controls the stretching profile and therefore the increasing strain rate and cycling frequency for a given applied strain.

The release process is the release of the cantilever beam substrate from the strained position, which restores the nanogenerator to its original state. The release process speed dominates electrical outputs more than the cycling frequency. Note also that the output impedance of the nanogenerator varies with frequency; as the cycling frequency increases from 2 to 4 Hz, the impedance drops from 900 to 300 MΩ (Figure 5.51). This leads to better impedance matching with the measurement system and results in higher electrical outputs. Output voltages with respect to time from the stretch-release frequency cycling experiments (Figure 5.57) exhibited high consistency with the current output results, controlled by the substrate rigidity and resulting in a higher strain rate.

The long-term stability of the PVDF piezoelectric nanogenerator is examined through operation over an extended period of time, exposing it to the ambient environment without protective packaging. For these experiments, the electrical outputs were relatively stable without noticeable degradation for a stretch-release cycle frequency of 0.5 Hz and an operation time of 100 min (Figure 5.58). Output voltages

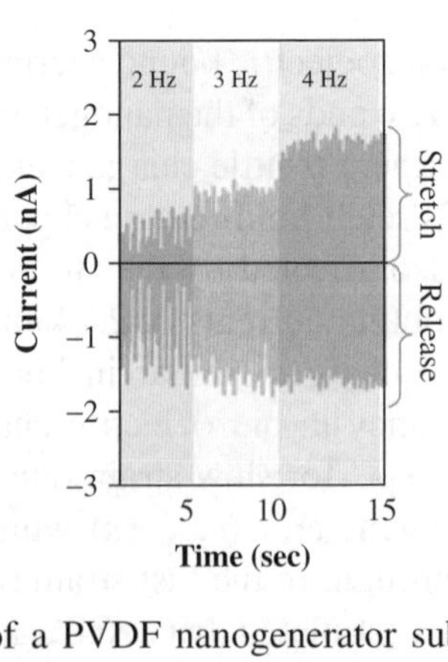

Figure 5.56 Output current of a PVDF nanogenerator subject to different stretch-release cycling frequencies of 2–4 Hz [26]

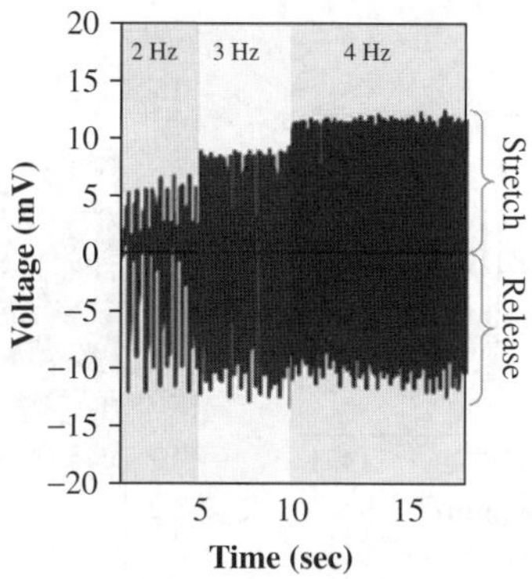

Figure 5.57 Output current of a PVDF nanogenerator subject to different stretch-release cycling frequencies of 2–4 Hz [26]

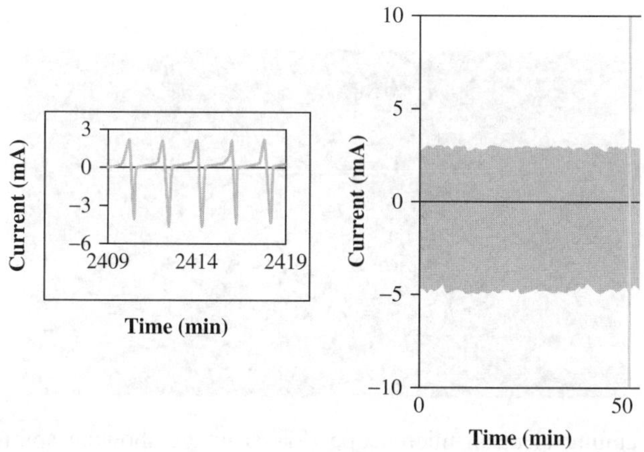

Figure 5.58 Output current of a PVDF nanogenerator under 0.5 Hz of continuous stretch-release for 100 min, demonstrating the stability of the nanogenerator. The inset shows the detailed profiles of the electrical outputs [26]

under long-term operation of 0.5 Hz for 100 min (Figure 5.59) also illustrated similar performance with the current outputs.

Limitations of Conventional Electrospinning

The conventional electrospinning process produces randomly oriented nanofibers such that the piezoelectric outputs mostly cancel out one another as shown in Figure 5.60. By designing different electrode patterns in conventional electrospinning [2], the electrostatic forces can guide PVDF nanofibers across the gap of two electrodes (Figure 5.61). However, the PVDF nanofiber arrays do not have controlled polarity. The polarities of these PVDF nanofibers cancel out one another such that the overall electric outputs were close to zero (Figure 5.62).

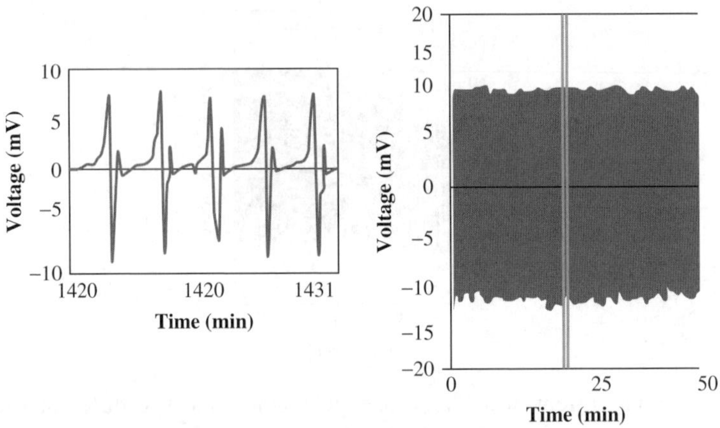

Figure 5.59 Output voltage of a PVDF nanogenerator under continuous stretch and release for 100 min [26]

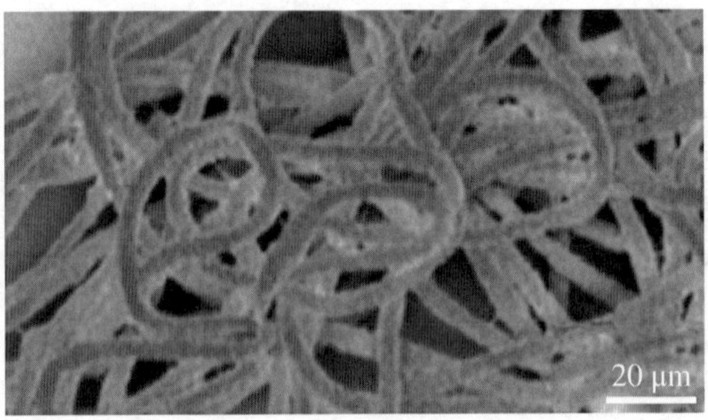

Figure 5.60 Scanning electron microscope (SEM) image showing spiraling and random patterns of as-spun PVDF nanofibers as the result of the conventional electrospinning process [26]

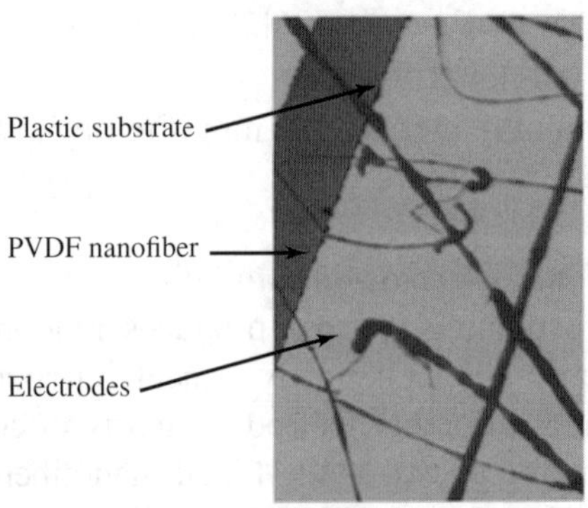

Figure 5.61 Optical image of PVDF nanofibers deposited between two metal electrodes by conventional electrospinning [26]

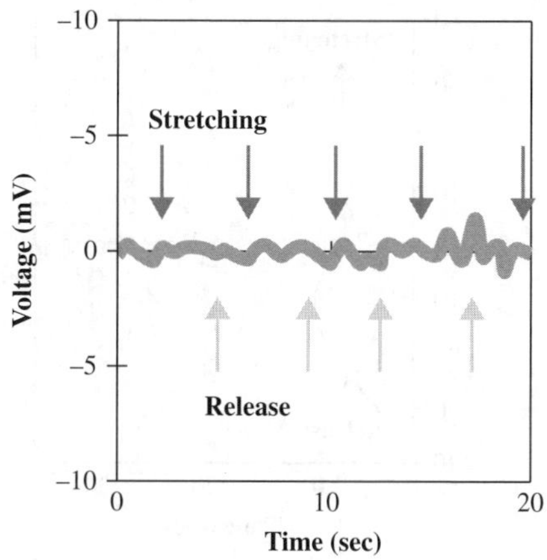

Figure 5.62 Voltage output of the random PVDF nanofibers subject to continuous stretch and release [26]

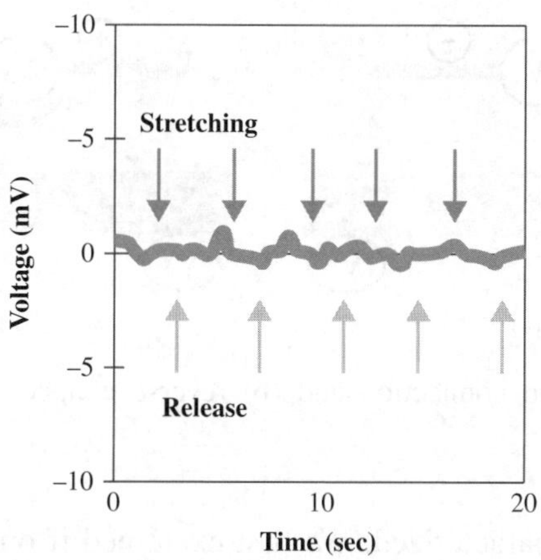

Figure 5.63 Voltage output of a single PEO nanofiber subject to continuous stretch and release [26]

Validation of Piezoelectric Response

To verify that the electrical outputs were generated from piezoelectric properties of PVDF nanofibers, we performed the same experiments but change the nanofiber material from PVDF to the non-piezoelectric polymer, PEO. The output voltage (Figure 5.63) showed only noise with no visible peaks, indicating that PEO nanofibers did not produce piezoelectric behavior in the same way as the PVDF nanofiber (Figure 5.64). The data precluded the possibility of residual charges, friction, or contact potential as the cause of the observed electric outputs for the PVDF nanofibers.

To confirm the validity of the recorded piezoelectric responses, various combinations of probe connections and experiments on stretching/compressing of the

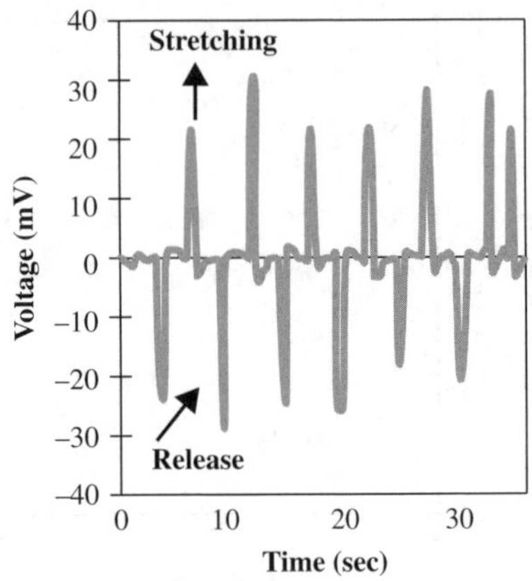

Figure 5.64 Voltage output of a single PVDF nanofiber subject to continuous stretch and release [26]

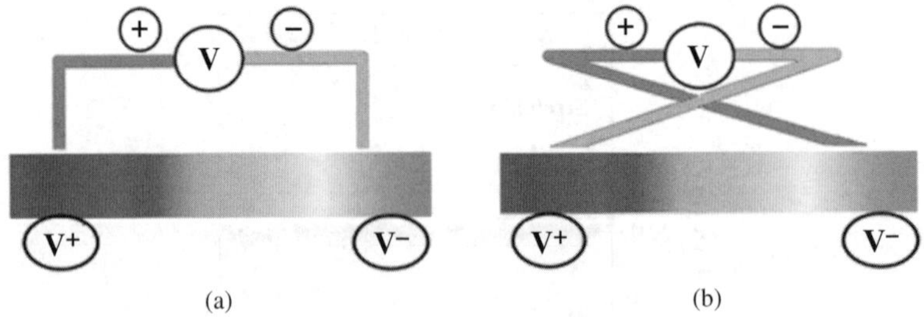

Figure 5.65 (a) Forward connection and (b) reverse connection for switching polarity tests [26]

nanogenerators were characterized. We first examined if our PVDF nanofibers met the 'switching polarity' criterion, which would confirm that the measured outputs were indeed generated by the piezoelectric response [80]. In forward connections, the positive and negative probes were connected to the positive and negative potential of the PVDF nanofiber, respectively. In the reverse connection, this connection is reversed as shown in Figure 5.65. Since the polarity of the PVDF nanofiber was fixed, this switching polarity test generated electrical outputs (Figure 5.66).

In our experiments, compressing the PVDF nanofiber generated electrical outputs (Figure 5.67) with polarities opposite to those of the stretch experiments. This further confirmed the piezoelectric properties of our PVDF nanofiber. We also performed polarity switching tests in compression experiments to verify the piezoelectric response. The polarity of electrical outputs in the forward and reverse connections had opposite polarities as shown in Figures 5.67 and 5.68.

It was also found that the output voltage and current of the PVDF nanogenerator could be enhanced by serial and parallel connections, respectively (Figure 5.69).

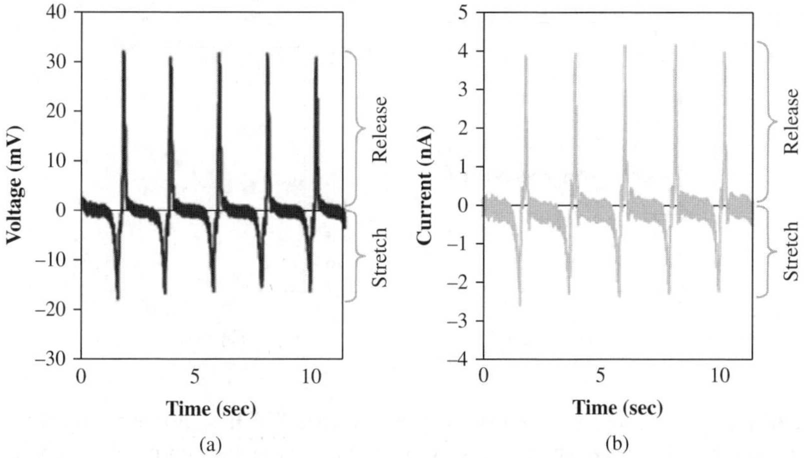

Figure 5.66 Electrical outputs of a PVDF nanogenerator subject to continuous stretch and release when reversely connected to the measurement system: (a) output voltage and (b) output current [26]

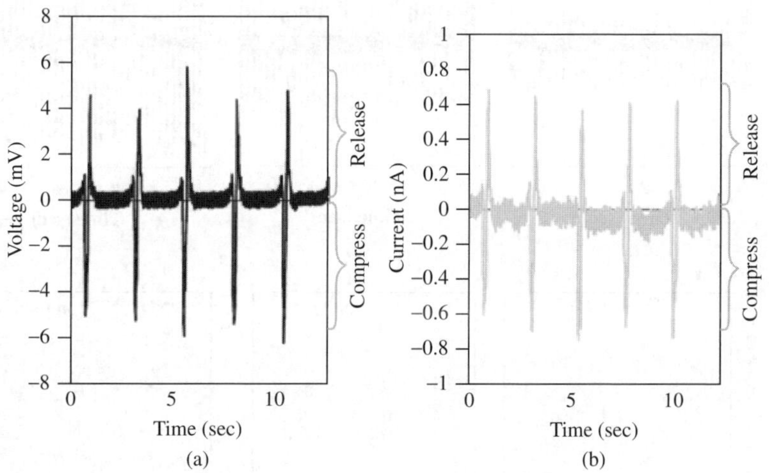

Figure 5.67 Electrical outputs of a single piezoelectric PVDF nanofiber subject to continuous compress and release: (a) output voltage and (b) output current [26]

The electrical outputs of serial and parallel connections are approximately the sum of the two individual nanogenerators, which satisfies the 'linear superposition' criterion proposed by Yang *et al.* [80] to confirm the true piezoelectric responses. This is achieved by the direct-write characteristics of NFES, as the placement and polarity of the PVDF nanofibers can be manipulated such that nanofibers are either in serial or parallel connections during the fabrication process.

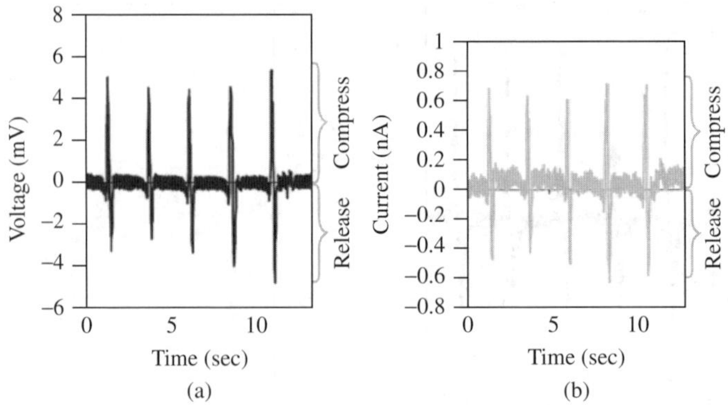

Figure 5.68 Electrical outputs of a single piezoelectric PVDF nanofiber subject to continuous compression and release when reversely connected to the measurement system: (a) output voltage and (b) output current [26]

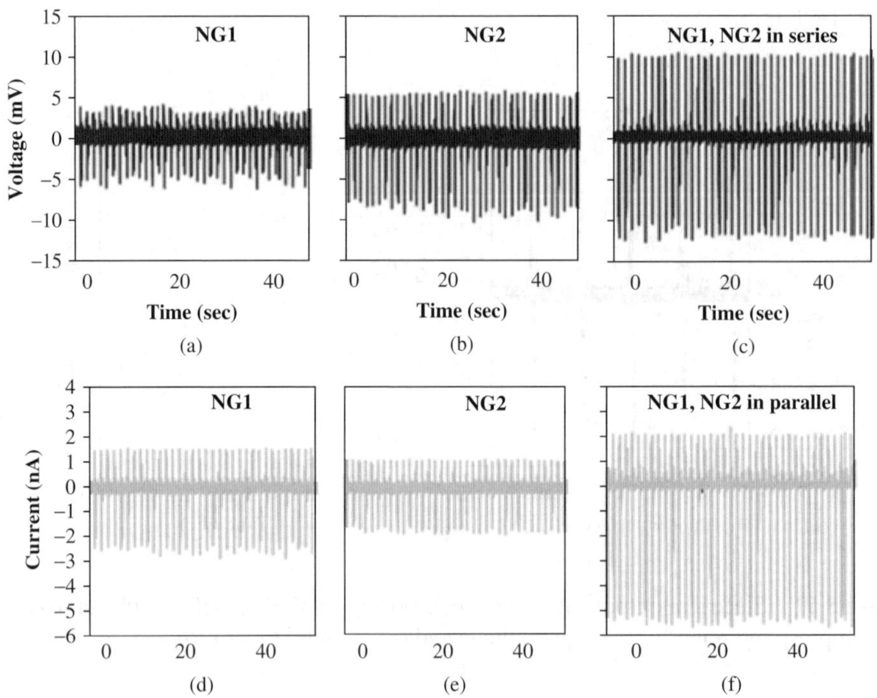

Figure 5.69 Output voltages of (a) nanogenerator #1 and (b) nanogenerator #2 subject to continuous stretch and release. (c) Output voltages constructively add when two nanogenerators are in serial connection. Output currents of (d) nanogenerator #1 and (e) nanogenerator #2 subject to continuous stretch and release. (f) Output currents constructively add when two nanogenerators are in parallel connection [26]

Energy Conversion Efficiency

The ECE is defined

$$\text{ECE} \equiv \frac{\text{Electric energy generated}}{\text{Mechanical energy applied}} = \kappa^2 \tag{5.2}$$

where κ is the electromechanical coupling coefficient of the piezoelectric material. The electric energy W_e was estimated by integrating the product of output voltage and current of the PVDF nanogenerator when stretched. The elastic strain energy W_s was estimated from

$$W_s = \frac{1}{2} E A \varepsilon^2 L_0 \tag{5.3}$$

where E is Young's modulus of the material, A is cross-sectional area, ε is the strain applied on the material, and L_0 is the length of the material. The dimensions of the PVDF nanofibers were measured with SEM after the energy generation experiments. There were more than 45 PVDF nanogenerators used in our experiments to estimate the ECE. Among all samples in the ECE experiments, the nanofibers were 600 nm to 6.5 μm in diameter, 100–600 μm in length, and Young's modulus was estimated as 1.4 GPa [101].

By analyzing 45 samples, the ECE of our nanogenerators was found to be as high as 21.8% with an average of 12.5%, a value that is much greater than typical power generators made from experimental piezoelectric PVDF thin films (0.5–4%) [4–6] and commercial PVDF thin films (0.5–2.6%) tested under the same conditions in this work (Figure 5.70).

The efficiency of the PVDF thin film was estimated using the same method as the PVDF nanogenerators. However, d_{31} mode was utilized in PVDF thin film instead of d_{33} mode to accommodate for the strain loading set-up. In the experiments, the output voltage and current of the PVDF thin film were *c.* 60 mV and 60 μA, resulting in a ECE

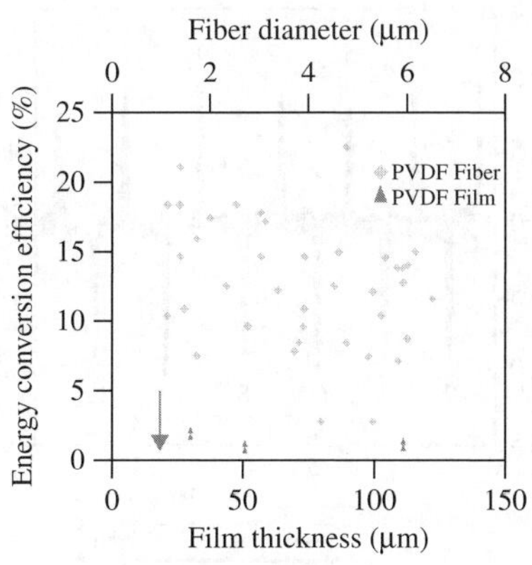

Figure 5.70 Plots of measured energy conversion efficiency of PVDF nanogenerators and thin films with different feature sizes [26]

of 0.5–2.6% (average 1.3%). We realized that the ECE of d_{31} mode would be smaller than that of d_{33} mode ($d_{33} = 30\,\mathrm{pC\,N^{-1}}$ and $d_{31} = 22\,\mathrm{pC\,N^{-1}}$ from the company's data sheet [98]). If taken into account, the efficiency would be enhanced by a factor of *c.* 1.8, resulting in an average efficiency of 2.3%. However, it is noted that the real strain applied on the PVDF thin film would be larger than the strain on the top surface of the plastic substrate, which was measured by a strain gauge. The ECE of the PVDF thin films would therefore become smaller if we used the real strains on the PVDF thin films to calculate the efficiency. The ECE of PVDF thin film provided by the company was 1.96%, which was close to the estimation from our experiment. This demonstrated that the approach we used in calculating the ECE was reasonable.

First, we estimated the electrical energy error by calculating the average and standard deviation of electric energy generated from the same PVDF nanogenerator under the same operating conditions (Figure 5.71). The calculated electric energy had an average of 4.07×10^{-13} J and a standard deviation of 8.74×10^{-14} J, corresponding to an error ($\Delta W_e/W_e$) of 21.5%. The elastic strain energy error mainly came from the mechanical strain variations, which was caused by the strain loading machine and plastic substrate.

We estimated this error by continuously bending the plastic substrate by the same amount and measuring the strains to obtain the average and standard deviation. The measured strain had an average value of 0.085% and a standard deviation of 0.0013%, corresponding to a strain error ($\Delta\varepsilon/\varepsilon$) of 1.5% and an elastic strain energy error ($\Delta W_s/W_s$) of 3.0%. The total error of the energy conversion efficiency (ΔECE/ECE) was therefore *c.* 21.71%.

The general trend in Figure 5.70 indicates that nanogenerators with smaller diameters exhibit higher ECEs, even with variable piezoelectric properties resulting from slight changes in processing conditions. Several possible factors could have

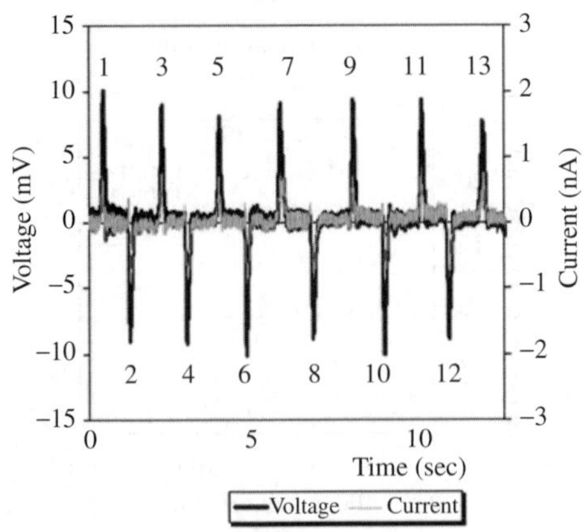

Figure 5.71 Output voltage and current of a PVDF nanogenerator subject to continuous stretch and release under the same strain [26]

contributed to the observed enhanced electromechanical response. For example, previous reports have demonstrated a drop in the piezoelectric coefficient due to internal defects in ultra-thin lead zirconium titanate (PZT) films [102], while nanogenerators made by the NFES process could possibly have fewer defects than the PVDF thin film due to a higher degree of crystallinity and chain orientation [103]. The difference in elastic boundary conditions (thin film/bulk samples have metal electrodes on both ends as constraints) [104] and the physical size of the nanogenerators could promote size-dependent piezoelectricity such as 'flexoelectricity', which is caused by strain gradients that locally break inversion symmetry and induce polarization [105].

Furthermore, piezoelectric thin films could have a significant fraction of the measured responses coming from extrinsic contributions collectively known as 'domain wall motion' [106]. PVDF thin-film material has a domain wall motion barrier at *c.* 0.3% strain [107], under which the piezoelectric response results from the mostly linear intrinsic response (the responses of single domains). Once the applied strain increases beyond the barrier, the piezoelectric response is dominated by the remarkably large non-linear extrinsic responses due to domain wall motion [107].

5.5.7 Summary

In summary, we have demonstrated PVDF nanogenerators that are directly written onto flexible plastic substrate using NFES. The piezoelectric responses of single PVDF nanofibers were measured and multiple nanofibers were arranged to enhance the electrical outputs. The reported nanogenerator has several advantages over other macro-, micro-, and nanogenerators, including high ECE, manufacturability, and the capability of integration with other micro-/nanofabrication processes. The principle and the nanogenerator demonstrated could be the basis for integrated power sources for nanodevices and wireless sensors or new self-powered textile by direct-writing nanofibers onto fabric.

5.6 Materials, Structure, and Operation of Nanogenerator with Future Prospects

5.6.1 Material and Structural Characteristics

The applications of piezoelectric nanofibers strongly rely on good piezoelectric properties; the formation of the β-phase is key to enhancing piezoelectricity for PVDF nanofibers. Piezoelectric analysis techniques and tools are therefore necessary to characterize crystal/molecular structure and inherent piezoelectric properties of electrospun nanofibers for optimal parameters during the electrospinning processes. Techniques such as XRD analysis, FTIR, piezoresponse force microscopy (PFM), second-harmonic generation (SHG), and Raman spectroscopy have all been utilized for this purpose and are briefly discussed in the following sections.

5.6.1.1 XRD

XRD has been commonly used for the analysis of the crystalline structure of materials. Baji *et al.* [108] analyzed PVDF fibers fabricated by far-field electrospinning using XRD. It was found that the β-phase is most abundant in electrospun PVDF nanofibers while other crystalline forms also existed. Electrospun nanofibers with smaller diameters were found to have higher contents of β-phase, probably due to the strong stretching effect. Other groups utilized XRD to analyze the structures of electrospun polar β-glycine nanofibers and polycrystalline barium titanate nanofibers [109]. For PZT fiber-based nanogenerators, XRD can be used to confirm the presence of the pervoskite phase of the material. This phase is crucial for the piezoelectric effect. Using a MOD (metallo-organics decomposition) method, the pervoskite phase is formed during the fiber annealing process at a temperature of 600–850 °C [87]. Chen *et al.* conducted a similar experiment using a PZT sol-gel, polyvinyl pyrrolidone and alcohol mixture [110]; the XRD results confirmed the pure pervoskite phase of the fibers with an annealing temperature at 650 °C.

5.6.1.2 FTIR

FTIR can be used to characterize both the dipole orientation and crystallographic structure of nanofibers based on the sensitivity of CF_2 structure orientation changes. For example, Mandal *et al.* used FTIR to examine the dipole orientation of electrospun P(VDF-TrFE) nanofibers [71]. They showed that the dipoles were aligned in the direction of the electrical field during the electrospinning process [71]. A comparison between the FTIR spectra of an as-spun fiber and a heat-treated electrospun fiber (heated above the Curie temperature in order to ensure random dipole alignments) resulted in a difference in absorbance of perpendicular polarized light for CF2-sensitive wavelengths. Baji *et al.* also confirmed the presence of the crystal β-phase using FTIR in good correspondence to the results obtained using XRD [108]. Similarly, Wang and Santiago-Aviles used FTIR to examine the PZT fiber to identify the decomposition temperature of the organic groups as well as the appearance of the pervoskite phase [111].

5.6.1.3 PFM

PFM measurements are based on the detection of a voltage-induced deformation of the piezoelectric material. The obtained data allows quantification of the degree of the polarization as well as the polarization direction of the tested sample [112]. The PFM set-up is comparable to AFM; a micro tip is brought into contact with the surface while applying a sinusoidal bias voltage. In the case of a piezoelectric material with a poling direction parallel to the direction of the PFM bias voltage, this leads to a detectable deformation of the material due to the piezoelectric response (referred to as the 'in-phase' response). For an opposite poling direction at the contact area, the observed phase would shift by 180° (as the 'out-of-phase' response). Lateral deformations can also be sensed by the detection of the torsional responses of this same tip. PFM is particularly well suited for the analysis of individual nanofibers in contrast to other characterization tools because it allows local piezoelectric measurements. For example, researchers used PFM to generate images of TGS (triglycine sulfate) nanocrystals embedded in PEO nanofibers [113].

PFM has also been used for the characterization of PZT fibers. For example, Santiago-Aviles *et al.* characterized the polarization domains of electrospun PZT fibers [114] and found that the domains of original fibers were randomly oriented in directions of upward, downward, or parallel to the substrate on which the fiber is deposited. A post-poling process was necessary for nanogenerator applications.

5.6.1.4 SHG

The polarity of a piezoelectric material can also be characterized by SHG as demonstrated by Isakov *et al.* [115] analyzed a 532-nm second-harmonic spectrum from a Nd:YAG (neodymium-doped yttrium aluminum garnet) laser on β-glycine and γ-glycine nanofibers [116]. The non-linear optical responses were attributed to the embedded nanocrystals with their polar axis oriented parallel to the fiber mat plane. The same technology has been used to measure the piezoelectric property of other nanofibers such as electrospun PVDF and α-helical poly(α-amino acid) fibers [108].

5.6.1.5 Raman Spectroscopy

Raman spectroscopy can be used for the detection of various low-frequency modes in a system. It has been used to characterize material properties such as crystallographic orientation, including the detection of the formation of PZT structures [117].

5.6.2 Operation of Nanogenerator

Many prototype nanofiber nanogenerators have been described in the literature. Several issues are discussed in the following sections including measurement artifacts, linear superposition criterion, artifacts due to contact electrodes, and piezoelectric effects in nanofibers.

5.6.2.1 Measurement Artifacts

In order to test the performance of the nanogenerators, electrical outputs generated by mechanical inputs are often conducted for nanofibers made of different materials. These signals are often very small and are difficult to measure due to the nature of the small size of the nanofibers. The background noise or artifacts can easily overshadow the true signals. It is especially problematic when dealing with single nanofiber structures. For example, the capacitance changes between the wires and the electrodes and possible electrical coupling of the measurement instruments could surpass the real nanofiber signals. It is therefore important to filter out or reject noise generated from the surrounding experimental environment. Some of these experimental validation conditions have been previously proposed for semiconductor-based or ceramic-based nanogenerators such as ZnO nanowires, while PVDF nanogenerators do not apply to some of the requirements such as the Schottky behavior test.

5.6.2.2 Linear Superposition Criterion

When two devices are connected in series, the voltage response is the sum of the two; the current response is sum of the two when connected in parallel. Single PVDF nanofiber nanogenerators have been used in previous demonstrations. All data are measured using the single PVDF nanofiber nanogenerator operated under the same strain, strain rate, and frequency. Similarly, connecting two identical devices in series with opposite polarity should result in an almost-zero voltage response.

5.6.2.3 Artifacts due to Contact Electrodes

In order to measure the electrical outputs from the nanogenerators, it is necessary to have at least two electrical contact pads (or multiple electrodes) typically placed on top of a substrate with nanofibers deposited on top of them, as illustrated in Figure 5.72a. The mechanical actuation of the substrate nanogenerator could result in a capacitance change due to the varying distance of adjacent contact electrodes and add to the true output signals as artifacts. Specifically, the total capacitance of the device due to contact pads is the sum of the capacitance of the gap between adjacent electrodes and that of the overlapping area, i.e.

$$C_{\text{contact\_pad}} = C_{\text{gap}} + C_{\text{overlap}} \tag{5.4}$$

Since the magnitude of capacitance is proportional to the area and inversely proportional to the gap distance, a small contact electrode area and a large gap distance are preferable to reduce the impact of artifact due to the contact pads. Under small actuation as shown in the left side of Figure 5.72a, C_{overlap} could be insignificant; under large actuation as shown in the right side of Figure 5.72a however, C_{overlap} could dominate the total responses. By switching the actuation direction upward or downward, the output responses due to the electrode gap will have opposite effects as C_{gap} between the two electrodes decreases or increases, respectively. On the other hand, C_{overlap} should always increase due to the increased overlapping area under either upward or downward actuation. Figure 5.72b depicts experimental results based on upward actuation and releasing to the original flat position under small actuation (left) and large actuation (right), respectively. Figure 5.72c are results from the same tests based on

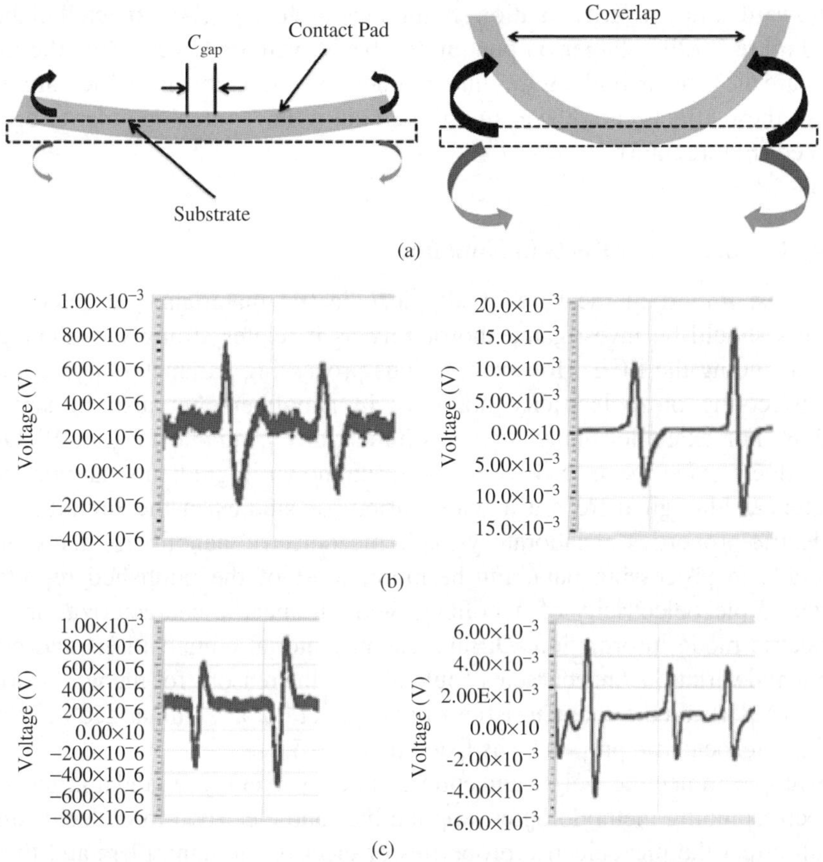

Figure 5.72 (a) Schematic diagram showing the possible artifacts of nanogenerator outputs due to the contact electrodes from the sources of C_{gap} and $C_{overlap}$ under small (left) and large (right) substrate actuation tests. (b, c) Experimental voltage outputs from upward and downward actuation tests, respectively, with small (left figures) and large (right figures) substrate actuation. Under small actuation, C_{gap} could be the dominating factor to change the polarity of the output signals of upward and downward actuation as shown. Under large actuation, $C_{overlap}$ could be the dominating factor and no polarity change is found due to upward or downward actuation as $C_{overlap}$ increases in both actuation directions [118]

downward actuation and releasing to the original flat position. The specimen used in these tests has an electrode area of 20 mm × 20 mm and cross-sectional area in the gap of 50 nm × 20 mm without depositing any nanofibers. The output signals from small actuation (left side of Figure 5.72b,c) are less than 1 mV, which can be considered as noise when compared to the typical voltage outputs of a single nanofiber at tens of mV [5]. The clear change in the polarities of the output signals suggests the outputs come from the changes in C_{gap}. On the other hand, under large actuation the contribution of C_{gap} is small and the changes of $C_{overlap}$ dominate the output signals;

either upward or downward actuation should increase $C_{overlap}$. As expected in the right side of Figure 5.72b,c, larger outputs up to 10 mV were observed while the polarity changes are the same in both figures under either upward or downward actuation. Nevertheless, these effects could be easily filtered out if the 'switching polarity criterion' is checked (Figure 5.65).

5.6.2.4 Piezoelectric Effects in Nanofibers

As the most important measure of all piezoelectric materials, piezoelectricity in nanofibers should be investigated thoroughly as it relates strongly to their performance, including the ECE. However, various processing parameters and conditions easily affect the piezoelectricity not only in nanofibers but also in large-scale thin films. For example, many of the commercially available PZT or PVDF thin films/products have been processed with unknown ingredients and/or process parameters. Although there are a wide variety of studies in the literature on the piezoelectric properties of laboratory-made piezoelectric nanofibers, many of them lack details in processing data. Furthermore, most of the published reports only characterize piezoelectricity of nanofibers without specific characterizations of the piezoelectric mode information. Differences in material compositions and possible experimental variations/inaccuracies could also be the reasons for the wide variations in various reports on the characteristics of nanogenerator as summarized in Table 5.1, as well as piezoelectric properties as listed in Table 5.2.

For example, molecular weight and concentration, solvent types and mixing percentage, electrospinning methodologies, applied bias and electrode-to-collector distance could all affect the piezoelectric properties of electrospun nanofibers and there is a wide range of these parameters used by different groups. In the work of Yee *et al.* [119], it is concluded that electrospinning of PVDF from DMF/acetone solution promotet the formation of β-phase and aligned electrospinning using a rotation disk results in the c-axis of the β-phase crystallites being oriented along the fiber axis. They also suggest that the formation of the β-phase is likely by the electric field instead of the mechanical shear force. Using a modified rotating disk collector, Yee *et al.* [119] further suggest that effective stretching by the rotation disk could help the formation of highly oriented β-phase which is quite different from their previous report. The study by Zhong *et al.* on PVDF/polyacrylonitrile and PVDF/polysulfone nanofibers had different conclusions [120]; they found that mechanical stretching is more effective than electric poling to induce ferroelectric phase. Clearly, these recent reports indicate the lack of uniform and optimal processing parameters and a lack of fundamental understanding of the piezoelectric effects of nanofibers.

In several different approaches, researchers have worked on enhancing the piezoelectric effects of nanofibers by adjusting solvents or adding other nanomaterials. For example, Andrew and Clarke [121] demonstrated that β-phase PVDF fibers can be

Table 5.1 Calculated electric energy of PVDF nanogenerator

No. of peaks	Electric energy (J)
1	3.23×10^{-3}
2	5.10×10^{-3}
3	3.28×10^{-3}
4	5.06×10^{-3}
5	3.03×10^{-3}
6	5.47×10^{-3}
7	4.76×10^{-3}
8	4.56×10^{-3}
9	3.09×10^{-3}
10	5.38×10^{-3}
11	3.93×10^{-3}
12	4.83×10^{-3}
13	4.06×10^{-3}
14	3.86×10^{-3}
15	3.03×10^{-3}
16	4.21×10^{-3}
17	2.64×10^{-3}
18	4.51×10^{-3}
19	2.89×10^{-3}
20	4.39×10^{-3}
Average	$\mathbf{4.07 \times 10^{-3}}$
Standard deviation	**8.74-14**

Table 5.2 Summary of nanofiber nanogenerators

Material	Synthesis method	Diameter (nm)	Strain/ strain rate	Peak current	Peak voltage	Ref.
PZT	FFES (M)	60	12%	N/A	1.63 V	[127]
	FFES (M)	100	N/A	N/A	0.4 mV	[128]
	FFES (M)	50–150	0.5%	N/A	0.17 V	[129]
PVDF	FFES (M)	187	34 mm/s (at 5 Hz)	4 μA/2 cm^2	2.21 V	[130]
	NFES (M)	500–6500	0.085%	0.5–3 nA	5.30 mV	[123]
	NFES (M)	1000–2000	N/A	30 nA	0.2 mV	[131]
	FFES (M)	600	0.05%, 1.67%/s	0.3 nA	20 mV	[125]
P(VDF-TrFE)	FFES (M)	60–120	N/A	N/A	400 mV	[71]

FFES: far-field electrospinning, NFES: near-field electrospinning, (S): singe fiber, (M): multiple fibers.

electrospun directly from DMF solution with a maximum fraction of β-phase of 0.75, and the addition of well-dispersed ferrite ($Ni_{0.5}Zn_{0.5}Fe_2O_4$) nanoparticles can form nanofibers with high crystalline fraction of ferroelectric α and β phases [122]. Huang *et al.* found that adding a low concentration of SWCNTs in the electrospinning process can induce highly oriented-form crystallites [123]. The studies by Isakov *et al.* concluded that low evaporation rate favors the β-phase PVDF while high evaporation rate preferentially produces the α-phase PVDF [115].

These are just some of the examples of recent publications on the study and possible enhancement of β-phase PVDF nanofibers. Note that many of these studies involved different experimental set-ups such as different solvent types, concentrations, and electrical bias voltages. It is therefore possible that contradictory conclusions could have been drawn on specific processing parameters or materials. Having more uniform and optimized processing parameters with consistent characterizations of the piezoelectric effects of electrospun nanofibers is highly desirable.

5.6.3 *Summary and Future Prospects*

Clearly, current and future stand-alone devices could all take advantage of self-powering energy harvesters and piezoelectric nanofibers. For the production of cost-effective piezoelectric nanofibers, electrospinning processes are good candidates as various materials and nanoparticles have been successfully fabricated as nanofibers and/or nanofibers with embedded nanoparticles. Conventional electrospinning processes have been modified for good alignment and position controllability as well as the feasibility to embed nanoparticles. Piezoelectric nanofibers, such as PVDF, are highly flexible and easy to fabricate for possible integration in implantable and/or flexible devices, as well as textile applications such as electric clothing. Several groups have demonstrated the first prototypes of nanofiber nanogenerators. Further development of fiber-based nanogenerators, beyond the scope of academic laboratories, are necessary for practical applications.

5.7 Case Study: Large-Array Electrospun PVDF Nanogenerators on a Flexible Substrate

5.7.1 *Introduction and Background*

This section presents the use of PVDF-based electrospun nanofibers as nanogenerators in series and/or in parallel to amplify harvested energy due to mechanical strain on a flexible substrate. A chemically resistive flexible polymer substrate has been used with patterned comb-shape gold electrodes by means of a conventional lithography process. A total of 500 parallel nanofibers have been fabricated and connected to amplify current outputs under repeated mechanical straining tests. A peak current of 35 nA has been collected with 0.2 mV of peak voltage.

Various approaches in scavenging energy from the environment have been pursued in the global energy crisis. There are many forms of energy harvesters ranging from human scale [77] to large scale such as ocean waves and waterfalls [75]. Energy harvesting devices based on one-dimensional nanostructures have achieved much recent progress toward potential practical applications. The most common yet most important method is piezoelectric based. For example, nanogenerators using large amounts (1400–1500 mm^{-2}) of ZnO nanowires have been shown to generate decent values of energy outputs in which LED lights are powered and up to 1 V of voltage is generated [78, 80, 124]. However, these nanowires have short structural lengths (a few micrometers) which could limit their energy-harvesting capacity. Furthermore, high processing temperature and difficulty in controlling the orientation of individual wires could complicate their manufacturing process and cost. Polymer-based piezoelectric materials could be the alternative approach. For example, single PVDF nanogenerators have been demonstrated by using NFES [125] with good control over the deposition location and theoretically infinite length for possible applications such as electric clothing [70, 126]. This work advanced the current state-of-the-art technologies in which parallel connections of multiple nanofibers are used to enhance the current outputs. The prototype device is fabricated on a flexible substrate through conventional lithography processes such that direct integration with surfaces of arbitrary shape could be feasible.

5.7.2 Working Principle

Piezoelectric materials have been widely studied in the micro- to nanoscale as energy harvesters because of their low cost and higher piezoelectric constants. Of the piezoelectric polymers, PVDF has been widely used for its good piezoelectric constant, superior mechanical properties and biocompatibility [26, 75]. Specifically, PVDF powders can be easily dissolved in solvents and electrospun on various substrates. Figure 5.73 shows the working principle of the prototype device. A total of 10 pairs of comb-shaped gold electrodes have been designed by using 100-m-wide gaps as the working distance and 1-mm-wide gaps as the separation distance. A total of 50 parallel nanofibers have been electrospun on top of the electrodes such that there are 500 active working contacts to collect charges generated from these PVDF nanofibers. Under axial strain by bending the bottom plastic substrate, aligned dipoles can generate electrical potentials. Since the connections are in parallel, output currents are amplified.

5.7.3 Device Fabrication

Device fabrication process is based on conventional lithography and the NFES process. First, the flexible substrate (Thermanox®) is cleaned with isopropyl alcohol (IPA) and nitrogen dried several times prior to the lithography process. A G-line

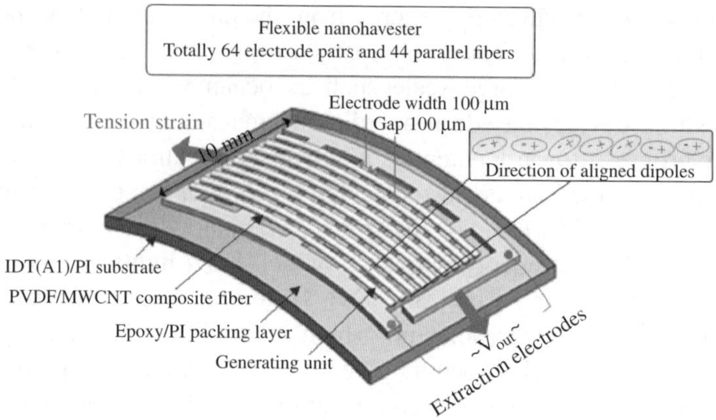

Figure 5.73 Working principle of the large array energy harvesting device by electrospun nanofibers. A total of 50 parallel nanofibers and 10 electrode pairs have been designed in the device on top of a flexible substrate [118]

photo resist is spun on the polymer substrate followed by pattering the photoresist. A 15-nm-thick silicon dioxide layer is deposited by the e-beam evaporation to enhance electrical insulation and promote adhesion. A 10-nm-thick chromium is evaporated followed by the deposition of 150-nm-thick gold layer. After that, the lift-off process defines the electrode area and NFES is performed to deposit PVDF nanofibers. These nanofibers are deposited under a computer-controlled x–y stage with continuous loops such that the deposition direction of nanofibers on top of the electrodes is always unchanged. A high potential is applied afterwards as part of the poling process to further assist alignment of dipoles in PVDF nanofibers. The breakdown voltage of air ($c.$ 60 kV cm^{-1}) is too low for the poling process for PVDF, so a 2-mm-thick polydimethylsiloxane (PDMS) (breakdown voltage $c.$ 250 kV cm^{-1}) layer is applied to cover the electrodes and cured at 70 °C for 30 minutes. Electrical poling of 200 kV cm^{-1} for 30 min [127] on a 70 °C hot plate is applied to further enhance the piezoelectricity of PDVF nanofibers in the lateral direction, as a higher temperature helps the active movement of diploes. Figure 5.74a,b shows optical images of a fabricated device with electrospun nanofibers on top of the plastic substrate before PDMS is applied. The typical diameters of these nanofibers are $c.$ 1–2 μm. The magnified optical photos on the right-hand side of the figure illustrate working gaps of $c.$ 100 μm across these well-aligned nanofibers. The diameter of these nanofibers can be controlled via applied voltage and mixture ratio of the polymer solutions. Generally, smaller applied voltage results in a thinner diameter of nanofibers.

For simulations, electrostatic fields between both 100 μm and 1 mm gaps are used during the electrical poling process. Results show the electrical potential streamlines when one electrode is grounded and the neighboring electrode is applied at 2 kV. As expected, a stronger electrical field occurred in the 100 μm gap ($c.$ 2×10^7 V m^{-1}) as compared with the 1 mm gap (2×10^6 V m^{-1}). The typical electrical poling for PVDF

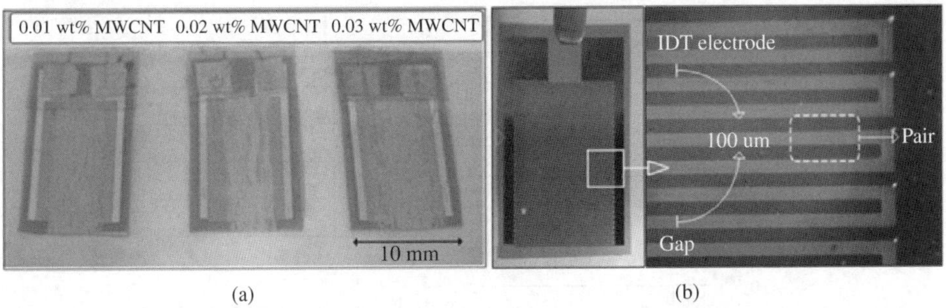

Figure 5.74 (a) Optical images of fabricated device with well-aligned nanofibers. (b) The working gap between two electrodes is 100 μm. Nanofibers have been deposited by electrospinning continuously without stopping controlled by an x–y stage and the deposition direction for nanofibers was maintained [26]

should be higher than 2×10^7 V m^{-1}. The local concentration effects could enhance the local electrical field to be higher than the average values. It is therefore expected that some of the local areas in the 100 μm gap had undergone successful electrical poling process, while the area in the 1 mm gap probably did not experience enough electrical poling to have meaningful piezoelectric properties.

5.7.4 Experimental Results

The fabricated devices have been characterized under repeated external strains by using a DC motor at 1 Hz to control the deformation magnitude of the plastic substrate. As the dipoles in nanofibers are aligned parallel to the longitudinal direction of the nanofibers, d_{33} mode has the strongest piezoelectric effect. The estimated strain from the bending of the substrate from the motor is $0.06 \pm 0.02\%$. Current and voltage generated from the device were measured inside a Faraday cage to block out external noise. Figure 5.75 shows the current measurement results for the forward and reverse connections. This test is important to validate that results were due to the piezoelectric responses and not artificial effects. If the signal is coming from the noise or other forms instead of piezoelectric responses, the shape of the response should remain the same even if the polarity of the contacts is changed. Since the induced piezoelectric response has its own polarity depending on the electrical poling direction, the electrical measurements should have reversed responses when the polarity of the contacts was reversed. The bottom plots show the magnified pattern of a single bending and release process. The peak current in the forward direction measurement was *c.* 35 nA while the reverse connection shows *c.* 2 nA. This observed reduced response after polarity change is not clearly understood. One reason could be inherent current flowing inside the measurement equipment. If that is the case, the real current generated from the device is the average of two responses, in which inherent current should be subtracted from the forward direction data and should be added to the reverse direction

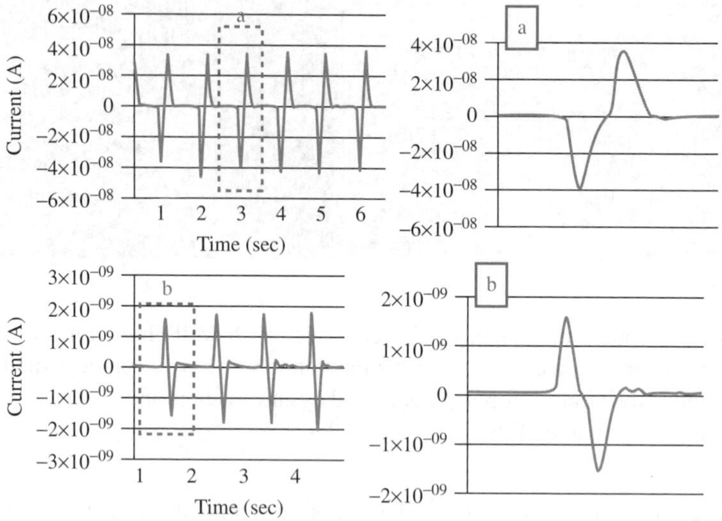

Figure 5.75 (a, b) Current generated from device is monitored under repeated application of strains. The response shape is flipped when the measurement polarity is switched. Peak current is measured at *c*. 35 nA in the forward connection and 2 nA in reverse connection. The small current flowing inside the measurement device may have resulted in the unsymmetrical response [26]

data. The voltage measurement result is plotted in Figure 5.76. Similar to the current measurement, the forward and reverse direction measurement was conducted and the shape of response was flipped as expected. The bottom plots in Figure 5.76 show the voltage measurement of a single bending and release event. The peak voltage was *c*. 0.2 mV in both the forward and reverse connections. A significant decrease in the voltage measurement was not observed as the voltage measurement is not affected by current flow but by potential difference.

It is noted that both current and voltage outputs of these PVDF nanogenerators are about at least one order of magnitude less than the previous report on a single PVDF nanofiber [6]. Possible explanations are: (1) the *in situ* poling process could be more effective to induce stronger piezoelectric effects than the post-poling process in this work; (2) partially reversed energy generation effects from the opposite neighboring electrodes generate counter voltages and currents; and (3) encapsulation of the nanofiber in the current work could increase the loss of charges to the package material. These factors will all be further investigated.

5.7.5 Summary

Piezoelectric energy harvesters based on electrospun nanofibers with current amplifying design have been demonstrated on flexible substrate. A total of 500 active parallel

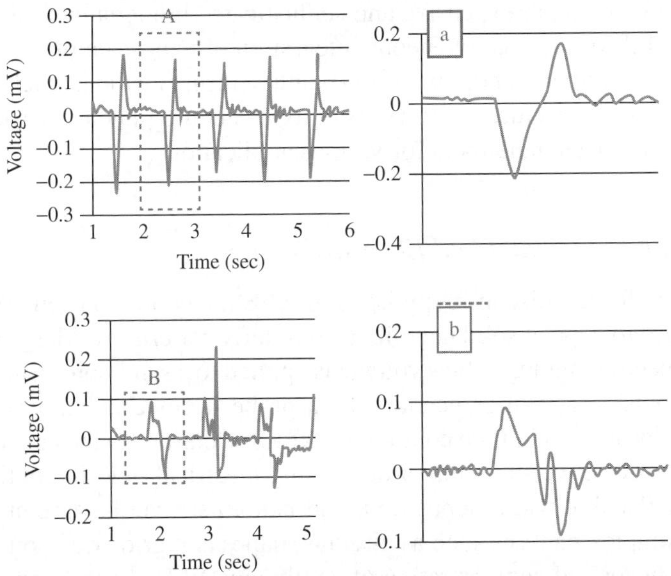

Figure 5.76 (a, b) Voltage generated from the device is measured in both forward and reverse connections. The response shape is flipped when the measurement polarity is switched. Peak voltage is *c.* 0.2 mV and no significant reduction of signal is monitored in the reverse connection measurement. The output is small under the parallel connection of nanogenerators [26]

connecting points have been made between the gold electrodes and PVDF nanofibers. Under a repeated maximum strain of 0.05%, a peak current of 35 nA and peak voltage of 0.2 mV were measured. Switching the measurement polarity resulted in flipped responses of the nanogenerators, which validates the piezoelectric responses. As such, power generation from these PVDF energy harvesters can be further enhanced by various electrodes designs to boost either current or voltage.

5.8 Conclusion

5.8.1 Near-Field Electrospinning

A NFES process has been developed to deposit solid nanofibers in a direct and controllable manner. Specifically, a tungsten electrode with tip diameter of 25 μm is used to construct nanofibers of 50–500 nm line width on silicon-based collectors while the liquid polymer solution is supplied in a manner analogous to that of a DPN. To achieve position controllable depositions, the minimum applied bias voltage is 600 V and minimum electrode-to-collector distance is 500 μm. Charged nanofibers can be orderly collected, making NFES a potential tool in direct-write nanofabrication for a variety of materials. NFES is a simple yet powerful method for direct-write deposition of nanofibers with unprecedented controllability at resolutions comparable to those achieved with much more expensive and sophisticated lithography tools. It should be especially useful for the heterogeneous integration of nanoscale materials to devices prepared by conventional lithographic and manufacturing methods, such as microelectronics and MEMS structures. Furthermore, NFES could play a key role in building up ordered non-woven nanofibers for various applications.

5.8.2 Continuous Near-Field Electrospinning

To compensate for the discontinuity issue of NFES, a continuous NFES process has been developed to deposit solid nanofibers in orderly patterns over large areas. Before the onset of electrospinning, a bias voltage is applied to a semi-spherical-shaped polymer droplet outside a syringe needle, and a probe tip mechanically draws a single fiber from the droplet to initiate continuous NFES. Contrary to the conventional electrospinning process, we show that a decreasing electrical field in continuous NFES results in smaller line-width deposition, and nanofibers can be assembled into controlled and complex patterns such as circular shapes and grid arrays on large and flat areas. The concept and demonstration of continuous NFES enables applications that were difficult to achieve by conventional electrospinning. We believe that the ability to deposit polymeric materials at precise locations with specific patterns can open up opportunities for cost-effective heterogeneous integration of a variety of materials.

5.8.3 Direct-Write Piezoelectric PVDF

Nanogenerators capable of converting energy from mechanical sources to electricity with high effective efficiency using low-cost non-semiconducting organic nanomaterials are attractive for many applications, including energy harvesters. In this chapter, NFES is used to direct-write PVDF nanofibers with *in situ* mechanical stretch and electrical poling characteristics to produce piezoelectric properties. Under mechanical stretching, nanogenerators have shown repeatable and consistent electrical outputs with ECE an order of magnitude higher than those made of PVDF thin films. The early onset of the non-linear domain wall motions behavior has been identified as one mechanism responsible for the apparent high piezoelectricity in nanofibers, rendering them potentially advantageous for sensing and actuation applications. The reported nanogenerator has several advantages over other maco-, micro-, and nanogenerators, including high ECE, manufacturability, and the capability of integration with other micro-/nanofabrication processes. The principle and the nanogenerator demonstrated could be the basis for integrated power sources in nanodevices and wireless sensors, or new self-powered textiles by direct-writing nanofibers onto a large area of cloth to boost the total power output for portable electronics.

5.9 Future Directions

The concept and initial demonstrations of NFES make possible new applications that are not possible by conventional electrospinning. In addition to the nanogenerator demonstrated in this chapter, there are many other interesting directions that can be implemented by NFES. The following sections give several examples.

5.9.1 NFES Integrated Nanofiber Sensors

The ability to deposit polymeric materials at specific locations provides opportunities for heterogeneous integration of sensing materials with microelectronics. Figure 5.77 shows the schematic diagram of a simple way to complete the heterogeneous integration of nanofiber materials with prefabricated contact electrodes for possible sensing applications. The chaotic nature of conventional electrospinning prevents the possibility of site-specific deposition. The SEM microphoto shows preliminary results. A nanofiber of 100 nm in diameter made of polypyrrole-conducting polymer was

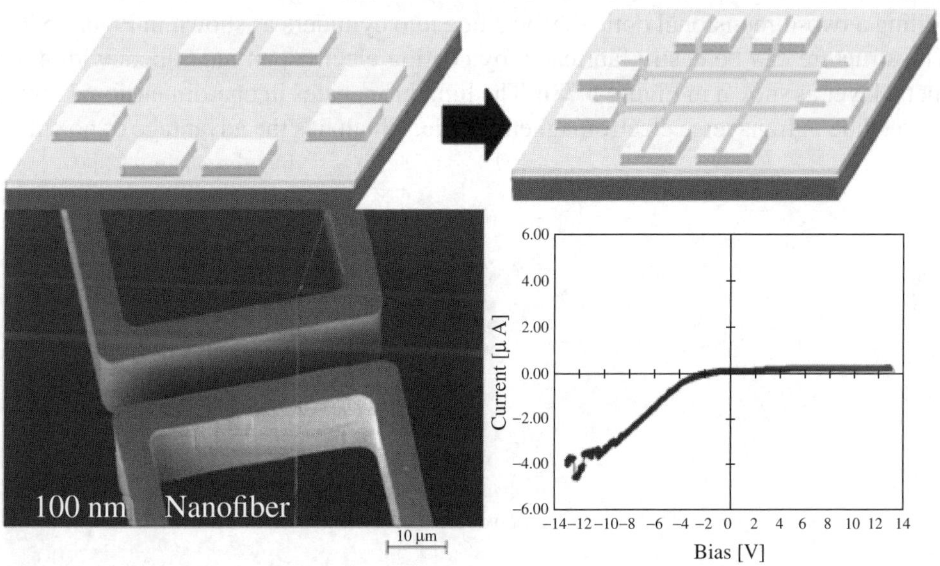

Figure 5.77 Schematic diagram showing a simple way to complete the heterogeneous integration of nanofiber materials with prefabricated contact electrodes for possible sensing applications. The SEM microphoto shows that the 100-nm-diameter polypyrrole-conducting polymer was deposited by NFES over two suspended microbridges made of heavily doped polysilicon. This single nanofiber was deposited with good control over a 10-μm-wide air gap. Initial I–V characteristics show rectifying behavior under negative voltage input [26]

deposited by NFES in a suspended fashion over two micro-bridges made of heavily doped polysilicon. The nanofiber was deposited with good control over a 10-μm-wide air gap as shown. Initial I–V characteristics show rectifying behavior under negative voltage input, which requires further investigation. Nevertheless, the possible expansion of the proposed architecture with integrated circuitry underneath the contact pads for transduction, readout, processing, and communications can be forecast. Furthermore, there are a wide variety of materials for the electrospinning process such that different kinds of sensors can be constructed by the proposed method.

5.9.2 NFES One-Dimensional Sub-Wavelength Waveguide

The smallest diameter of an optical beam propagating in dielectric waveguides is of the order of the wavelength because of optical diffraction limits. This limits the size of integrated optical devices and the spatial resolution of the optical measurements. If there were a way to decrease the smallest beam diameter beyond diffraction limit, it would make a great contribution to optical devices in the nanometer range. The structure of the proposed one-dimensional sub-wavelength waveguide is similar to rolling a two-dimensional optical waveguide into cylinders as shown in Figure 5.78a. The structure can be easily fabricated by coating electrospun nanofibers with a thin metal layer as shown in Figure 5.78b. The light propagates in one-dimensional optical waveguide in transverse electromagnetic mode, which has the advantage of no cut-off

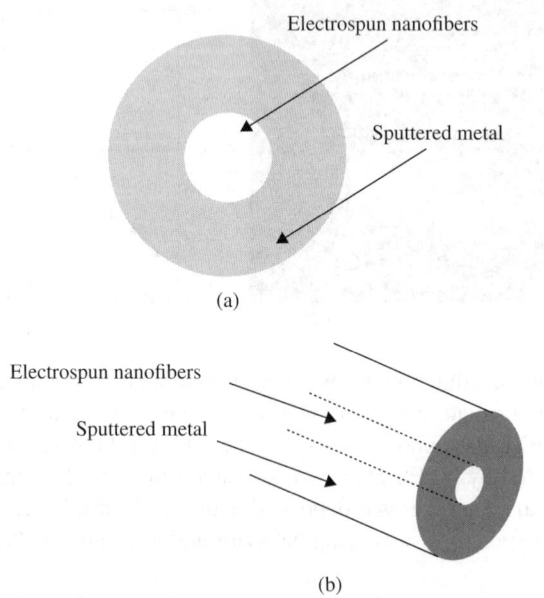

Figure 5.78 (a) Illustration of a 1D sub-wavelength optical waveguide and (b) 3D drawing [26]

frequency. The attenuation is dominated by resistive loss in metal and light can be guided around sharp corners.

5.9.3 NFES Biological Applications

NFES can enable bioengineers and biologists to create highly tailored and functionalized culture substrates quickly and at low cost by varying the concentration of factors such as biological compounds (e.g. growth factors/inhibitors, enzymes, and proteins), conductivity, porosity, and surface chemistry (e.g. hydrophobicity) in polymeric solutions. Possible applications include drug tests, neuron guidance, stem cell differentiation, proliferation, apoptosis, focal adhesion and migration, cell growth and cessation, and tissue engineering. In recent years, artificial tissue and organs have been the focus of much attention due to the drastic mismatch between the numbers of tissue and organ donors and patients requiring transplants. For example, conventional electrospinning [54] has been used to create scaffolds to mimic ECMs for tissue engineering to study cell behavior such as attachment and proliferation. However, these techniques are not ideal for tissue engineering because the highly ordered and specialized structures found in real tissues and organs [128] are unachievable. NFES has the potential to not only create ordered structures through automated processes, but to also specifically functionalize different parts of these structures through the use different materials. The ability to develop and customize ordered 3D structures (Figure 5.79) would enable bioengineers to create 3D scaffolds *in vitro* that are highly

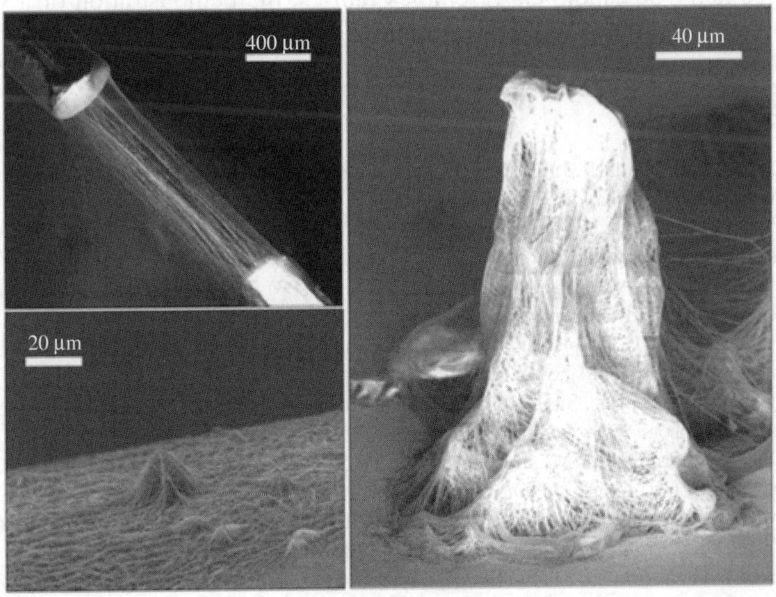

Figure 5.79 SEM microphoto showing preliminary result of 3D structures fabricated by continuous NFES, demonstrating potential for bioscaffolds/substrates applications [26]

representative of real ECMs. Such scaffolds could extend current tissue engineering techniques to develop previously unachievable organ types and increase the viability of these tissues.

5.9.4 Direct-Write Piezoelectric PVDF Nanogenerators

The current experimental results in this chapter lead to several directions that can be exploited in the future. First, what are the geometrical and material effects on the energy generation? These are complex issues as fiber diameter control, material properties, and manufacturing parameters are related to each other and should be investigated at the same time. Processing parameters such as electrical field strength, viscosity of the polymer solution, and environmental conditions will all affect the fiber diameter and piezoelectric properties. Second, piezoelectric nanofibers have never been studied or demonstrated before and it could present interesting potentials in various applications beyond power scavenging, including piezoelectric sensing and actuation.

In future studies we aim to investigate manufacturing protocols, material properties and optimizations. A prototype energy harvester demonstration using the proposed electrospun nanofibers is definitely needed for its practical applications. For example, can a single nanofiber itself be a power source for nanosensors that may require low power? Can a bundle of these nanofibers be assembled for larger power output? Future research should extend the investigations to large-scale demonstrations for several possible applications including buildings, automobiles, and bridges. Because the natural geometry of nanofibers resembles fabrics, the demonstration of a wearable nanogenerator on clothes could be a starting point.

References

1. Greiner, A. and Wendorff, J.H. (2007) Electrospinning: a fascinating method for the preparation of ultrathin fibers. *Angewandte Chemie International Edition*, **46** (30), 5670–5703.
2. Agarwal, S., Wendorff, J.H. and Greiner, A. (2008) Use of electrospinning technique for biomedical applications. *Polymer*, **49** (26), 5603–5621.
3. Pinto, N.J., Johnson, A.T. and MacDiarmid, A.G. (2003) Electrospun polyaniline/polyethylene oxide nanofiber field-effect transistor. *Applied Physics Letters*, **83** (20), 4244–4246.
4. Liu, H., Kameoka, J., Czaplewski, D.A. and Craighead, H.G. (2004) Polymeric nanowire chemical sensor. *Nano Letters*, **4** (4), 671–675.
5. Wang, X., Drew, C., Lee, S.-H. *et al.* (2002) Electrospun nanofibrous membranes for highly sensitive optical sensors. *Nano Letters*, **2** (11), 1273–1275.
6. Takahashi, T., Taniguchi, M. and Kawai, T. (2005) Fabrication of DNA nanofibers on a planar surface by electrospinning. *Japanese Journal of Applied Physics*, **44** (27), L860–L862.
7. Theron, A., Zussman, E. and Yarin, A.L. (2001) Electrostatic field-assisted alignment of electrospun nanofibers. *Nanotechnology*, **12** (3), 384–390.
8. Reneker, D.H., Yarin, A.L., Fong, H. and Koombhongse, S. (2000) Bending instability of electrical charged liquid jets of polymer solutions in electrospinning. *Journal of Applied Physics*, **87** (9), 4531–4547.

9. Chew, S.Y., Wen, J., Yim, E.K.F. and Leong, K.W. (2005) Sustained release of proteins from electrospun biodegradable fibers. *Biomacromolecules*, **6** (4), 2017–2024.
10. Fennessey, S.F. and Farris, R.J. (2004) Fabrication of aligned and molecularly oriented electrospun polyacrylonitrile nanofibers and the mechanical behavior of their twisted yarns. *Polymer*, **45** (12), 4217–4225.
11. Kim, K.W., Lee, K.H., Khil, M.S. *et al.* (2004) The effect of molecular weight and the linear velocity of drum surface on the properties of electrospun poly(ethylene terephthalate) nonwovens. *Fibers and Polymers*, **5** (2), 122–127.
12. Matthews, J.A., Wnek, G.E., Simpson, D.G. and Bowlin, G.L. (2002) Electrospinning of collagen nanofibers. *Biomacromolecules*, **3** (2), 232–238.
13. Li, D., Wang, Y. and Xia, Y. (2003) Electrospinning of polymeric and ceramic nanofibers as uniaxially aligned arrays. *Nano Letters*, **3** (8), 1167–1171.
14. Li, D., Wang, Y. and Xia, Y. (2004) Electrospinning nanofibers as uniaxially aligned arrays and layer-by-layer stacked films. *Advanced Materials*, **16** (4), 361–366.
15. Theron, A., Zussman, E. and Yarin, A.L. (2003) Experimental investigation of the governing parameters in the electrospinning of polymer solutions. *Polymer*, **45** (6), 2017–2030.
16. Kowalewski, T.A., Blonski, S. and Barral, S. (2005) Experiments and modelling of electrospinning process. *Bulletin of the Polish Academy of Sciences*, **53** (4), 385–394.
17. White, B.E. (2008) Energy-harvesting devices: beyond the battery. *Nature Nanotechnology*, **3**, 71–72.
18. Paradiso, J.A. and Starner, T. (2005) Energy scavenging for mobile and wireless electronics. *IEEE Pervasive Computing*, **4** (1), 18–27.
19. Churchill, D.L., Hamel, M.J., Townsend, C.P. and Arms, S.W. (2003) Strain energy harvesting for wireless sensor networks. *Proceedings of the SPIE – Smart Structures and Materials*, **5005**, 319–327.
20. C.G. Triplett. (1985) Vehicular mounted piezoelectric generator. US Patent No. 4,504,761.
21. Shenck, N.S. and Paradiso, J.A. (2001) Energy scavenging with shoe-mounted piezoelectrics. *IEEE Micro*, **21** (3), 30–42.
22. Niu, P., Chapman, P., Riemer, R. and Zhang, X. (2004) Evaluation of motions and actuation methods for biomechanical energy harvesting. IEEE Power Electronics Specialists Conference, pp. 2100–2106.
23. Qin, Y., Wang, X. and Wang, Z.L. (2008) Microfibre-nanowire hybrid structure for energy scavenging. *Nature*, **451**, 809–813.
24. Starner, T. (1996) Human powered wearable computing. *IBM Systems Journal*, **35** (3–4), 618–629.
25. Kymissis, J., Kendall, C., Paradiso, J. and Gershenfeld, N. (1998) Parasitic power harvesting in shoes. 2nd IEEE International Symposium on Wearable Computers, pp. 132–139.
26. C. Chang. (2009) DirectWrite piezoelectric nanogenerator by near-field electrospinning. PhD Thesis, University of California, Berkeley.
27. Definition of *Textile*. Available at http://www.answers.com/topic/textile (accessed January 2014).
28. Teo, W.E. and Ramakrishna, S. (2006) A review on electrospinning design and nanofibre assemblies. *Nanotechnology*, **17** (14), R89–R106.
29. Taylor, G. (1964) Disintegration of water drops in an electric field. *Proceedings of the Royal Society of London Series A*, **280** (1382), 383–397.
30. Larrondo, L. and John Manley, R.S. (1981) Electrostatic fiber spinning from polymer melts. I. Experimental observations on fiber formation and properties. *Journal of Polymer Science: Polymer Physics Edition*, **19** (6), 909–920.
31. Reznik, S.N., Yarin, A.L., Theron, A. and Zussman, E. (2004) Transient and steady shapes of droplets attached to a surface in a strong electric field. *Journal of Fluid Mechanics*, **516**, 349–377.
32. Boudriot, U., Dersch, R., Goetz, B. *et al.* (2004) Electrospun poly-l-lactide nanofibres as scaffolds for tissue engineering. *Biomedical Engineering*, **49** (9), 242–247.

33. Inai, R., Kotaki, M. and Ramakrishna, S. (2001) Structure and properties of electrospun PLLA single nanofibres. *Nanotechnology*, **16** (2), 208–213.
34. Xu, C.Y., Inai, R., Kotaki, M. and Ramakrishna, S. (2004) Aligned biodegradable nanofibrous structure: a potential scaffold for blood vessel engineering. *Biomaterials*, **25** (5), 877–886.
35. Katta, P., Alessandro, M., Ramsier, R.D. and Chase, G.G. (2004) Continuous electrospinning of aligned polymer nanofibers onto a wire drum collector. *Nano Letters*, **4** (11), 2215–2218.
36. Bhattaraia, N., Edmondsona, D., Veiseha, O. *et al.* (2005) Electrospun chitosan-based nanofibers and their cellular compatibility. *Biomaterials*, **26** (31), 6176–6184.
37. Pan, H., Li, L., Hu, L. and Cui, X. (2005) Continuous aligned polymer fibers produced by a modified electrospinning method. *Polymer*, **47** (14), 4901–4904.
38. Teo, W.E., Kotaki, M., Mo, X.M. and Ramakrishna, S. (2005) Porous tubular structures with controlled fibre orientation using a modified electrospinning method. *Nanotechnology*, **16** (6), 918–924.
39. W. E. Teo and S. Ramakrishna. (2005) Electrospun fibre bundle made of aligned nanofibres over two fixed points. *Nanotechnology*, **16** (9), 1878–1884.
40. Boland, E.D., Wnek, G.E., Simpson, D.G. *et al.* (2001) Tailoring tissue engineering scaffolds using electrostatic processing techniques: a study of poly(glycolic acid) electrospinning. *Journal of Macromolecular Science Pure and Applied Chemistry*, **A38** (12), 1231–1243.
41. Buttafoco, L., Kolkman, N.G., Engbers-Buijtenhuijs, P. *et al.* (2006) Electrospinning of collagen and elastin for tissue engineering applications. *Biomaterials*, **27** (5), 724–734.
42. Deitzel, J.M., Kleinmeyer, J.D., Hirvonen, J.K. and Beck Tan, N.C. (2001) Controlled deposition of electrospun poly(ethylene oxide) fibers. *Polymer*, **42** (19), 8163–8170.
43. Kim, G.H. (2006) Electrospinning process using field-controllable electrodes. *Journal of Polymer Science Part B: Polymer Physics*, **44** (10), 1426–1433.
44. Stankus, J., Guan, J. and Wagner, W. (2004) Fabrication of biodegradable elastomeric scaffolds with sub-micron morphologies. *Journal of Biomedical Materials Research Part A*, **70A** (4), 603–614.
45. Kameoka, J., Orth, R., Yang, Y. *et al.* (2003) A scanning tip electrospinning source for deposition of oriented nanofibres. *Nanotechnology*, **14** (10), 1124–1129.
46. Rayleigh, F.R.S. (1882) On the equilibrium of liquid conducting masses charged with electricity. *London Edinburgh Dublin Philosophical Magazine and Journal*, **44**, 184–186.
47. Zeleny, J. (1917) Instability of electrified liquid surfaces. *Physical Review*, **10**, 1–6.
48. A. Formhals. (1934) Process and apparatus for preparing artificial threads. US Patent No. 1,975,504.
49. Reneker, D.H. and Chun, I. (1996) Nanometre diameter fibres of polymer, produced by electrospinning. *Nanotechnology*, **7** (3), 216–223.
50. Zhou, Y., Freitag, M., Hone, J. *et al.* (2003) Fabrication and electrical characterization of polyaniline-based nanofibers with diameter below 30 nm. *Applied Physics Letters*, **83** (18), 3800–3802.
51. Dai, H., Gong, J., Kim, H. and Lee, D. (2002) A novel method for preparing ultra-fine alumina-borate oxide fibres via an electrospinning technique. *Nanotechnology*, **13** (5), 674–677.
52. Bergshoef, M.M. and Vancso, G.J. (1999) Transparent nanocomposites with ultrathin, electrospun nylon-4,6 fiber reinforcement. *Advanced Materials*, **11** (16), 1362–1365.
53. Ko, F., Gogotsi, Y., Ali, A. *et al.* (2003) Electrospinning of continuous carbon nanotube-filled nanofiber yarns. *Advanced Materials*, **15** (14), 1161–1165.
54. Li, W.-J., Laurencin, C.T., Caterson, E.J. *et al.* (2002) Electrospun nanofibrous structure: a novel scaffold for tissue engineering. *Journal of Biomedical Materials Research*, **60** (4), 613–621.
55. Min, B.-M., Lee, G., Kim, S.H. *et al.* (2004) Electrospinning of silk fibroin nanofibers and its effect on the adhesion and spreading of normal human keratinocytes and fibroblasts in vitro. *Biomaterials*, **25** (7–8), 1289–1297.
56. Gibson, P., Schreuder-Gibson, H. and Riven, D. (2001) Transport properties of porous membranes based on electrospun nanofibers. *Colloids and Surfaces A: Physicochemical and Engineering Aspects*, **187–188**, 469–481.

57. Wannatong, L., Sirivat, A. and Supaphol, P. (2004) Effects of solvents on electrospun polymeric fibers: preliminary study on polystyrene. *Polymer International*, **53** (11), 1851–1859.
58. Fong, H., Liu, W.-D., Wang, C.-S. and Vaia, R.A. (2002) Generation of electrospun fibers of nylon 6 and nylon 6-montmorillonite nanocomposite. *Polymer*, **43** (3), 775–780.
59. Dersch, R., Liu, T., Schaper, A.K. *et al.* (2003) Electrospun nanofibers: internal structure and intrinsic orientation. *Journal of Polymer Science Part A: Polymer Chemistry*, **41** (4), 545–553.
60. Zussman, E., Theron, A. and Yarin, A.L. (2003) Formation of nanofiber crossbars in electrospinning. *Applied Physics Letters*, **82** (6), 973–975.
61. Han, T., Reneker, D.H. and Yarin, A.L. (2007) Buckling of jets in electrospinning. *Polymer*, **48** (20), 6064–6076.
62. Frenot, A. and Chronakis, I.S. (2003) Polymer nanofibers assembled by electrospinning. *Current Opinion in Colloid and Interface Science*, **8** (5), 64–75.
63. Son, W.K., Youk, J.H., Lee, T.S. and Park, W.H. (2004) The effects of solution properties and polyelectrolyte on electrospinning of ultrafine poly(ethylene oxide) fibers. *Polymer*, **45** (9), 2959–2966.
64. Piner, R.D., Zhu, J., Xu, F. *et al.* (1999) Dip-pen nanolithography. *Science*, **283** (5402), 661–663.
65. Sirringhaus, H., Kawase, T., Friend, R.H. *et al.* (2000) High-resolution inkjet printing of all-polymer transistor circuits. *Science*, **290** (5499), 2123–2126.
66. Pawlowski, K.J., Belvin, H.L., Raney, D.L. *et al.* (2003) Electrospinning of a micro-air vehicle wing skin. *Polymer*, **44** (4), 1309–1314.
67. Dzenis, Y. and Wen, Y. (2002) Continuous carbon nanofibers for nanofiber composites. *Materials Research Society Symposium Proceedings*, **702**, 173–178.
68. Ye, H., Lam, H., Titchenal, N. *et al.* (2004) Reinforcement and rupture behavior of carbon nanotubes–polymer nanofibers. *Applied Physics Letters*, **85** (10), 1775–1777.
69. Yeo, I.-S., Oh, J.-E., Jeong, L. *et al.* (2008) Collagen-based biomimetic nanofibrous scaffolds: preparation and characterization of collagen/silk fibroin bicomponent nanofibrous structures. *Biomacromolecules*, **9** (4), 1106–1116.
70. Chang, C., Limkrailassiri, K. and Lin, L. (2008) Continuous near-field electrospinning for arge area deposition of orderly nanofiber patterns. *Applied Physics Letters*, **93**, 123111.
71. Mandal, D., Yoon, S. and Kim, K.J. (2011) Origin of piezoelectricity in an electrospun poly (vinylidene fluoride-trifluoroethylene) nanofiber web-based nanogenerator and nano-pressure sensor. *Macromolecular Rapid Communications*, **32**, 831–837.
72. Lee, J.S., Choi, K.H., Do Ghim, H. *et al.* (2004) Role of molecular weight of atactic poly(vinyl alcohol) (PVA) in the structure and properties of PVA nanofabric prepared by electrospinning. *Journal of Applied Polymer Science*, **93** (4), 1638–1646.
73. Buchko, C.J., Chen, L.C., Shen, Y. and Martin, D.C. (2003) Processing and microstructural characterization of porous biocompatible protein polymer thin films. *Polymer*, **40** (26), 7397–7407.
74. Megelski, S., Stephens, J.S., Chase, D.B. and Rabolt, J.F. (2002) Micro-and nanostructured surface morphology on electrospun polymer fibers. *Macromolecules*, **35** (22), 8456–8466.
75. Lu, X., McElroy, M.B. and Kiviluoma, J. (2009) Global potential for wind-generated electricity. *Proceedings of the National Academy of Sciences of the United States of America*, **106**, 10933–10938.
76. Scruggs, J. and Jacob, P. (2009) Harvesting ocean wave energy. *Science*, **323** (5918), 1176–1178.
77. Donelan, J.M., Li, Q., Naing, V. *et al.* (2008) Biomechanical energy harvesting: generating electricity during walking with minimal user effort. *Science*, **319** (5864), 807–810.
78. Wang, Z.L. and Song, J. (2006) Piezoelectric nanogenerators based on zinc oxide nanowire arrays. *Science*, **312** (5771), 242–246.
79. Wang, X., Song, J., Liu, J. and Wang, Z.L. (2007) Direct-current nanogenerator driven by ultrasonic. *Science*, **316** (5821), 102–105.
80. Yang, R., Qin, L., Dai, L. and Wang, Z.L. (2009) Power generation with laterally packaged piezoelectric fine wires. *Nature Nanotechnology*, **4**, 34–39.

81. Ye, Y., Jiang, Y., Wu, Z. and Zeng, H. (2006) Phase transitions of poly(vinylidene fluoride) under electric fields. *Integrated Ferroelectrics*, **80** (1), 245–251.
82. Sajkiewicz, P., Wasiak, A. and Golclowski, Z. (1999) Phase transitions during stretching of poly(vinylidene fluoride). *European Polymer Journal*, **35** (3), 423–429.
83. Salimi, A. and Yousefi, A.A. (2003) FTIR studies of β-phase formation in PVDF stretched films. *Polymer Testing*, **22** (6), 699–704.
84. Neidhöfer, M., Beaume, F., Ibos, L. *et al.* (2004) Structural evolution of PVDF during storage or annealing. *Polymer*, **45** (5), 1679–1688.
85. Gregorio, R. and Chaves Pereira de Souza Nociti, N. (1995) Effect of PMMA addition on the solution crystallization of the alpha and beta phases of poly(vinylidene fluoride). *Journal of Physics D: Applied Physics*, **28** (2), 432–436.
86. Andrew, J.S., Mack, J.J. and Clarke, D.R. (2007) Electrospinning of polyvinylidene difluoride-based nanocomposite fibers. *Journal of Materials Research*, **23** (1), 105–114.
87. Andrew, J.S. and Clarke, D.R. (2008) Effect of electrospinning on the ferroelectric phase content of polyvinylidene difluoride fibers. *Langmuir*, **24** (3), 670–672.
88. Gao, K., Hu, X., Dai, C. and Yi, T. (2006) Crystal structures of electrospun PVDF membranes and its separator application for rechargeable lithium metal cells. *Materials Science and Engineering B*, **131** (1–3), 100–105.
89. Nasir, M., Matsumoto, H., Danno, T. *et al.* (2006) Control of diameter, morphology, and structure of PVDF nanofiber fabricated by electrospray deposition. *Journal of Polymer Science Part B: Polymer Physics*, **44** (5), 779–786.
90. Nasir, M., Matsumoto, H., Minagawa, M. *et al.* (2007) Formation of β-phase crystalline structure of PVDF nanofiber by electrospray deposition: additive effect of ionic fluorinated surfactant. *Polymer Journal*, **39** (7), 670–674.
91. Ren, X. and Dzenis, Y. (2006) Novel continuous poly(vinylidene fluoride) nanofibers. *Materials Research Society Symposium Proceedings*, **920**, 0920-S03-03.
92. Yee, W.A., Kotaki, M., Liu, Y. and Lu, X. (2006) Morphology, polymorphism behavior and molecular orientation of electrospun poly(vinylidene fluoride) fibers. *Polymer*, **48** (2), 512–521.
93. Yu, L. and Cebe, P. (2009) Crystal polymorphism in electrospun composite nanofibers of poly(vinylidene fluoride) with nanoclay. *Polymer*, **50** (9), 2133–2141.
94. Zhao, Z., Li, J., Yuan, X. *et al.* (2004) Preparation and properties of electrospun poly(vinylidene fluoride) membranes. *Journal of Applied Polymer Science*, **97** (2), 466–474.
95. Huang, Z.-M., Zhangb, Y.-Z., Kotakic, M. and Ramakrishna, S. (2003) A review on polymer nanofibers by electrospinning and their applications in nanocomposites. *Composites Science and Technology*, **63** (15), 2223–2253.
96. Morgret, L.D., Pawlowski, K.J. and Hinkley, J.A. (2005) Electrospinning of Polyvinylidene Fluoride and Polyetherimide From Mixed Solvents. NASA, Technical Report, NASA/TM-2005-213786.
97. Pham, Q.P., Sharma, U. and Mikos, A.G. (2006) Electrospinning of polymeric nanofibers for tissue engineering applications: A review. *Tissue Engineering*, **12** (5), 1197–1211.
98. Precision Acoustics Ltd. Available at http://acoustics.co.uk/products/piezo-electric-materials /properties-pvdf-film/ (accessed January 2014).
99. Alexe, M., Senz, S., Schubert, M.A. *et al.* (2008) Energy harvesting using nanowires. *Advanced Materials*, **20** (21), 4021–4026.
100. Sirohi, J. and Chopra, I. (2000) Fundamental understanding of piezoelectric strain sensors. *Journal of Intelligent Material Systems and Structures*, **11** (4), 246–257.
101. Vidhate, S., Shaito, A., Chung, J. and D'Souza, N.A. (2009) Crystallization, mechanical, and rheological behavior of polyvinylidene fluoride-carbon nanofiber composites. *Journal of Applied Polymer Science*, **112** (1), 254–260.

102. Dunn, S. (2003) Determination of cross sectional variation of ferroelectric properties for thin film (ca. 500 nm) PZT (30/70) via PFM. *Integrated Ferroelectrics*, **59** (1), 1505–1512.
103. Gu, S.-Y., Wu, Q.-L., Ren, J. and Vancso, G.J. (2005) Mechanical properties of a single electrospun fiber and its structures. *Macromolecular Rapid Communications*, **26** (9), 716–720.
104. Zhao, M.-H., Wang, Z.-L. and Mao, S.X. (2004) Piezoelectric characterization of individual zinc oxide nanobelt probed by piezoresponse force microscope. *Nano Letters*, **4** (4), 587–590.
105. Majdoub, M.S., Sharma, P. and Cagin, T. (2008) Enhanced size-dependent piezoelectricity and elasticity in nanostructures due to the flexoelectric effect. *Physical Review B*, **77** (12), 125424-1–125424-9.
106. Bassiri-Gharb, N., Fujii, I., Hong, E. *et al.* (2007) Domain wall contributions to the properties of piezoelectric thin films. *Journal of Electroceramics*, **19** (1), 47–67.
107. Wang, Y., Ren, K. and Zhang, Q.M. (2007) Direct piezoelectric response of piezopolymer polyvinylidene fluoride under high mechanical strain and stress. *Applied Physics Letters*, **91** (22), 222905-1–222905-3.
108. Baji, A., Mai, Y.-W., Li, Q. and Liu, Y. (2011) Electrospinning induced ferroelectricity in poly(vinylidene fluoride) fibers. *Nanoscale*, **3**, 3068–3071.
109. Farrar, D., Ren, K., Cheng, D. *et al.* (2011) Permanent polarity and piezoelectricity of electrospun α-helical poly(α-amino acid) fibers. *Advanced Materials*, **23**, 3954–3958.
110. Chen, X., Yao, N., Shi, Y. and Zang, L. (2011) *Energy Efficiency and Renewable Energy Through Nanotechnology*, Springer, London.
111. Wang, Y. and Santiago-Aviles, J.J. (2004) Synthesis of PZT nanofibers and the FTIR characterization of its MOD process. *Nanotechnology*, **15**, 32–36.
112. Kalinin, S. and Bonnell, D. (2002) Imaging mechanism of piezoresponse force microscopy of ferroelectric surfaces. *Physical Review B*, **65**, 1–11.
113. Isakov, D.V., Gomes, E.D.M., Almeida, B.G. *et al.* (2010) Ferroelectricity in electrospun nanobers with triglycine sulphate. *Journal of Applied Physics*, **108**, 042011.
114. Santiago-Aviles, J.J., Wang, Y., Furlan, R. and Ramos, I. (2004) Synthesis and characterization of micro/nanoscopic $Pb(Zr_{0.52}\,Ti_{0.48})O_3$ fibers by electrospinning. *Applied Physics A*, **78**, 1043–1047.
115. Isakov, D., de Matos Gomes, E., Bdikin, I. *et al.* (2011) Production of polar β-glycine nanofibers with enhanced nonlinear optical and piezoelectric properties. *Crystal Growth & Design*, **11**, 4288–4291.
116. Majdoub, M., Sharma, P. and Cagin, T. (2008) Erratum: size-dependent piezoelectricity and elasticity in nanostructures due to the flexoelectric effect. *Physical Review B*, **77**, 125424.
117. Wang, Y. and Santiago-Aviles, J.J. (2011) A review on synthesis and characterization of lead zirconate titanate nanofibers through electrospinning. *Integrated Ferroelectrics*, **126**, 60–76.
118. Chang, J., Dommer, M., Chang, C. and Lin, L. (2012) Piezoelectric nanofibers for energy scavenging applications. *Nanoenergy*, **1**, 356–371.
119. Yee, W.A., Nguyen, A.C., Lee, P.S. *et al.* (2008) Stress-induced structural changes in electrospun polyvinylidene difluoride nanofibers collected using a modified rotating disk. *Polymer*, **49**, 4196–4203.
120. Zhong, G., Zhang, L., Su, R. *et al.* (2011) Understanding polymorphism formation in electrospun fibers of immiscible poly(vinylidene fluoride) blends. *Polymer*, **151–162** (52), 2228–2237.
121. Andrew, J.S. and Clarke, D.R. (2008) Enhanced ferroelectic phase content of polyvinylidene difluoride fibers with the addition of magnetic nanoparticles. *Langmuir*, **24** (16), 8435–8438.
122. Andrew, J.S. and Clarke, D.R. (2008) Effect of electrospinning on the ferroelectric phase content of polyvinylidene difluoride fibers. *Langmuir*, **24** (3), 670–672.
123. Huang, S., Yee, W.A., Tjiu, W.C. *et al.* (2008) Electrospinning of polyvinylidene difluoride with carbon nanotubes: synergistic effects of extensional force and interfacial interaction on crystalline structures. *Langmuir*, **24** (3), 13621–13626.

124. Hu, Y., Zhang, Y., Xu, C. *et al.* (2010) High-output nanogenerator by rational unipolar assembly of conical nanowires and its application for driving a small liquid crystal display. *Nano Letters*, **10** (12), 5025–5031.
125. Chang, C., Tran, V.H., Wang, J. *et al.* (2010) Direct-write piezoelectric polymeric nanogenerator with high energy conversion efficiency. *Nano Letters*, **10**, 726–731.
126. Sun, D., Chang, C., Li, S. and Lin, L. (2006) Near-field electrospinning. *Nano Letters*, **6** (4), 839–842.
127. Hansen, B.J., Liu, Y., Yang, R. and Wang, Z.L. (2010) Hybrid nanogenerator for concurrently harvesting biomechanical and biochemical energy. *ACS Nano*, **4**, 3647–3652.
128. Lutolf, M.P. and Hubbell, J.A. (2005) Synthetic biomaterials as instructive extracellular microenvironments for morphogenesis in tissue engineering. *Nature Biotechnology*, **23** (1), 47–55.

Index

Design and Fabrication of Self-Powered Micro-Harvesters: Rotating and Vibrating Micro-Power Systems, First Edition.
C. T. Pan, Y. M. Hwang, Liwei Lin and Ying-Chung Chen.